KW-010-511

THE ADHESION MOLECULE

FactsBook

Second Edition

Clare M. Isacke
Imperial College of Science, Technology and Medicine, London, UK

Michael A. Horton
University College London, London, UK

ACADEMIC PRESS

A Harcourt Science and Technology Company

San Diego San Francisco New York Boston
London Sydney Tokyo

This book is printed on acid-free paper.

Copyright © 2000 by ACADEMIC PRESS

All Rights Reserved.
No part of this publication may be reproduced or transmitted in any form or by
any means, electronic or mechanical, including photocopying, recording, or any
information storage and retrieval system, without permission in writing from the
publisher.

Academic Press
A Harcourt Science and Technology Company
Harcourt Place, 32 Jamestown Road, London NW1 7BY
http://www.hbuk.co.uk/ap/

Academic Press
525 B Street, Suite 1900, San Diego, California 92101-4495, USA
http://www.apnet.com

ISBN 0-12-356505-7

Library of Congress Catalog Card Number: 00-104879

A catalogue for this book is available from the British Library

Typeset by Mackreth Media Services, Hemel Hempstead, UK
Printed in Great Britain by Redwood Books, Trowbridge, Wiltshire

00 01 02 03 04 05 RB 9 8 7 6 5 4 3 2 1

UNIVERSITY
LIBRARY
WITHDRAWN

THE
ADHESION
MOLECULE
FactsBook
Second Edition

Other books in the FactsBook Series:

Robin Callard and Andy Gearing
The Cytokine FactsBook

Steve Watson and Steve Arkinstall
The G-Protein Linked Receptor FactsBook

Shirley Ayad, Ray Boot-Handford, Martin J. Humphries, Karl E. Kadler
and C. Adrian Shuttleworth
The Extracellular Matrix FactsBook, 2nd edn

Grahame Hardie and Steven Hanks
The Protein Kinase FactsBook
The Protein Kinase FactsBook CD-Rom

Edward C. Conley
The Ion Channel FactsBook
I: Extracellular Ligand-Gated Channels

Edward C. Conley
The Ion Channel FactsBook
II: Intracellular Ligand-Gated Channels

Edward C. Conley and William J. Brammar
The Ion Channel FactsBook
IV: Voltage-gated Channels

Kris Vaddi, Margaret Keller and Robert Newton
The Chemokine FactsBook

Marion E. Reid and Christine Lomas-Francis
The Blood Group Antigen FactsBook

A. Neil Barclay, Marion H. Brown, S.K. Alex Law, Andrew J. McKnight,
Michael G. Tomlinson and P. Anton van der Merwe
The Leucocyte Antigen FactsBook, 2nd edn

Robin Hesketh
The Oncogene and Tumour Suppressor Gene FactsBook, 2nd edn

Jeffrey K. Griffith and Clare E. Sansom
The Transporter FactsBook

Tak W. Mak, Josef Penninger, John Rader, Janet Rossant
and Mary Saunders
The Gene Knockout FactsBook

Bernard J. Morley and Mark J. Walport
The Complement FactsBook

Steven G.E. Marsh, Peter Parham and Linda Barber
The HLA FactsBook

Hans G. Drexler
The Leukemia-Lymphoma Cell Line FactsBook

Contents

Preface

The authors acknowledge all those who have helped in the completion of this book. In particular we would like to thank Chau Chong for excellent secretarial help, our laboratories for their patience while the book was being written and members of the UK Adhesion Group who filled in questionnaires suggesting molecules that should be included. Many colleagues were invaluable in checking entries and providing unpublished information. We would particularly like to thank Ann Ager, Chris Buckley, Paul Crocker, Pat Doherty, Kurt Drickamer, Nancy Hogg, Gudrun Ihrke, David Jackson, Sirpa Jalkanen, Clive Landis, Tony Magee, Fiona Watt, Sue Watt and Lisa Williams. Writing this book was greatly helped by the recent publication of *The Leucocyte Antigen FactsBook*, and we would like to acknowledge Neil Barclay, Marion Brown, Alex Law, Andrew McKnight, Michael Tomlinson and Anton van der Merwe whose hard work made our task much easier.

With the published literature on adhesion molecules increasing by several thousand papers per annum and the various mammalian genome projects sequencing apace, a book will inevitably contain omissions and inaccuracies, and the references have a danger of being superseded. For this we apologize. However, with access to electronic databases on the World Wide Web (see Chapter 3), this is relatively easy to remedy and the reader is asked to bear with us and make their own updates relating to a favourite molecule.

Abbreviations

AIDS	Acquired immunodeficiency syndrome
B	B lymphocyte
BCR	B cell antigen receptor
CCP	Complement control protein
CD	Cluster of differentiation
CNR	Cadherin-related neuronal receptor
CS	Chondroitin sulphate
ECM	Extracellular matrix
EGF	Epidermal growth factor
ERM	Ezrin/radixin/moesin
ES	Embryonic stem
EST	Expressed sequence tag
FAK	Focal adhesion kinase
FGF	Fibroblast growth factor
GAG	Glycosaminoglycan
GalNAc	N-Acetylgalactosamine
GlcNAc	N-Acetylglucosamine
GPI	Glycosyl-phosphatidylinositol
GSK	Glycogen synthase kinase
HEV	High endothelial venule
IFN	Interferon
Ig	Immunoglobulin
IgSF	Immunoglobulin superfamily
IL	Interleukin
ITIM	Immunoreceptor tyrosine-based inhibitory motif
kb	Kilobase pairs
kDa	Kilodalton
LAD	Leucocyte adhesion deficiency
LPS	Lipopolysaccharide
LRR	Leucine-rich repeat
MAPK	Mitogen-activated protein kinase
MGI	Mouse Genome Informatics
MHC	Major histocompatibility complex
NK	Natural killer
NMR	Nuclear magnetic resonance
OMIM	Online Mendelian Inheritance in Man
PI	Phosphatidylinositol
SDS-PAGE	Polyacrylamide gel electrophoresis in sodium dodecyl sulphate
SH2	Src homology domain 2
SH3	Src homology domain 3
sLe^A	Sialyl-Lewis^A
sLe^x	Sialyl-Lewis^x
T	T lymphocyte
TM	Transmembrane

TM4SF	Tetraspan superfamily
TNF	Tumour necrosis factor
Wg	Wingless
WWW	World Wide Web

Type		Size (approximate amino acids)	Type		Size (approximate amino acids)
Cadherin repeat	Cd	110	Lectin C-type	CL	120
Collagenous domain		Variable length	Galectin or Lectin S-type	G	140
Complement control protein (CCP)	C	60	LRR repeats		24
			Link	Lk	90
Desmoglein repeat	D	30	Ly-6	L6	70–90
Epidermal growth factor (EGF)	E	40	Scavenger receptor	Sc	110

OTHER SYMBOLS USED

Type		Size (approximate amino acids)		
Fibronectin type III (Fn3)	F3	100	N-glycosylation sites	
			O-linked glycosylation	
			Glycosoaminoglycan	GAG
Immunoglobulin (Ig) V set	V S S	110	GPI anchor in lipid bilayer	
Immunoglobulin (Ig) C2 set	C2 S S	90–100	Site of propeptide cleavage (e.g. in cadherins)	
			Unknown or unclear molecular conformation or membrane attachment/linkage	?

Figure 1 *Icons used for the protein domains and repeats that are present in adhesion molecules.*

THE INTRODUCTORY CHAPTERS

1 Introduction

AIMS OF THE BOOK

The primary aim of this book is to provide a compendium of the major cell surface adhesion molecules. The entries have been designed to allow the reader to establish the main structural and functional features of a molecule quickly and to provide sufficient information for accessing further information. A detailed explanation of the organization of the entries is included in this chapter. In addition there are two further chapters preceding the individual entries. Chapter 2 provides a background to the main adhesion molecule families and Chapter 3 provides information as to how to access information on adhesion molecules on the World Wide Web.

CRITERIA FOR SELECTION

Deciding on which molecules should be included in this book was a difficult task and many researchers may be offended to find the subject of their work missing. The difficulties arise in several areas. First, the majority of adhesion receptors belong to a family group based on recognizable structural features. With the rapid advances in cloning and proteomics, new molecules which contain these structural features are frequently appearing in the literature and databases. However, for many of these new molecules there is no information as to whether they function as adhesion molecules. Unless such information has been obtained, these molecules have not been given individual entries but they are discussed in the relevant sections in Chapter 2.

Second, some molecules have been shown to be bona fide adhesion molecules but have been omitted unless a mammalian orthologue has been established. In the majority of cases these molecules have been identified in *Drosophila melanogaster* or in *Caenorhabditis elegans*. Both these organisms have extensive databases associated with them that can be accessed for further information (see Chapter 3 and *C. elegans*: http://www.sanger.ac.uk/Projects/C_elegans/; http://www.wormbase.org/; *D. melanogaster*: http://flybase.bio.indiana.edu/).

Third, molecules which function in cell recognition rather than cell adhesion have been omitted. These include components of the immune recognition complexes such as the T and B cell receptors, receptor:ligand pairs such as Notch and Jagged/Serrate. Although not easy to assess, recognition molecules are judged as those that play a crucial role in mediating cell:cell interactions but do not alone mediate adhesion. Similarly, in the past few years, the distinction between adhesion molecules and signalling receptors has become increasingly blurred, and indeed many adhesion and signalling receptors act in concert to mediate their biological effects. Again, unless such receptors have been demonstrated themselves to have adhesive properties, they have been omitted. Finally, there is an increasing number of transmembrane molecules that associate with adhesion molecules and in many cases modulate their function. However, these association molecules, such as members of the tetraspan superfamily (TM4SF) do not alone mediate adhesion. Similarly, intracellular components which associate with adhesion molecules have not been included.

ORGANIZATION OF THE DATA

Name

The most commonly used name for each molecule has been used and other names that are currently still in use are given also. Many of the adhesion molecules originally were given different names but gradually a consensus of nomenclature has arisen. As will be discussed in Chapter 2, cadherins, integrins, selectins and syndecans have family names. An equivalent situation does not exist for the immunoglobulin superfamily (IgSF), which reflects the diversity of function within this family. However, within the IgSFs, nomenclature for subfamilies, such as the siglecs and ICAMs, is becoming more common. Many of the adhesion molecules are expressed by leucocytes and consequently they have a CD (cluster of differentiation) nomenclature (see *Leucocyte Antigen FactsBook*[1]). In some cases this is the most commonly used name, especially for those molecules that were originally characterized in different laboratories and hence given different names. In other cases, molecules have a CD nomenclature but an earlier consensus for another name means that the CD number is not commonly used.

Family

The family to which the adhesion molecule belongs is given. Some families are further divided and, in these cases, the subfamily is also provided. Further information into the organization of families and subfamilies is given in Chapter 2. The 'other molecule' section contains those proteins that we decided to include on the basis of functional information but which again do not fit into the major adhesion molecule families.

Molecular diagram

Many of the adhesion molecules are expressed by leucocytes and are therefore listed in the *Leucocyte Antigen FactsBook*[1]. To aid cross-referencing, we have used the same icons for the protein domains and repeats. In addition, new icons have been used to represent domains not found in leucocyte antigens. The aim of the diagram is to provide a visual representation of the molecule indicating the subunit composition, how it is attached to the membrane, its orientation, the number and type of protein domains, the degree and type of glycosylation and the position of any proteolytic cleavage sites. The icons and symbols are shown in Fig. 1. The cytoplasmic domains of the majority of adhesion molecules do not contain recognizable protein domains and therefore these are represented by squiggly lines, the length of which is proportional to the sequence length. Nuclear magnetic resonance (NMR) and crystallographic data are increasingly available for domains present within some of the families of adhesion molecules (e.g. IgSF, cadherins, integrins). We have not attempted to discuss or illustrate detailed atomic structural data for any of the molecules; we do though discuss and reference this in general terms in Chapter 2 and the individual molecule entries.

Size of the receptor

The number of amino acids and calculated relative molecular weight (M_r) of the entire polypeptide backbone (including the signal sequence) is given. The M_r of the fully processed molecule obtained from sodium dodecyl sulphate-polyacrylamide gel electrophoresis (SDS-PAGE) under reducing conditions is given in kilodaltons (kDa). Where relevant, sizes under non-reducing conditions are also listed. As all the adhesion molecules are processed through the rough endoplasmic reticulum and Golgi apparatus, this value reflects the size of the glycosylated protein. In the case of the type I proteins this reflects the loss of the N-terminal signal sequence and for the classical cadherins, desmocollins and desmogleins, the additional loss of the propeptide sequence.

Glycosylation

All of the adhesion molecules are glycosylated, but in the majority of cases the extent and nature of the carbohydrate modification is not known.

N-linked carbohydrates are attached to Asn residues within the consensus motif Asn-X-Ser or Asn-X-Thr, where X is any amino acid except Pro. The number of these motifs present in the polypeptide backbone is given and the approximate positions shown in the diagram. It should be noted that not all potential N-linked sites are necessarily occupied.

O-linked sugars are attached to Ser or Thr residues but there is no consensus sequence for this modification. In most molecules, O-linked glycosylation is found in stretches of sequence with a preponderance of Ser, Thr and Pro amino acids, where specificity depends on a combination of sequence context, secondary structure and accessibility. It should be noted that the acceptor sequence context for O-glycosylation of serine is different from that of threonine. A programme for the prediction of O-glycosylation sites is available at the web site: http://www.cbs.dtu.dk/services/.

The approximate predicted positions of the O-linked sugars are indicated in the diagrams and the following terms are used to describe the extent of O-glycosylation:

No entry	There are no data from sequence or biochemical studies to indicate the level of O-glycosylation.
0	The absence of O-glycosylation is indicated by biochemical analysis.
+ to +++	Biochemical data indicate that O-glycosylation occurs a low (+), moderate (++) or high (+++) level.

Glycosaminoglycans (GAG) have been identified on syndecans 1–4, CD44 and dystroglycan. With the exception of keratan sulphate, which is not found on any of the adhesion molecules listed here, GAGs are attached via a xylose residue linkage to specific serine residues in the protein core. Attachment requires the presence of a C-terminal Gly; however, not all Ser-Gly sites are GAG modified.

Gene location and size

The chromosome location of the human gene is given together with the gene size and number of exons, where known. As indicated in the text, where the genomic structure has been established in non-human species, the gene size and number of exons for that species is given. Due to the rapidly changing information on the 'human genome' we have not attempted to provide a gene structure diagram which inevitably is likely to be superceded.

Alternative forms

Many of the adhesion molecules are subject to alternative splicing and other modifications and this section summarizes what is known about the different isoforms. Where relevant, the position of sequences encoded by alternatively spliced exons or the position of exon insertion is indicated in the amino acid sequence.

Structure

Biochemical and structural data are summarized in this section.

Ligands

This section describes the ligands characterized for each adhesion molecule. In many cases the 'ligand' is another membrane-bound molecule and therefore may be more properly described as a 'counter receptor'. In some cases a physiological ligand has not been identified but molecules have still been included in the book where their structure or function strongly suggests that they are adhesion molecules.

Function

Functional data are summarized in this section. In some cases a physiological function can only be predicted.

Distribution

The tissue and cellular distribution of the adhesion molecule is described in this section. It should be noted that in many cases a full analysis of tissue distribution has not been undertaken and therefore may be wider than described. Where relevant the subcellular distribution of the molecule is also discussed.

Disease association

The OMIM (Online Mendelian Inheritance in Man) accession number for each molecule is given (information as to how to access information from the OMIM databases is given in Chapter 3). Where relevant the association of a molecule with a particular disease is described. In addition, the use of reagents directed against these molecules for disease prediction or in therapy is also indicated.

Knockout

The MGI (Mouse Genome Informatics) accession number is given. As described in Chapter 3, accessing knockout mouse databases provides a large amount of information about the function of the murine homologue of the adhesion molecule. Where a knockout has been generated, the phenotype is described.

5

Amino acid sequence

Amino acid sequences are shown in the single-letter code (Table 1). If it is known, the human amino acid sequence is given. The initiating methionine is amino acid 1. The predicted signal sequence is show in bold and the predicted transmembrane domain is underlined. Any further information is described for individual entries.

Table 1. *Single-letter and three-letter amino acid codes*

Amino acid	Single-letter code	Three-letter code
Alanine	A	Ala
Arginine	R	Arg
Asparagine	N	Asn
Aspartic acid	D	Asp
Cysteine	C	Cys
Glutamic acid	E	Glu
Glutamine	Q	Gln
Glycine	G	Gly
Histidine	H	His
Isoleucine	I	Ile
Leucine	L	Leu
Lysine	K	Lys
Methionine	M	Met
Phenylalanine	F	Phe
Proline	P	Pro
Serine	S	Ser
Threonine	T	Thr
Tryptophan	W	Trp
Tyrosine	Y	Tyr
Valine	V	Val

Database accession

The accession numbers are given for the nucleic acid sequences in the GenBank/EMBL databases and for the amino acid sequences in SwissProt databases. Where there is no SwissProt entry, the TrEMBL database accession number is given. Unless otherwise stated, these accession numbers refer to the human molecules. Information as to how to retrieve information from the databases is given in Chapter 3.

References

It is not feasible to give a comprehensive list of references. The references given are recent ones, which, together with the databases accession information described in Chapter 3, will allow access to the rest of the literature.

References

[1] Barclay, A.N. et al. (1997) The Leucocyte Antigen FactsBook, 2nd edn. Academic Press, London.

2 Adhesion Molecule Families

The majority of adhesion molecules can be grouped into families. In the case of cadherins, integrins, selectins and syndecans, this grouping is based both on structural and functional similarities of the family members. The immunoglobulin superfamily (IgSF) represents a more diverse group of proteins that have structural similarity but a variety of different functions, and consequently subfamilies with functional similarity are arising. Finally, a large number of adhesion molecules do not fit into any of these families and these have been listed as 'others'. Some of the 'other' adhesion molecules can be divided into subfamilies. It is anticipated that many of the remaining 'other' molecules are the first identified members of a family, and that other members will be isolated and characterized within the next few years.

CADHERINS

Introduction

The cadherin proteins now comprise a very large superfamily which can be divided into two subfamilies, the classic cadherins and the protocadherins, each of the subfamilies is further subdivided (Fig. 1). Only the classic cadherins are given individual entries in this book but, as is discussed in this section, it is anticipated that in this rapidly moving field many of the protocadherins will prove to mediate cell adhesion *in vivo*.

Structure

Classic cadherins

The classic cadherins can be subdivided into four subfamilies, the type I classic cadherins, the type II classic cadherins, the desmosomal cadherins and the modified or other classic cadherins.

The first cadherin to be identified was E-cadherin (cadherin-1), which acts as the prototype for the type I classic cadherins, namely N-cadherin (cadherin-2), P-cadherin (cadherin-3) and R-cadherin (cadherin-4). In each case the letter indicates the tissue in which the cadherin was first identified.

A number of features characterize the type I classic cadherins. They are synthesized as precursor polypeptides with an N-terminal propeptide sequence being proteolytically removed intracellularly. This event is a necessary step in the activation of cadherins to generate adhesion-competent molecules. In their extracellular domains, the type I classic cadherins have five tandemly arranged cadherin repeats of approximately 110 amino acids in length, also called extracellular cadherin (EC) domains, in between which are located four calcium binding pockets. Structural analysis of the N-terminal cadherin repeat of E-cadherin and N-cadherin show a barrel-like structure composed of seven antiparallel β-strands with the N- and C-termini located at the top and bottom, respectively. Although there is no sequence similarity, this structure is similar to the immunoglobulin domains found in the immunoglobulin superfamily (IgSF) members (see

Classic cadherins

Type I

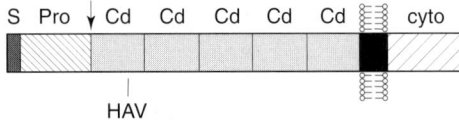

Type II

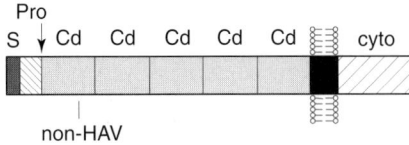

Desmocollins

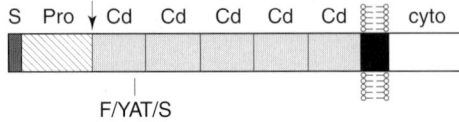

Desmogleins

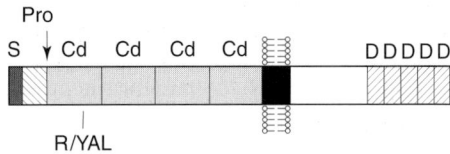

Protocadherins

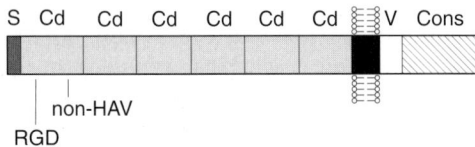

Figure 1 *Structure of the cadherin family members. As described in the text, cadherins can be divided into classic cadherins and protocadherins. Within the classic cadherins there is a further subdivision into type I classic cadherins, type II classic cadherins, desmosomal cadherins, and other or modified classic cadherins. The desmosomal cadherins fall into two groups, the desmocollins and the desmogleins. The characteristics of each cadherin type are described in the text. (S, signal sequence; Pro, propeptide sequence; Cd, extracellular cadherin repeats; cyto, cytoplasmic domain; D, desmoglein repeats; V, variable cytoplasmic domain sequence; Cons, conserved cytoplasmic domain sequence). Arrow indicates position of propeptide cleavage to generate the mature form of the classic cadherins. Amino acid motifs within the N-terminal cadherin repeat indicate the adhesion recognition sequence. Note (a) other or modified classic cadherins: the three other/modified classic cadherins covered in this book are not shown in this diagram as these are described individually in the text and in*

'Immunoglobulin superfamily'). One major difference is that the IgSF are not calcium binding proteins, whereas in the cadherins the five domains assume a rod-shape that is stabilized by the binding of calcium ions between successive repeat domains. Located within the N-terminal domain of the classic type I cadherins is the conserved HAV motif which is part of the active site for interaction with cadherins on neighbouring cells. Other residues within the active site provide the specificity for homophilic interaction. The exceptions are N- and R-cadherin which have almost identical amino acids at the homophilic binding interface and can interact with each other. In addition, a tryptophan residue in the +2 position after removal of the propeptide sequence is required for this *trans* interaction. Within domain 1, on the opposite side to the adhesive or *trans*-face, is the *cis* face which mediates homophilic association with cadherins in the plane of the membrane. This *cis* interaction is dependent on calcium being bound between domains 1 and 2, and is necessary for the subsequent *trans* interaction and hence cell:cell adhesion[1-3a].

The cytoplasmic domain of the type I class cadherins is 150–160 amino acids in length and is required for their adhesive function as it is this domain that mediates the association with the actin cytoskeleton. Binding to actin is not direct but via linker proteins known as catenins. The arrangement of the catenin binding is as follows: cadherin–β-catenin–α-catenin–actin. Originally three catenins were identified but it is now established that the desmosomal component plakoglobin (previously called γ-catenin), like β-catenin, can bind to both cadherins and α-catenin, and that the inclusion of β-catenin and plakoglobin in the cadherin/catenin complex is mutually exclusive. However, as evidenced by the different phenotypes associated with the knockout of β-catenin (MGI:88276) and plakoglobin (MGI:96650) *in vivo*[4-6], these two components have different physiological roles and are not functionally redundant. As discussed below, β-catenin is considered to have a dual role in that it is not only required for cadherin association with the cytoskeleton but it also has an important signalling function.

The type II classic cadherins include VE-cadherin (cadherin-5), K-cadherin (cadherin-6), cadherin-8, OB-cadherin (cadherin-11), cadherin-14 and M-cadherin (cadherin-15) and are structurally similar to the type I subfamily in that they have a propeptide sequence, five cadherin repeats and, where it has been examined, they have been shown to mediate homophilic binding and to associate intracellularly with catenins. The notable differences are that the propeptide sequence tends to be

---

their separate entries. (b) Desmogleins: the structure shown here corresponds to desmoglein 1. Desmoglein 2 and desmoglein 3 have six and two desmoglein repeats, respectively, in their cytoplasmic domains. (c) Protocadherins: protocadherins are shown with six extracellular domain repeats. However, as described in the text, protocadherins with seven repeats have been isolated. The position of the variable and conserved protocadherin cytoplasmic sequences are shown. As described in the text, the conserved sequences are conserved between members of a protocadherin subfamily but not between members of different subfamilies. In the CNR protocadherins (a subset of the protocadherinα subfamily) there is an RGD motif in the N-terminal cadherin repeat. This is not conserved between all protocadherins. Finally, it is anticipated that more protocadherins will be identified and it remains to be determined whether they all conform to the protocadherin structure as shown here.

shorter and the HAV motif on the adhesive face of the N-terminal cadherin domain is not conserved. It is known that additional type II cadherins exist and it is anticipated that these will be cloned and further characterized in the near future. For example, full length cadherin-7 (GenBank AB035301), cadherin-9 (GenBank AB035302) and cadherin-10 (GenBank AF039747)[7] have recently been isolated but as there are no functional data available on these molecules, they have not been given individual entries.

Included in this book are three modified or other classical type cadherins; H-cadherin (T-cadherin, cadherin-13), Ksp-cadherin (cadherin-16) and LI-cadherin (cadherin 17). H-Cadherin has not conserved the HAV motif but has retained a long propeptide sequence. However, the major difference between this cadherin and the type I/type II classic cadherins is that it does not have a cytoplasmic domain, rather it is held in the membrane via a glycosyl-phosphatidylinositol (GPI) anchor. Ksp- and LI-cadherin are the most diverged in that they have no propeptide sequence, the extracellular domain contains seven cadherin repeats, the HAV motif is not conserved and the cytoplasmic domain is much shorter.

The desmosomal cadherins, the desmocollins and desmogleins, mediate adhesion in desmosomes, which are intercellular contacts that are linked on their cytoplasmic face to intermediate filaments[8–11]. The three desmocollins show overall homology to the classic type I cadherins with a long N-terminal propeptide sequence and five extracellular cadherin repeats. Unlike the type I cadherins, the adhesion recognition sequence is F/YAT/S rather than HAV. The three desmogleins differ from type I classic cadherins in having a short, approximately 29-amino-acid, propeptide sequence, only four cadherin repeat domains and a R/YAL rather than HAV adhesion motif. The intracellular domains of the desmogleins have diverged substantially from other cadherins and contain between two and six, 28–30-amino-acid, desmoglein repeat sequences. Although these desmoglein repeats are smaller, they show some homology to cadherin repeats. The cytoplasmic domains of desmosomal cadherins interact with intermediate filaments (keratin in skin and desmin in heart) via plakoglobin (γ-catenin) and other desmosome-associated proteins, such as desmoplakins and plectin.

Protocadherins
Structurally the protocadherins differ from the classic cadherins in that they do not have propeptide sequences and the extracellular domains contain more than five cadherin repeats (usually six or seven)[12,13]. Recently it has become clear that there are a very large number of protocadherins. This first became apparent with the identification of a family of approximately 20 murine genes in a screen for receptors coupled to the Fyn signalling pathway in the brain[14]. These cadherins were named the CNR (cadherin-related neuronal receptor) cadherins. Examination of eight members of this subfamily revealed six extracellular cadherin repeats with conserved calcium binding sequences. Most notable about the structure was the presence in the N-terminal cadherin repeat of a conserved RGD motif and within the 220–230-amino-acid cytoplasmic domains, the presence of four PXXP motifs and the complete conservation of the C-terminal 153 amino acids[14]. The presence of these 3' identical sequences led to the suggestion that the CNR protocadherins might arise from the joining of diverse 5' exons to a common set of 3' exons. Support for this hypothesis has come from the identification of 52 human protocadherin genes arranged in three tandemly linked clusters in the 5q31 region

of chromosome 5. These three clusters have been named protocadherin (*Pcdh*)α, *Pcdh*β and *Pcdh*γ with the *Pcdh*α genes being the probable human orthologues of the murine CNR protocadherins. Strikingly, within each cluster there is a tandem array of large exons each of which encodes for the extracellular, transmembrane and membrane proximal cytoplasmic domain of the different protocadherins. To generate a functional protocadherin, one of the 'variable' 5' exons must be joined to the conserved C-terminal cytoplasmic domain portion which is encoded for by three 'constant' 3' exons. In this manner protocadherins with differing extracellular domains (and hence presumably differing binding specificity) and a common C-terminal cytoplasmic domain are generated in a manner that can be regarded as a simplified version of the generation of antibodies and T cell receptors[15,16]. Cytoplasmic domain components associated with the *Pcdh*β and *Pcdh*γ gene products have not been identified but, given that there is no cytoplasmic domain conservation between the *Pcdh* protocadherin subfamilies, it is unlikely that the *Pcdh*β and *Pcdh*γ receptors will associate with Fyn. It should be noted that these three gene clusters do not encode all of the protocadherins. Human protocadherins with a gene structure similar to that of the *Pcdh*α, *Pcdh*β and *Pcdh*γ protocadherins, have been identified. These include the *Pcdh*8 gene (chromosome 13)[17] and the *Pcdh*12 gene (chromosome 5q31). In addition, protocadherins with different gene and predicted protein structures such as two homologues of the Drosophila *fmi* gene, *hFlamingo*1 (chromosome3) and *hFlamingo*2 (chromosome 22q13),[17a] and a human homologue of the Drosophila *fat* gene (*hFat*2, chromosome 5q33) have been described. Although these latter protocadherins differ in their number of cadherin and transmembrane domains and the presence of additional extracellular domain motifs, they contain the typical protocadherin feature of a very large first exon encoding multiple cadherin domains[17a].

Ligands and regulation of ligand binding

The characteristic feature of the classic cadherins is their ability to mediate Ca^{2+}-dependent homophilic adhesion with specificity being generated by sequence differences at the adhesive face of the N-terminal domain. As discussed above, structural studies combined with mutagenesis suggest that binding can be regulated by the cadherins switching between non-binding monomers and adhesive competent dimers in the plane of the membrane. However, this is probably not the only form of binding regulation and there is increasing evidence for a regulatory role both for the cytoplasmic domain[5] and by the *cis*-association of the cadherins with other receptors[18,19]. It should be noted that although a robust cell:cell adhesive function has been demonstrated for the type I classic cadherins, a similar role for many of the other classic cadherins has yet to be demonstrated[20]. In addition, the demonstration that E-cadherin can bind the $\alpha_E\beta_7$ and $\alpha_2\beta_1$ integrins[21,22] raises the possibility that other classic cadherins may have additional ligands.

Less is known about the ligand binding properties of the protocadherins, however, despite the differences in extracellular domain organization, at least some of the protocadherins can mediate Ca^{2+}-dependent homophilic cell adhesion[12,13,23]. In addition the CNR protocadherins contain an RGD motif in their N-terminal cadherin repeat, which is predicted to be located on a protruding loop thereby implicating this subclass of *Pcdh*α protocadherins as counter-receptors for integrins on neighbouring cells[14,24].

Signal transduction

As with the majority of adhesion molecules, there is evidence that the cadherins act in concert with signal transduction pathways to provide a coordinated regulation of cell behaviour and gene expression. In the case of the type I classic cadherins, this linking of signalling and adhesive function is particularly strong owing to the dual role of β-catenin as both a structural component of the adhesive complexes and as a potent transcriptional cofactor. β-Catenin is a key component in the Wnt/wingless (Wg) pathway. In the absence of Wnt/Wg signalling, free cytoplasmic β-catenin is phosphorylated by the cytoplasmic enzyme glycogen synthase kinase 3β (GSK) and in its phosphorylated state is recognized by the ubiquitin-proteosome system and targeted for degradation. Activation of the Wnt/Wg pathways results in inhibition of GSK, accumulation of stable non-phosphorylated β-catenin, which is able to associate with the lymphoid enhancer binding factor (LEF)/T cell specific factor (TCF) transcription factor and activate gene expression[5,6,25]. This pathway is conserved throughout evolution; however, it is now clear that it much more complex than the simplified scheme outlined here. For example, the interaction of GSK and β-catenin involves a number of other components, including the tumour suppressor gene product APC (adenomatous polyposis coli) and the scaffolding protein axin/conductin, and these play a key role in regulating β-catenin signalling. In addition, the relationship between β-catenin in the cadherin adhesive complexes and β-catenin as a signalling molecule has yet to be fully elucidated[5,6,25,26]. More recently, evidence has been accumulating that the *cis*-association of cadherins with growth factor receptors plays an important role in regulating N-cadherin-mediated neural outgrowth[18,19]. Whether similar growth factor receptor interactions are found in non-neural tissues with other classic cadherins has yet to be determined.

Little is known about putative signal transduction pathways associated with the protocadherins except that at least a subclass of the protocadherins, the CNR cadherins, were identified on the basis of their interaction with the non-receptor tyrosine kinase, Fyn. The demonstration that subfamilies of *Pcdh*α, β and γ protocadherins contain a common 3′ cytoplasmic domain indicates that they will activate common signalling pathways, although it should be noted that differences in the membrane proximal portions of their cytoplasmic domains, and indeed in their transmembrane and extracellular domains, allows the possibility of individual protocadherins to activate individual pathways in addition to common pathways[14–16].

Function

The first identified cadherins, the type I classic cadherins, were characterized by their ability to mediate Ca^{2+}-dependent homophilic adhesion and by their differential cell type distribution. It is these features that underlie the ability of cadherins to regulate the cell type specific adhesion required for the arrangement and rearrangement of cells during morphogenesis[27]. At least in the case of E-cadherin, which is localized to the adherens junctions in epithelial cells, and the desmosomal cadherins, cadherins also play a crucial role in establishing the intercellular junctions[5]. The finding that desmocollins and desmogleins are linked to intermediate filaments in desmosomes suggests a functional role in maintaining tissue integrity at sites undergoing mechanical stress as would be found, for example, in the heart and skin. In addition, the discovery that components of the type I classic cadherin adhesive complex have a signalling role in the cell suggests that these cadherins function as more than just molecular glue, and that they also serve to regulate cell:cell interactions and provide information to the cell as to its extracellular milieu. Currently the most exciting development in cadherin research is the recent demonstration of the large subfamilies of protocadherins. The unique genomic arrangement of these protocadherins, which allows variable extracellular domains to be joined to a constant cytoplasmic domain portion suggests a mechanism for tightly controlling cell:cell interactions by cell-specific expression of adhesion molecules which would activate a common signalling pathway. Consequently it has been speculated that protocadherins may provide the molecule code for specifying neuronal connections in the brain. Whether this is indeed a physiological function of protocadherins, and whether they have additional roles both in neural and non-neural tissues remains to be determined.

IMMUNOGLOBULIN SUPERFAMILY

Introduction

The immunoglobulin superfamily (IgSF) represents a large group of proteins with the common feature that they all contain one or more extracellular Ig domains. All cells in the body express such molecules and in leucocytes, where the cell surface proteins have been most thoroughly characterized, 34% of the transmembrane and membrane-associated proteins are IgSF members. However, it should be noted that not all IgSF members function as adhesion molecules and therefore the leucocyte IgSF molecules given entries in this book only represent a subset of those found in *The Leucocyte Antigen FactsBook*.[28]

Structure

General

Ig domains are 70–110 amino acids in length with a distinct overall structure referred to as the Ig fold. This fold consists of two β sheets each made up of short antiparallel β strands. In the majority of Ig domains, the two β sheets are joined by a disulphide linkage. Sequence and structural comparisons have shown that there is highest conservation in the in-pointing β sheet residues, i.e. those that are packed face:face and variation at the out-pointing β sheet residues and the loops that connect the β strands. Originally Ig domains were characterized as V-type or C-type based on features shared in common with the Ig and T cell receptor V-domains and the Ig, T cell receptor and major histocompatibility complex (MHC) C-domains, respectively. In V-type domains the two β sheets are made up of three and four β strands (the ABED and GFC strands), and lying between these sheets are the C' and C" additional strands. These two additional strands are missing in the C-type domains. Subsequently, sequence and structural analysis of more IgSF members revealed a group of domains which were similar to C-type domains but with some sequence patterns found in V-type domains. This new group are referred to as C2 domains, and are characterized by the lack of the D strand but the presence of the C' strand. The original C-type domains, which contain only the ABED and GFC strands, have been renamed C1 domains[29,30]. Further structural analysis has demonstrated that the C2-type domains in fact represent two distinct sets, which have been termed C2 and I domains[30,31]. However, as structural information is not available for many of the IgSF members described in this book, the term C2 will be used here for both C2 and I domains.

As discussed under 'Cadherins', structural analysis of cadherin domains gave the surprising finding that, despite no sequence similarity, these repeats showed an overall structural arrangement similar to that found in the Ig domains, particularly I-type members of the C2 domains. It has been proposed that this similarly results from convergent evolution owing to the favourable stability of protein folds generated by two β sheets[32]. In support of this theory is the fact that similar arrangements are found in other protein domains. For example, fibronectin type III (Fn3) repeats,[33,34] such as those found in the IgSF members L1 and NCAM (see individual entries), again have a fold generated by two β sheets.[31,35,36]

As illustrated by the entries in this book, the number of Ig domains within the IgSF members is variable and, in addition, it should be noted that many IgSF members also contain non-Ig domains. The only common features are that, with the exception of CD147, in molecules with V-type domains, these domains are N-terminal to the C2-type domains and the Ig domains are N-terminal to non-Ig domains.

Subfamilies

The IgSF can be grouped into subfamilies but in many cases this represents an artificial grouping as individual molecules can belong to more than one subfamily. For example, the five ICAMs can be grouped together on the basis of sequence similarity and their ability to bind integrins. However, VCAM, MAdCAM-1 and CD31, which are not closely related to the ICAMs, have functional overlap in that they also act as integrin counter-receptors. Consequently we have not divided the IgSF entries into subgroups but a few relevant issues will be discussed here.

Siglecs

The siglecs are functionally characterized by their ability to bind sialic acid. The now agreed nomenclature of this group of molecules is sialoadhesin (siglec-1), CD22 (siglec-2), CD33 (siglec-3), MAG (myelin-associated glycoprotein; siglec-4a) and SMP (Schwann cell myelin protein; siglec-4b).[37] All the siglecs are transmembrane proteins with extracellular domains consisting of an unusual N-terminal V-type Ig domain followed by a variable number of C2-type domains. It is now clear that these five molecules do not represent the full siglec subfamily. Cloning of siglec-5[38], siglec-6[39], siglec-7[40] and siglec-8[40b] has recently been reported, and it is likely that other siglecs will be identified.

Neural IgSF members

In this book individual entries are included for the following IgSF members found in neural tissues: ALCAM, contactin-1, ICAM-5, MAG, L1, NCAM, P(0) and TAG-1. This discussion will concentrate on contactin-1, L1, NCAM and TAG-1. NCAM and L1 are type I transmembrane adhesion molecules whereas contactin-1 and TAG-1 are GPI anchored. One of the difficulties in classifying these four molecules is that, over the past decade, a large number of related molecules have been cloned from different species and consequently there is some controversy as to whether different molecules are species orthologues of each other or different subfamily members. In general it is agreed that NCAM is the mammalian orthologue of *Drosophila* fasciclin II. TAG-1, originally cloned in mice, is the product of the human TAX-1 gene and is the orthologue of chicken axonin-1. Human contactin-1 is the orthologue of rodent F3 and chicken F11. The situation for L1 is more controversial. L1 (previously known as NILE) shows sequence similarity to two chicken adhesion molecules, NgCAM and NrCAM. NgCAM is considered to be a probable L1 orthologue, whereas NrCAM together with the murine CHL1 (close homologue of L1) and neurofascin are regarded as vertebrate L1 family members. It has yet to be determined whether *Drosophila* neuroglian and leech tractin represent invertebrate orthologues or other L1 family members[41]. The second area of controversy is in the ligands for these four molecules and this will be discussed below. Finally, there is an increasing number of other IgSF molecule that are predicted to function as neural adhesion molecules. Owing to the limited functional data on these molecules or their ligands, or the fact that they have only been characterized in invertebrates, they have not been given individual entries in this book. Of particular note are DCC (deleted in colorectal carcinoma; SwissProt P43146; OMIM 120470), which is probably the mammalian orthologue of *Drosophila* frazzled and *Caenorhabditis elegans* Unc40 genes, and the DCC-related protein neogenin (OMIM 601907). It is now well established that DCC and neogenin are structurally related type I transmembrane proteins which bind to members of the netrin family. Netrins play a key role in axon guidance but, as they are secreted molecules, they and their receptors have not been included in this book. However, it is likely that the netrins are in some way tethered to the extracellular matrix and it remains to be determined whether binding of the netrins to their receptors represents an 'adhesive' interaction[42].

Ligands and signal transduction

Unlike the cadherins, which essentially mediate homophilic interactions via the antiparallel binding of the N-terminal cadherin repeats with the equivalent domain on neighbouring cells, the IgSF members show a broad range of ligand interactions. These include integrins (e.g. ICAMs), sialic acid (members of the siglec subfamily), extracellular matrix components (e.g. contactin-1) and homophilic interactions (e.g. CD31, P(0), CEACAM family). In some cases, more than one ligand has been identified. For example, in addition to binding integrins, ICAMs act as receptors for various viruses and parasites, CD31 can also mediate homophilic interactions and MAdCAM additionally binds L-selectin. The IgSF ligands are discussed in the individual entries. However, of particular interest in the past few years has been the lessons learnt from the neural IgSF members. One of the curiosities in this field was how molecules such as TAG-1 and contactin might mediate adhesive interactions when they are linked to the membrane via GPI anchors. Some clues as the mechanisms operating have come from the demonstration that these GPI-anchored molecules can associate in the plane of the membrane with other neural IgSF members. For example, both contactin and TAG-1 bind in *cis* to L1 family members. Furthermore, both L1 and NCAM can interact in *cis* with each other and growth factor receptors. Importantly these different interactions can modulate ligand binding and downstream signal transduction pathways[41,43-45]. It has yet to be determined whether similar *cis* interactions regulate non-neural IgSF members.

INTEGRINS

Cell adhesion molecules of focal adhesion contacts

Integrins are so called because they 'integrate' the extracellular matrix with the intracellular cytoskeleton[46] and hence deliver 'outside-in' signals from the external environment to the cell to modify cellular structure and functions: cell adhesion, motility, proliferation, apoptosis, induction of gene transcription and differentiation. To achieve this, integrins are recruited into complex structures, focal adhesion contacts (or related structures at cell:cell contacts, which can involve integrins interacting with counter-receptors, such as $\alpha_E\beta_7$ and $\alpha_2\beta_1$ with E-cadherin – see entries). The function of the integrin receptor is itself modulated by signals from within the cell delivered via other receptor systems ('inside-out' signalling) through complex cytoskeletal interactions with their cytoplasmic tails[47-49]. The intracellular proteins that interact with integrins at focal adhesion contacts[50-52] include cytoskeletal elements and link proteins, such as α-actinin, talin and paxillin, and associated non-receptor tyrosine kinases as c-Src and focal adhesion kinase (FAK). Other transmembrane proteins (e.g. syndecan-4 and dystroglycan; see individual entries) and cell junction-specific components (such as dystrophin in neuromuscular junctions; see dystroglycan entry) are also concentrated in these adhesion complexes[52].

Integrin structure

A wide range of other proteins have been shown from biochemical, immunological and cDNA and genomic cloning approaches in a large number of species from all phyla[53,54] to be structurally and functionally related. The original integrin family included the α and β subunits of platelet glycoprotein $\alpha_{IIb}\beta_3$ (also called gpIIbIIIa, which shares its β_3 chain with the $\alpha_v\beta_3$ vitronectin receptor), the β_1 matrix receptors and the leucocyte α_L (LFA-1), α_M (Mac-1) and α_X (p150/95) molecules, which share a common β_2 chain. Seventeen α and 8 β chains have now been characterized. We have excluded those 'integrins' that have been only defined as EST sequences or which exist only in non-mammalian species. An additional collagen-binding integrin alpha chain, $\alpha11$[54a,54b], has recently been cloned. Integrins are heterodimers and form 23 $\alpha\beta$ associations as shown in Fig. 2. Some chains associate widely (e.g. the β_1 and α_v subunits having several partners) whereas others show no promiscuity (e.g. $\alpha_E\beta_7$, $\alpha_{IIb}\beta_3$). Additional complexity is generated by several subunits undergoing mRNA splicing[55]; some of these isoforms differ in their tissue distribution, developmental regulation and ligand binding specificity.

Biochemical, molecular and physical analyses have revealed the protein structure of the subunits (see Fig. 3). The α and β subunits are both transmembrane, N-glycosylated glycoproteins with large extracellular domains, a single hydrophobic transmembrane region, and short cytoplasmic domain, apart from β_4. Rotary shadowing electron microscopy of a number of purified integrin receptors has revealed that they are extended structures of $\sim 10 \times 20$ nm, with an N-terminal globular 'head', formed by the association of the two subunits, linked to the membrane by two extended 'rod' structures[56].

α Subunits are between 120 and 180 kDa and comprise a large N-terminal extracellular domain, a 20–30 amino acid transmembrane region and a short,

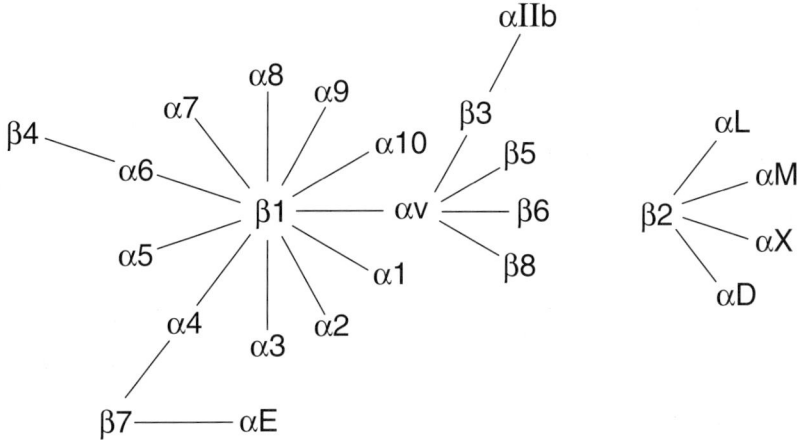

Figure 2 *The integrin family is composed of a series of α/β dimers. The various combinations that have been observed are illustrated. Thus, for example, β_1 chains are associated with a large number of a chains, whereas α_{IIb} only combines with β_3.*

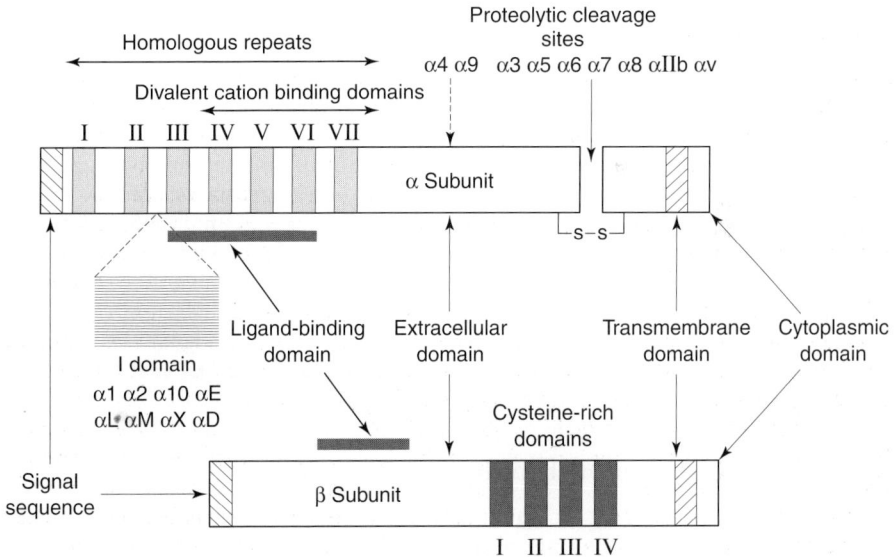

Figure 3 *Integrin subunit structures. The general features of an integrin heterodimer are shown. In addition, RNA splice variants exist (see text) that modify the structure of integrin subunits α_6, α_7, α_{IIb} and β_3 in their extracellular domains, and the cytoplasmic tails of α_3, α_6, α_7, β_1, β_3 and β_5. The cytoplasmic tail of the β chain is shown as an 'average' for all β subunits except β_4, which has a unique, approximate 1000 amino acid long, intracellular extension (see entry for integrin β_4). Glycosylation sites vary between individual integrins and are omitted.*

usually hydrophilic, cytoplasmic tail. All α chains contain seven homologous, approximately 60 amino acids long, tandem repeat sequences; the C-proximal three or four contain putative divalent cation binding sites, which are critical in ligand binding and subunit association.

Some integrin α chains contain an inserted, or 'I', domain of ~200 amino acids between the second and third repeats. These include the four α subunit partners of the β2 integrins and α_1, α_2, α_{10} and α_E. The I domain has sequence homology with several molecules, including cartilage matrix protein, type VI collagen and the von Willebrand factor 'A' domain[57], all of which can all interact with collagen, and is involved in ligand binding[58]. The $\alpha_1\beta_1$ and $\alpha_2\beta_1$ integrins have been shown to be collagen receptors; however, none of the β_2 heterodimers bind collagen suggesting that the I domain here has other roles. The other α subunits, without I domains, are post-translationally cleaved near to the transmembrane domain, the resultant fragments being disuphide bonded. The α_4 and α_8 subunits are, however, cleaved within the middle of the chain into two non-disulphide bonded components; the different forms of the $\alpha_4\beta_1$ dimer have altered ligand binding (see entry for α_4). The cytoplasmic domains of the α subunits show little sequence similarity with the exception of the conserved amino-acid motif GFFKR, which is involved in the transmission of signals into the cell, possibly via its ability to bind calreticulin[47–49,52].

The β subunits are usually smaller than α subunits and are between 90–110 kDa, apart from the 210 kDa β_4 chain. The C-terminal halves of all β subunits have a high cysteine content (e.g. 56 in the β_3 chain), grouped into four, 40 amino acid cysteine-rich regions, which are internally disulphide bonded. The cytoplasmic tails are usually short (40–50 amino acids), although the β_4 is unique with a 1018 amino acid long cytoplasmic domain, which contains four fibronectin type III repeats. A three amino acid sequence (TTT) is found within the cytoplasmic tail of several β subunits and in $\alpha_L\beta_2$ it is essential for ligand binding. Further functional domains have been defined[47–49,52], frequently by mutational analysis. These include the binding site for α–actinin in the cytoplasmic tail of β_1 which mediates linkage between the integrin and actin cytoskeleton and focal contact localization sequences in β_1 (via the 'Cyto-1' motif – REFAKFAKEK – and other sequences). More globally, integrin cytoplasmic tails are key to the establishment of the two-way linkage between the 'outside-in' and 'inside-out' signal transduction processes; however, conserved linear amino-acid sequence motifs that mediate the signalling processes have not been identified[59–64].

Three-dimensional models of integrins

Various structural models for integrin α-subunit domains have been proposed based upon sequence modelling of the α and β chains, and the solution of the crystal structure of several examples of 'I' domains[65–68]. To date, studies have been insufficient to take account of the impact of the known effects of cation and ligand binding upon protein conformation[65,69–71]; this will have to await analysis of the integrin dimer in its membrane environment when this becomes technically feasible.

The seven-fold repeats seen in the N-terminal extracellular part of all the α chains of integrins have been predicted to take up a radial arrangement, giving a 'β-propellar fold'[72]. Divalent cation-binding domains, containing the consensus sequence DXD/NXDGXXD (which is similar to the EF-hand loop structure of calmodulin) are found within the C-proximal three to four repeats (numbers IV, V, VI and VII); different binding locations are predicted for calcium and magnesium ions, the latter being positioned such that it can co-ordinate with a donated negative charge (Glu or Asp) from a bound protein ligand[72-74].

The crystal structures of several 'I' domains have now been solved (αL, αM, α1, α2; see individual entries) with different protein conformations observed in the presence of co-crystallized divalent cations. The I domain also contains a divalent cation binding site (metal ion-dependent adhesion site or 'MIDAS'), which has homology to the dinucleotide-binding (Rossmann) fold first identified in lactic dehydrogenase. It consists of a conserved DxSxS motif together with distally placed Thr and Asp residues; this leaves one metal coordination site free and this is proposed to be involved in ligand binding, possibly by accepting the acidic LDV sequence ligand motif[75].

Finally, an analysis of the β chain has been performed using the conserved DxSxS motif seen in the α chain as a probe sequence. An 'I domain-like' structure has been identified at its N-terminus and this is suggested to serve similar cation- and ligand-binding functions[76,77].

Alternative forms of integrins – RNA splice variants

An increasing number of RNA splice variants of both integrin subunits are being identified[55]; these increase further the structural diversity of integrins and thus their functional complexity.

Alternate extracellular domains have been identified for two β_1 integrins, α_6 and α_7, and α_{IIb}. α_6 and α_7 isoforms are created by mutually exclusive utilization of alternative exons between the N-terminal repeats III and IV, whereas variation in the light chain of α_{IIb} is due to variable exon splicing. A truncated, soluble 60 kDa form of β_3 with a unique C-terminus is due to use of a cryptic donor splice site.

Alternate forms of cytoplasmic tails also exist for both α and β integrin subunits. Variable C-termini have been described for α_3, α_6 and α_7, and there is good evidence for variation in tissue distribution and regulation in embryonic development for the α_6 and α_7 isoforms (see entries). Likewise, several forms of β_1 (five isoforms), β_3 and β_5 have been identified. Again there is clear evidence for functional importance in such variation; only the β_{1A} subunit is localized in focal adhesion contacts, and the β_3C variant has altered adhesion properties (see entries).

Integrin ligand specificity

Physicochemical analysis of integrins in conjunction with cross-linking studies with radioactively labelled RGD and other peptide probes has revealed that the ligand binding site of the functional integrin heterodimer resides in its globular head and includes the cation-binding region of the α subunit and the N-terminal portion of the β subunit[78]. Other studies have shown that the I domain of the α chains of β_2 integrins[58], and its homologous region in the N-terminal part of the β chain, also are invloved in interaction with ligands. Thus, ligand binding by integrins, and hence their specificity, is likely to depend on the particular $\alpha\beta$ subunit combination utilized. However, the influence of receptor activation via conformational changes following signalling cannot be underestimated, for example, β_2 and β_3 integrins generally require activation before they can recognize ligand, whereas β_1 receptors are usually constitutively activated[66,69-71].

In the main, integrins act as cell surface receptors for ECM proteins; the range of ligands recognized by individual integrin dimers are given in the individual molecule entries. Some integrins are promiscuous and can bind a number of ligands (e.g. $\alpha_V\beta_3$ can recognize vitronectin, fibrinogen, fibronectin, denatured collagen and other proteins), whereas others show a more restricted binding pattern (e.g. $\alpha_5\beta_1$ and $\alpha_2\beta_1$). Conversely, several ECM proteins are recognized by a number of different integrin receptors. Thus, laminin is bound by $\alpha_1\beta_1$, $\alpha_2\beta_1$, $\alpha_3\beta_1$, $\alpha_6\beta_1$, $\alpha_7\beta_1$, $\alpha_V\beta_3$ and $\alpha_6\beta_4$. Added complexity is produced by different integrins recognizing different regions of the same molecule (e.g. $\alpha_V\beta_3$ and $\alpha_{IIb}\beta_3$ bind distinct sites on fibrinogen[79]) and splice variants (e.g. β_3, α_3 and α_6) show differing ligand affinities. Integrins can also serve as cell:cell adhesion molecules by recognizing counter-receptors on other cells, for example, $\alpha_4\beta_1$ for VCAM-1, the β_2 integrins with ICAMs and E-cadherin for some β_1 integrins. The recognition of microbial coat proteins by the β_2 subfamily probably reflects ancestral functions of the integrin genes together with molecular mimicry and selection by prokaryotic pathogens.

Important insights into the interaction between integrins and their ligands will come from an understanding of the three-dimensional structure of recognition sequences in ligands, in addition to knowledge of binding pockets of the integrins themselves (as discussed above). Thus, the structure of RGD sequence containing fibronectin type III repeats (FNIII) from fibronectin and tenascin have been solved. The RGD tripeptide is presented at the end of a flexible loop[80-82], whereas the binding surface for the IgSF ligand ICAM-2[83-85] is flat, lacking the protruding loop seen with VCAM-1.

Integrin-associated proteins

Integrins are associated with intracellular and cell surface molecules, 'integrin-associated proteins'[86,87]. Early observations were that the platelet integrin $\alpha_{IIb}\beta_3$ interacted with CD9, a tetraspan family member (TM4SF)[88], and CD47 with β_3 integrins[89].

The cytoplasmic tails of integrins have been shown to interact directly both with general intracellular signalling proteins (e.g. PI3 kinase) and also with those exhibiting specific integrin-linked functions – integrin-linked kinase (ILK), focal adhesion kinase (FAK), β_3 endonexin and ICAP-1[59-64,90].

There is also an ever-expanding list of membrane proteins that have been shown to associate directly with integrins at the cell surface[86,87]. These modify the activation status of integrins, and hence their adhesive functions, and also participate in integrin-mediated signal transduction by activating a different set of signalling molecules (e.g. activation of PI4 kinase via TM4SF proteins) to those associated with cytoskeletal proteins (e.g. FAK and MAP kinases in association with the actin cytoskeleton).

These associations include: β_1 integrins (α_3, α_4, α_6), $\alpha_{IIb}\beta_3$ and $\alpha_4\beta_7$ interacting with members of the TM4SF (including CD9, CD53, CD63, CD81, CD151); $\alpha_V\beta_3$ with growth factor receptors; and CD87/urokinase type plasminogen activator receptor (uPAR) with β_2 and some β_1 integrins (details of the cited CD molecules not found in this *FactsBook* can be found in *The Leucocyte Antigen FactsBook*[91]). Finally, there is also evidence supporting the functional interaction between

CD147 (see individual entry for details) and some β_1 integrins (α_3 and α_6)[92], between $\alpha_{IIb}\beta_3$ and CD36 (see entry)[93], and between CD98 (see entry) and β_1 integrins[94].

Some integrins interact with IgSF proteins. This has been best characterized between β_3 integrins and CD47, an ubiquitous hybrid protein predicted to have five transmembrane domains and a single extracellular Ig domain[89]. It modifies $\alpha_V\beta_3$ integrin adhesive function in some cell-types and directly transmits signals via G proteins; recent evidence suggests that it may also bind thrombospondin directly[95].

Integrin peptide recognition motifs and the clinical use of integrin antagonists

The first integrin binding site defined was the Arg-Gly-Asp (RGD) sequence[96], identified as the minimal binding site in fibronectin capable of supporting cell adhesion[97] (summarized in Table 1). Several hundred RGD sequences exist in protein and DNA databases; the majority of extracellular matrix proteins isolated to date seem to contain RGD or homologous sequences, although only a minority have been shown to be biologically active and many are so only after 'denaturation'. Progressive truncation and mutation of parent matrix molecules, and adhesion inhibition and competition studies with various synthetic peptides, have confirmed that many integrins recognize the RGD motif in their ligands. Further well-characterized motifs include the [HHLGGA]KQAGDV from fibrinogen (for $\alpha_{IIb}\beta_3$) and the LDV sequence from fibronectin (for $\alpha_4\beta_1$) (Table 1).

Identification of these peptide sequences has formed the basis for the development of adhesion-based therapeutics for use in a wide range of clinical indications and some of these are mentioned in the individual molecule entries. These include modification of the function of $\alpha_V\beta_3$, $\alpha_{IIb}\beta_3$ and $\alpha_4\beta_1$ in, for example, osteoporosis, vascular disorders, cancer, transplant rejection, autoimmune disorders and infection, by the use of modified peptides, non-peptidic chemicals and function-blocking antireceptor antibodies[98–104,104a,104b,104c].

Table 1. *Some well-characterized peptide recognition sequences identified for integrins*

RGD	KQAGDV	LDV, REDV	KRLDGS	DGEA, RKK
$\alpha_5\beta_1$	$\alpha_5\beta_1$			
$\alpha_V\beta_1$				
$\alpha_V\beta_3$				
$\alpha_V\beta_5$				
$\alpha_V\beta_6$				
$\alpha_{IIb}\beta_3$	$\alpha_{IIb}\beta_3$			
$\alpha_2\beta_1$				$\alpha_2\beta_1$
$\alpha_3\beta_1$				
$\alpha_4\beta_1$		$\alpha_4\beta_1$		
	$\alpha_M\beta_2$		$\alpha_M\beta_2$	

SELECTINS

Introduction

The three selectins, E-selectin (CD62E), L-selectin (CD62L) and P-selectin (CD62P), are related both structurally and functionally. They have been extensively studied due to their role in the inflammatory response and their potential use as therapeutic targets. These three molecules have been the subject of recent reviews[105–108], which should be consulted for greater detail.

Structure

All three selectins are type I transmembrane proteins with an N-terminal C-type lectin domain followed by an EGF repeat and a variable number of complement control protein (CCP) domains. The human E-selectin, L-selectin and P-selectin have six, two and nine CCP domains, respectively. However, the number of CCP domains in E- and P-selectin varies from four to six and from six to nine, respectively, in other species.

The C-type lectin domain is so called because it requires Ca^{2+} to bind carbohydrate[107,109] and it is this domain in the selectins which mediates ligand binding. Within the C-type lectin domain, all three selectins contain four correctly spaced cysteine residues required for nested disulphide bond formation, key residues for Ca^{2+} binding and other conserved residues necessary for packing of the hydrophobic core[109,110]. Structural analysis of this domain from L-selectin, together with the structural analysis of other C-type lectin domains engineered to introduce selectin–ligand interactions, has identified residues involved in cation coordination and oligosaccharide binding[106,111]. EGF domains[112] are found in a wide variety of proteins and are characterized by the presence of two antiparallel β strands with connecting loops. X-ray crystallography has shown that the EGF domain in E-selectin has a similar structure to EGF domains found in other molecules[111]. The precise role of this domain in the functioning of selectins is not known but certainly it is necessary as its deletion results in loss of E-selectin-mediated adhesion[113]. Most probably, the EGF domain is required as part of the selectin binding site by interacting with protein components on the ligand and/or for the correct spacing or orientation of the C-type lectin domain. CCP domains are so named because they are commonly found in proteins that control the complement cascade. By NMR analysis of the complement control protein Factor H, CCP domains have been shown to consist of two segments of antiparallel β sheet and a short triple-stranded β sheet[114]. The function of the selectin CCP domains is not known but it has been speculated that they may optimize the interaction with ligand by extending the C-type lectin/EGF domains away from the cell membrane and increasing the flexibility of the molecule.

Ligands and regulation of ligand binding

Selectins, as their names indicate, bind carbohydrates with all three selectins binding to the tetrasaccharide sialyl-Lewis^x (sLe^x) and its stereoisomer, sLe^A, in which the positions of the linkages of fucose and galactose are exchanged. In addition, both E- and P-selectin can bind sulpho-Le^X and sulpho-Le^A and L-selectin binds with high efficiency to sulphated sLe^X epitopes in which additional sulphate residues are attached to the galactose and *N*-acetylglucosamine residues (Fig. 4). In all cases these ligands contain fucosylated core trisaccharides with a negative charge due to the addition of a sialic acid or sulphate residue(s), and are generally found at the termini of O-linked sugars (Fig. 4[107,108]). To date the main identified physiological selectin ligands are as follows: E-selectin, which is predominantly expressed on endothelial cells can bind ESL-1, PSGL-1, and L-selectin on leucocytes. P-selectin expressed on endothelial cells is more restricted in its interaction, showing a preference for leucocyte PSGL-1 as a ligand. Leucocyte L-selectin in turn recognizes three main endothelial cell ligands, GlyCAM-1, CD34 and MAdCAM-1 and leucocyte PSGL-1 (see individual entries for details). It should be noted that although these represent the best-characterized selectin ligands, a number of other glycoprotein and glycolipid ligands have been identified. The identification of selectin ligands raises the issue as to how binding selectivity is achieved. For example, PSGL-1, which by a variety of criteria represents the major leucocyte P-selectin ligand *in vivo*, carries less than 1% of the cell surface sLe^x. It is now clear that the molecules on which the carbohydrate is presented, other modifications to the carbohydrate ligand and to the core protein, and the cell type in which the ligand is expressed all play an important role in determining selectin–ligand interactions. The biochemical nature of the selectin ligands has been extensively reviewed elsewhere[108] and will not be further discussed here.

Function

Both *in vivo* and *in vitro* studies have clearly established that selectins play an important role in mediating the rolling of leucocytes on endothelium and that this is a necessary component of leucocyte extravasation during recirculation and entry into inflamed tissues[115–118]. Leucocyte extravasation requires multiple molecular interactions and it is the function of the selectins to provide their initial tethering to the endothelial wall. As the leucocytes are under flow, this tethering causes them to roll along the endothelium, which allows them to sense their local environment. This in turn leads to activation of the leucocyte integrins, which then mediate firm adhesion and subsequent extravasation. Previously it was thought that this integrin activation was mediated by chemokines released from the endothelium; however, there is increasing evidence that the selectins and their ligands are an additional important component in this event. In the case of the leucocyte selectin, L-selectin, its interaction with GlyCAM-1 has been demonstrated to result in the activation of the β_2 leucocyte integrin[119]. The mechanism by which this is achieved is less clear. Although the cytoplasmic domain of L-selectin is very short, this domain is required for leucocyte rolling[120]. Ligation of L-selectin stimulates a signalling cascade via the non-receptor tyrosine kinase p56^{lck}, which results in the activation of mitogen-activated protein kinase (MAPK[121,122]). It has yet to be determined whether this signalling pathway is directly responsible for β_2 integrin activation and how L-selectin might interact with p56^{lck}. In the case of P-selectin and its ligand PSGL-1, it is PSGL-1 on leucocytes which can activate the β_2 integrins. Again the mechanism by which this occurs is not clear. In mouse neutrophils there is evidence that this process is direct and does not require co-stimulatory components[123], but, in human neutrophils, ligation of PSGL-1 alone is not sufficient for integrin activation. The selectin-mediated signal transduction pathways have been reviewed recently[106] and will not be discussed further here.

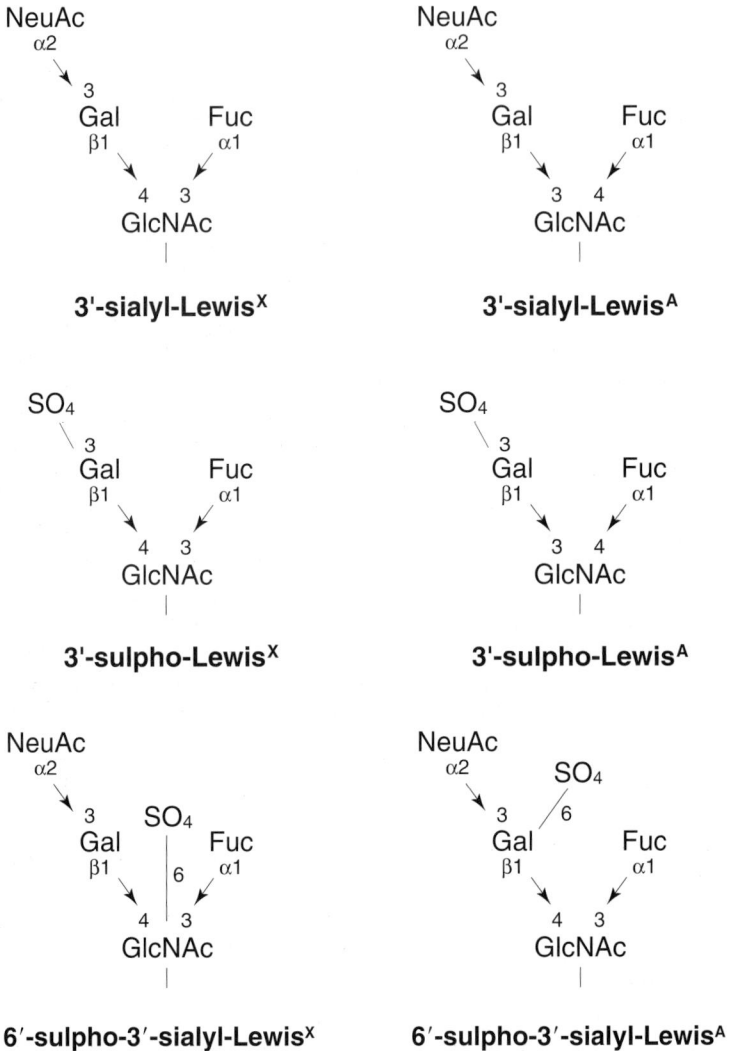

Figure 4 *Oligosaccharide ligands for the selectins. Structure of the core trisaccharide ligands for the selectins (Fuc, fucose; Gal, galactose; GlcNAc, N-acetylglucosamine; NeuAc, N-acetylneuraminic acid).*

SYNDECANS

Introduction

The syndecans are a group of heparan sulphate proteoglycans that participate in a range of cellular events including cellular signalling, cell:matrix and cell:cell adhesion.

Structure and ligand binding

Syndecans are all type I transmembrane molecules with the characteristic feature that all four family members are modified by the addition of heparan sulphate glycosaminoglycan (GAG) chains[124]. This modification occurs in the Golgi apparatus when a linkage tetrasaccharide is added to a serine residue which is N-terminal to a glycine (Fig. 5). The majority of syndecan Ser-Gly sequences that are heparan sulphate modified are surrounded by regions that contain acidic residues and are devoid of basic amino acids. In addition, many of the sites reside in Ser-Gly repeats[125]. An N-acetylglucosamine (GlcNAc) is then added to the tetrasaccharide and heparan sulphate chain elongation proceeds by the sequential addition of alternating glucuronic acid and GlcNAc sugars (Fig. 5). Importantly, some of the sugars are then further modified. First individual GlcNAc residues are N-deactylated and N-sulphated, then adjacent glucuronic acid residues are epimerized to form iduronic acid and finally O-sulphation occurs at various sites. These three reactions are carried out sequentially and result in blocks of modified sugars within the GAG chain that are required for high-affinity binding of ligands (see below). In addition to heparan sulphate, syndecan-1 also contains chondroitin sulphate chains, which are added to the membrane proximal GAG acceptor sites[126]. Although syndecan-4 can be modified by chondroitin sulphate, the attachment sites are the same as those for heparan sulphate and, in the majority of cells, heparan sulphate is the dominant GAG[127]. At least for syndecan-1, the extent and type of GAG modification is dependent on cell type[128] and this variability almost certainly has functional consequences. It should be noted that syndecans are not the only GAG-modified

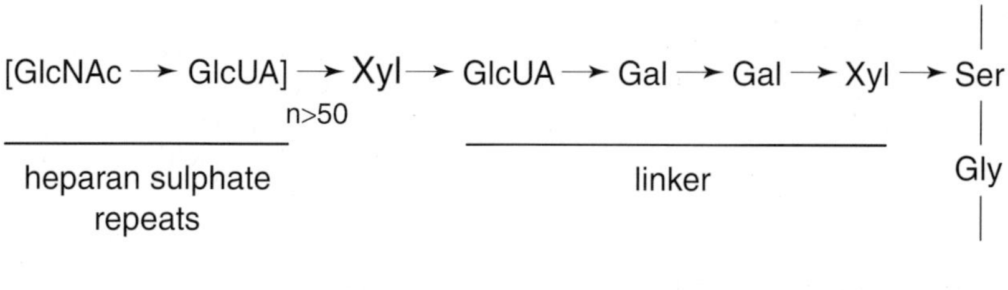

Figure 5 *Heparan sulphate modification of syndecans. Linkage of heparan sulphate to core syndecan protein occurs via a tetrasaccharide linker, GlcUA-Gal-Gal-Xyl. In the Golgi apparatus, the heparan sulphate sugars are further modified as described in the text. (Gal, galactose; GlcNAc, N-acetylglucosamine; GlcUA, glucuronic acid; Xyl, xylose).*

adhesion molecules, and the modification of CD44 and dystroglycan (see individual entries) by heparan and chondroitin sulphate occurs in a similar manner as described here.

Syndecans have been reported to bind a wide variety of different ligands. As these have recently been extensively reviewed[124], they will only be briefly summarized here. Key to the functioning of syndecans is their ability to bind extracellular ligands via their heparan sulphate chains and to date a very large number of heparan sulphate-binding proteins have been identified. These include polypeptide growth factors, enzymes, extracellular matrix proteins, cell adhesion molecules, lipid-binding proteins, blood coagulation factors, viruses and toxins. Given the array of potential ligands, it is important to consider how binding specificity might be generated. Clearly, there are cell-type-specific modifications of the heparan sulphate chains in terms of the extent and positioning of the O-sulphation. Thus, the number and type of GAG chains decorating the syndecans can vary. These differences can modulate both ligand specificity and binding efficiency.

Apart from the sites required for heparan sulphate addition, the extracellular domains of the syndecans show little sequence similarity. This is in contrast to their transmembrane and cytoplasmic domains, which are highly conserved both between family members and between syndecans isolated from mammals, chicken, *Xenopus*, *Drosophila* and *C. elegans*[129-131]. Examination of the cytoplasmic domains show two regions of exact conservation, termed conserved regions 1 and 2 (C1 and C2) separated by a variable (V) region.

Signal transduction

Given the high conservation in the cytoplasmic domains of all syndecans, it is unsurprising that they have a common function in mediating the association of syndecans with the cytoskeleton and downstream signal transduction pathways[124,130-132]. The sequence of the C-terminal C2 domain of syndecans predicts that is will act as a binding site for proteins with type II PDZ domains. Recently, association of two such proteins, syntenin and CASK, has been demonstrated[133-135]. CASK additionally contains a binding site for band 4.1 proteins[134], which include the ERM (ezrin, radixin, moesin) subfamily and these proteins are known to act as membrane:cytoskeletal linkers by interacting between transmembrane proteins and the actin cytoskeleton. Consequently, the association of such PDZ-containing proteins with the syndecans potentially provides a mechanism for linkage to the cytoskeleton, the recruitment of signalling components and, given the demonstrated interaction of ERM proteins with CD44 and ICAM-2 (see individual entries), the clustering with other adhesion molecules.

Less information is available on the role of the conserved membrane proximal C1 domain, although peptides from this region can compete with the ability of syndecan-3 to associate with both signalling and structural components, such as Src, Fyn, cortactin and tubulin[136]. Unlike the other family members, the V region in the syndecan-4 cytoplasmic domain can bind activated protein kinase Cα (PKCα) and PIP_2[137,138]. By definition, the V domain in syndecan-4 is not conserved in other family members. It will therefore be of interest to determine whether the other syndecans bind unique cytoplasmic components to mediate individual functions.

Function

As described above, the syndecan ligands include a wide array of heparan sulphate-binding proteins. The role of syndecans in binding to growth factors has been extensively studied and it is generally accepted the syndecans themselves do not act as growth factor receptors. Rather they are regarded as co-receptors and thus may play a variety of overlapping roles. First, they may sequester growth factors at the cell surface, thereby increasing their local concentration or bringing them into closer proximity to their signalling receptors. Second, binding to heparan sulphate can change the conformation of the growth factors which increases their affinity for their receptors. Finally, this interaction can promote ligand oligomerization which enhances the activation of the receptors[124]. Of particular interest with respect to this *FactsBook* is the role of syndecans in mediating cell adhesion. Cell:matrix adhesion is primarily mediated by the binding of integrins to extracellular matrix (ECM) components. However, many of these ECM molecules additionally bind heparan sulphate proteoglycans. In the case of fibronectin, it has been demonstrated that both integrin and heparan sulphate binding is required for cell adhesion[139] and cells which cannot synthesize heparan sulphate cannot form focal adhesions[140]. Owing to its widespread distribution and localization to focal adhesions, it has been proposed that syndecan-4 may play a role in the formation of these adhesive complexes[140a], possibly via its ability to recruit protein kinase C. In addition there is increasing evidence that other syndecan family members may have cell-type-specific roles both in cell:matrix and cell:cell adhesion[124,130–132]. In conclusion, syndecans appear to function as co-receptors both in growth factor signalling and in cell adhesion. It will be of interest to determine the molecular mechanisms that regulate these two functions and how interactions between these dual roles may act to co-ordinate cellular responses.

OTHER MOLECULES

Introduction

The 'other molecules' section contains a heterogeneous group of molecules that cannot be grouped solely by structural homology or by functional characteristics. The common factor connecting them is that there is good evidence for involvement in cell adhesion. Thus, we have included three enzymes with additional adhesive functions (AMOG, CD39, VAP-1), several ligands for the selectins not included elsewhere (ESL-1, GlyCAM-1, HSA, PSGL-1), two hyaluronan receptors (CD44 and LYVE-1), two scavenger receptors (CD6, CD36), four leucine-rich repeat domain containing proteins of the CD42 family, two lectins (CD23, galectin-3), six mucins (CD34, CD43, CD164, GlyCAM-1, PCLP1, PSGL-1) and three unclassified molecules (CD57, CD98, E48). We have also included entries for integral membrane proteins (claudins, occludin) associated with epithelial and endothelial 'tight junctions', and dystroglycan, a protein which was orignally found at neuromuscular junctions but which appears to have much broader adhesive functions. Clearly the 'other molecules' formed the most difficult section to compose in that the only selection criterion used was evidence for an adhesive function. Consequently, this is the part of the *FactsBook* in which errors of omission are most likely to have occurred; for this, we apologise.

Detailed discussion on these molecules is included in their individual entries. However, a brief review is given below on those adhesion molecules which belong to a larger family group of homologous proteins and where the majority of family members have no adhesive function.

Leucine-rich repeat (LRR) molecules

The LRR proteins form a diverse family that includes extracellular matrix, cytoplasmic and membrane molecules involved in signal transduction and protein–protein interactions in development, in addition to cell adhesion. They are defined by the presence of one or more 22–28 amino acid long 'flank–centre (leucine-rich region)–flank' sequence motifs (LRRs)[141], which were originally identified in leucine rich α2-glyprotein. Thus the LRR family includes: matrix glycoproteins and proteoglycans with structural roles (e.g. decorin, biglycan, fibromodulin)[142]; adhesive platelet glycoproteins of the CD42 family (first identified in the sequences of platelet gpIb and gpIX that bind von Willebrand factor)[143, 144]; hybrid membrane molecules, such as the neuronal tyrosine kinase receptor (trk)[145], mucins (e.g. densin-180)[146], and neural IgSF members LIG-1[147] and NLRR[148]; and a series of *Drosophila melanogaster* proteins involved in nervous system development – connectin[149], chaoptin[150] and TOLL[151].

CD36 family

The CD36 family of proteins consist of molecules that are class B1 scavenger receptors that bind native and modified lipoproteins[152] and are involved in the recognition and phagacytosis of apoptotic cells[153, 154]. As such, these macrophage receptors probably play an important role in, for example, the pathogenesis of atherosclerosis[155]. The gene family[156,157] encompasses CD36, which has additional adhesive functions as: a collagen/thrombospondin receptor (see entry); the human scavenger receptor, CLA-1 (CD36 and LIMP II Analogous-1)[158]; LIMP II, a lysosomal integral membrane protein of ubiticus distribution; and Croquemort, a *Drosophila melanogaster* receptor involved in macrophage recognition of apoptotic cells during development[159, 160].

References
Cadherins
1 Humphries, M. J. and Newham, P. (1998) Trends. Cell Biol. 8, 78–83.
2 Chothia, C. and Jones, E.Y. (1997) Annu. Rev. Biochem. 66, 823–862.
3 Tamura, K. et al. (1998) Neuron 20, 1153–1163.
3a Koch, A.W. et al. (1999) Curr. Op. Struct. Biol. 9, 275–281.
4 Aberle, H. et al. (1996) J. Cell. Biochem. 61, 514–523.
5 Yap, A.S. et al. (1997) Ann. Rev. Cell Dev. Biol. 13, 119–146.
6 Ben-Ze'ev, A. and Geiger, B. (1998) Curr. Opin. Cell Biol. 10, 629–639.
7 Kools, P. et al. (1999) FEBS Lett. 452, 328–334.

[8] Buxton, R.S. et al. (1992) J. Cell Biol. 121, 481–483.

[9] Garrod, D. et al. (1996) Curr. Opin. Cell Biol. 8, 670–678.

[10] Koch, P.J. and Franke, W.W. (1994) Curr. Opin. Cell Biol. 6, 682–687.

[11] Kowalczyk, A.P. et al. (1999) Int. Rev. Cytol. 185, 237–302.

[12] Sano, K. et al. (1993) EMBO J. 12, 2249–2256.

[13] Bradley, R.S. et al. (1998) Curr. Biol. 8, 325–334.

[14] Kohmura, N. et al. (1998) Neuron 20, 1137–1151.

[15] Wu, Q. and Maniatis, T. (1999) Cell 87, 779–790.

[16] Shapiro, L. and Colman, D.R. (1999) Neuron 23, 427–430.

[17] Strehl, S. et al. (1998) Genomics 153, 81–89.

[17a] Wu, Q. and Maniatis, T. (2000) Proc. Natl. Acad. Sci. USA 97, 3124–3129.

[18] Williams, E.J. et al. (1994) Neuron 13, 583–594.

[19] Saffell, J.L. et al. (1997) Neuron 18, 231–242.

[20] Suzuki, S.T. (1996) J. Cell. Biochem. 61, 531–542.

[21] Cepek, K.L. et al. (1994) Nature 372, 190–193.

[22] Karecla, P.I. et al. (1996) J. Biol. Chem. 271, 30909–30915.

[23] Obata, S. et al. (1995) J. Cell Sci. 108, 3765–3773.

[24] Shapiro, L. and Colman, D.R. (1998) Curr. Opin. Neurobiol. 8, 593–599.

[25] Gumbiner, B.M. (1995) Curr. Opin. Cell Biol. 7, 634–640.

[26] Huber, O. et al. (1996) Curr. Opin. Cell Biol. 8, 685–691.

[27] Takeichi, M. (1995) Curr. Opin. Cell Biol. 7, 619–627.

Immunoglobulin superfamily

[28] Barclay, A.N. et al. (1997) The Leucocyte Antigen FactsBook. Academic Press, London.

[29] Williams, A.F and Barclay, A.N. (1988) Annu. Rev. Immunol. 6, 381–405.

[30] Harpaz, Y. and Chothia, C. (1994) J. Mol. Biol. 238, 528–539.

[31] Chothia, C. and Jones, E.Y. (1997) Annu. Rev. Biochem. 66, 823–862.

[32] Shapiro, L. et al. (1995) Proc. Natl Acad. Sci. USA 92, 6793–6797.

[33] Leahy, D.L. et al. (1992) Science 258, 987–991.

[34] Main, A.L. et al. (1992) Cell 71, 671–678.

[35] Leahy, D.L. (1997) Annu. Rev. Cell Dev. Biol. 13, 363–393.

[36] Humphries, M. J. and Newham, P. (1998) Trends. Cell Biol. 8, 78–83.

[37] Crocker, P.R. et al. (1998) Glycobiology 8, v–vi.

[38] Crocker, P.R. et al. (1998) Blood 92, 2123–2132.

[39] Patel, N. et al. (1999) J. Biol. Chem. 274, 22729–22738.

[40] Nicoll, G. et al. (1999) J. Biol. Chem. 274, 34089–34095.

[40b] Floyd, H. et. al. (2000) J. Biol. Chem. 275, 861–866.

[41] Brummendorf, T. et al. (1998) Curr. Opin. Neurobiol. 8, 87–97.

[42] Cook, G. et al. (1998) Curr. Opin. Neurobiol. 8, 64–72.

[43] Buchstaller, A. et al. (1996) J. Cell Biol. 135, 1593–1607.

[44] Sakurai, T. et. al. (1997) J. Cell Biol. 136, 907–918.

[45] Walsh, F.S. and Doherty, P. (1997) Annu. Rev. Cell Dev. Biol. 13, 425–456.

Integrins

[46] Hynes, R.O. (1992) Cell 69, 11–25.

[47] Sastry, S.K. and Horwitz, A.F. (1993) Curr. Opin. Cell Biol. 5, 819–831.

[48] LaFlamme, S.E. et al. (1997) 16, 153–163.

[49] van Kooyk, Y. et al. (1998) Cell Adhes. Commun. 6, 247–254.

[50] Miyamoto, S. et al. (1995) J. Cell Biol. 131, 791–805.

[51] Burridge, K. et al. (1997) Trends Cell Biol. 7, 342–347.

[52] Burridge, K. and Chrzanowska-Wodnicka, M. (1996) Annu. Rev. Cell Dev. Biol. 12, 463–519.

[53] Brown, N.H. (1993) Bioessays 15, 383–390.

[54] Burke, R.D. (1999) Int. Rev. Cytol. 191, 257–284.

[54a] Lehnert, K. et al. (1999) Genomics 60, 179–187.

[54b] Velling, T. et al. (1999) J. Biol. Chem. 274, 25735–25742.

[55] de Melker, A.A. and Sonnenberg, A. (1999) Bioessays 21, 499–509.

[56] Carrell, N.A. et al. (1985) J. Biol. Chem. 260, 1743–1749.

[57] Collombatti, A. and Bonaldo, P. (1991) Blood 71, 2305–2325.

[58] Dickeson, S.K. and Santoro, S.A. (1998) Cell Mol. Life. Sci. 54, 556–566.

[59] Clark, E.A and Brugge, J.S. (1995) Science 268, 233–239.

[60] Ilic, D. et al. (1997) J. Cell Sci. 110, 401–407.

[61] Schlaepfer, D.D. and Hunter, T. (1998) Trends Cell Biol. 8, 151–157.

[62] Dedhar, S. et al. (1999) Trends Cell Biol. 9, 319–323.

[63] Aplin, A.E. et al. (1998) Pharmacol. Rev. 50, 197–263.

[64] Schlaepfer, D.D. et al. (1999) Prog. Biophys. Mol. Biol. 71, 435–478.

[65] Humphries, M.J. (1999) Biochem. Soc. Symp. 65, 63–78.

[66] Humphries, M.J. and Newham, P. (1998) Trends Cell Biol. 8, 78–83.

[67] Leahy, D.J. (1997) Annu. Rev. Cell Dev. Biol. 13, 363–393.

[68] Chothia, C. and Jones, E.Y. (1997) Annu. Rev. Biochem. 66, 823–862.

[69] Stewart, M. and Hogg, N. (1996) J. Cell Biochem. 61, 554–561.

[70] Diamond, M.S. and Springer, T.A. (1994) Curr. Biol. 4, 506–517.

[71] Hughes, P.E. and Pfaff, M. (1998) Trends Cell Biol. 8, 359–364.

[72] Springer, T.A. (1997) Proc. Natl Acad. Sci. 94, 65–72.

[73] Tuckwell, D.S. et al. (1994) Cell Adhes. Commun. 2, 385–402.

[74] Oxvig, C. et al. (1999) Proc. Natl Acad. Sci. USA 96, 2215–2220.

[75] Qu, A. and Leahy, D.J. (1996) Structure 4, 931–942.

[76] Tuckwell, D.S. and Humphries, M.J. (1997) FEBS Lett. 400, 297–303.

[76] Leahy, D.L. et al. (1992) Science 258, 987–991.

[77] Tozer, E.C. et al. (1996) J. Biol. Chem. 271, 21978–21984.

[78] Smith, J.W. and Cheresh, D.A. (1990) J. Biol. Chem. 265, 2168–2172.

[79] Hu, D.D. et al. (1999) J. Biol. Chem. 274, 4633–4639.

[81] Main, A.L. et al. (1992) Cell 71, 671–678.

[82] Aota, S. and Yamada, K.M. (1995) Adv. Enzymol. Relat. Areas Mol. Biol. 70, 1–21.

[83] Casasnovas, J.M. et al. (1997) Nature 387, 312–315.

[84] Wang, J. and Springer, T.A. (1998) Immunol. Rev. 163, 197–215.

[85] Shimizu, Y. et al. (1999) Adv. Immunol. 72, 325–380.

[86] Hemler, M.E. (1998) Curr. Biol. 10, 578–585.

[87] Porter, J.C. and Hogg, N. (1998) Trends Cell Biol. 8, 390–396.

[88] Slupsky, J.R. et al. (1989) J. Biol. Chem. 264, 12289–12293.

[89] Linberg, F.P. et al. (1993) J. Cell Biol. 123, 485–496.

[90] Danen, E.H. et al. (1998) Cell Adhes. Commun. 6, 217–224.

[91] Barclay, A.N. et al. (1997) The Leucocyte Antigen Factsbook, 2nd edn, Academic Press, London.

[92] Berditchevski, F. et al. (1997) J. Biol. Chem. 272, 29174–29180.

[93] Dorahy, D.J. et al. (1996) Biochem. Biophys. Res. Commun. 218, 575–581.

[94] Fenczik, C.A. et al. (1997) Nature 390, 81–85.

[95] Chung, J. et al. (1997) J. Biol. Chem. 272, 14740–14746.

[96] Ruoslahti, E. (1996) Ann. Rev. Cell Dev. Biol. 12, 697–715.

[97] Pierschbacher, M.D., Ruoslahti, E. (1984) Nature 309, 30–33.

[98] Horton, M.A. (1996a) Adhesion Receptors as Therapeutic Targets. CRC Press, Boca Raton, FL.

[99] Horton, M.A. (1996b) Moecular Biology of Cell Adhesion Molecules. John Wiley & Sons, Chichester.

[100] Moran, N. (1998) Br. J. Cardiol. 5, 413–423.

[101] Theroux, P. (1998) Am. Heart J. 135 (Suppl.), 107–112.

[102] Tcheng, J.E. (1996) Am. J. Cardiol. 78, 35–40.

[103] Collier, B.S. (1997) Thromb. Haemost. 78, 730–735.

[104] Lin, K.C. and Castro, A.C. (1998) Curr. Opin. Chem. Biol. 2, 453–457.

[104a] Ferguson, J.J. and Zaqqa, M. (1999) Drugs 58, 965–982.

[104b] Wang, W. et al. (2000) Curr. Med. Chem. 7, 437–453.

[104c] Jackson, D.Y. et al. (1997) J. Med. Chem. 40, 3359–3368.

Selectins

[105] Frenette, P.S. and Wagner, D.D. (1997) Thromb. Haemost. 78, 60–64.

[106] Crockett-Torabi, E. (1998) J. Leukocyte Biol. 63, 1–14.

[107] Weis, W.I. et al. (1998) Immunol. Rev. 163, 19–34.

[108] Vestweber, D. and Blanks, J.E. (1999) Physiol. Rev. 79, 181–213.

[109] Drickamer, K. (1993) Curr. Opin. Struct. Biol. 3, 393–400.

[110] Weis, W.I. et al. (1991) Science 254, 1608–1615.

[111] Graves, B.J. et al. (1994) Nature 367, 532–538.

[112] Campbell, I.D. and Bork, P. (1993) Curr. Opin. Struct. Biol. 3, 385–392.

[113] Pigott, R. et al. (1991) J. Immunol. 147, 130–135.

[114] Barlow, P.N. et al. (1993) J. Mol. Biol. 237, 268–284.

[115] Imhof, B.A. and Dunon, D. (1995) Adv. Immunol. 58, 345–416.

[116] Lasky, L.A. (1995) Ann. Rev. Biochem. 64, 113–139.

[117] McEver, R.P. et al. (1995) J. Biol. Chem. 270, 11025–11028.

[118] Tedder, T.F. et al. (1995) FASEB J. 9, 866–873.

[119] Hwang, S.T. et al. (1996) J. Exp. Med. 184, 1343–1348.

[120] Kansas, G.S. et al. (1993) J. Exp. Med. 177, 833–838.

[121] Waddell, T.K. et al. (1995) J. Biol. Chem. 270, 15403–15411.

[122] Brenner, B. et al. (1996) Proc. Natl Acad. Sci. USA 93, 15376–15381.

[123] Blanks, J.E. et al. (1998) Eur. J. Immunol. 28, 433–443.

Syndecans

[124] Carey, D.J. (1997) Biochem. J. 327, 1–16.

[125] Zhang, L. et al. (1995) J. Biol. Chem. 270, 27127–27136.

[126] Kokenyesi, R. and Bernfield, M. (1994) J. Biol. Chem. 269, 12304–12309.

[127] Shworak, N.W. et al. (1994) J. Biol. Chem. 269, 21204–21214.

[128] Sanderson, R.D. and Bernfield, M. (1988) Proc. Natl Acad. Sci. USA 85, 9562–9566.

[129] Rapraeger, A.C. and Ott, V.L. (1998) Curr. Opin. Cell Biol. 10, 620–628.

[130] Woods, A. and Couchman, J.R. (1998) Trends Cell Biol. 8, 189–192.

[131] Couchman, J.R. and Woods, A. (1996) J. Cell. Biochem. 61, 578–584.

[132] Zimmermann, P. and David, G. (1999) FASEB J. 13 (Suppl.), S91–S100.

[133] Grootjans, J.J. et al. (1997) Proc. Natl Acad. Sci. USA 94, 13683–13688.

[134] Cohen, A.R. et al. (1998) J. Cell Biol. 142, 129–138.

[135] Hsueh, Y.P. et al. (1998) J. Cell Biol. 142, 139–151.

[136] Kinnunen, T. et al. (1998) J. Biol. Chem. 273, 10702–10708.

[137] Oh, E.S. et al. (1998) J. Biol. Chem. 273, 10624–10629.

[138] Lee, D. et al. (1998) J. Biol. Chem. 273, 13022–13029.

[139] Woods, A. et al. (1993) Mol. Biol. Cell 4, 605–613.

[140] LeBaron, R.G. et al. (1988) J. Cell Biol. 106, 945–952.

[140a] Couchman, J.R. and Woods, A. (1999) J. Cell Sci. 112, 3415–3420.

Other molecules

[141] Kobe, B. and Deisenhofer, J. (1995) Nature 374, 183–186.

[142] Hocking, A.M. et al. (1998) Matrix Biol. 17, 1–19.

[143] Lopez, J.A. et al. (1987) Proc. Natl Acad. Sci. USA 84, 5615–5619.

[144] Hickey, M.J. et al. (1989) Proc. Natl Acad. Sci. USA 86, 6773–6777.

[145] Schneider, R. and Schweiger, M. (1991) Oncogene 6, 1807–1811.

[146] Apperson, M.L. et al. (1996) J. Neurosci. 16, 6839–6852.

[147] Suzuki, Y. et al. (1996) J. Biol. Chem. 271, 22522–22527.

[148] Taguchi, A. et al. (1996) Brain Res. Mol. Brain Res. 35, 31–40.

[149] Nose, A. et al. (1992) Cell 70, 553–567.

[150] Krantz, D.E. and Zipursky, S.L. (1990) EMBO J. 9, 1969–1977.

[151] Gay, N.J. et al. (1991) FEBS Lett. 291, 87–91.

[152] Calvo, D. et al. (1998) Lipid Res. 39, 777–788.

[153] Platt, N. et al. (1998) Trends Cell Biol. 9, 365–372.

[154] Yamada, Y. et al. (1998) Cell Mol. Life Sci. 7, 628–640.

[155] Hirano, K. et al. (1999) Circ. Res. 85, 108–116.

[156] Calvo, D. and Vega, M.A. (1993) J. Biol. Chem. 268, 18929–18935.

[157] Calvo, D. et al. (1995) Genomics 25, 100–106.

[158] Murao, K. et al. (1997) J. Biol. Chem. 272, 17551–17557.

[159] Franc, N.C. et al. (1999) Science 284, 1991–1994.

[160] Franc, N.C. et al. (1996) Immunity 4, 431–443.

3 Accessing information on the World Wide Web

We have made extensive use of the following World Wide Web (WWW)-based databases in the preparation of this book. For literature surveys, we utilized either PubMed (http://www.ncbi.nlm.nih.gov/PubMed/) or Medline on BioMedNet (http://www.biomednet.com/); the advantage of the former is that an increasing number of journal articles can be downloaded electronically as 'pdf files' without resource to the library. For certain molecules that have attained 'CD' status as leucocyte proteins, then an initial step was to examine the 'CD' database posted at NCBI (http://www.ncbi.nlm.nih.gov/prow/cd/index_molecule.htm). For the majority of 'CD' molecules this contains information on the individual protein and direct links to protein sequences and the Online Mendelian Inheritance of Man (OMIM) database (see later). Other information can be obtained from the human leucocyte differentiation antigen (HLDA) workshop database at: http://mol.genes.nig.ac.jp/hlda/. For our specific sequence entries, we used information obtained from the GenBank (hosted by NCBI: http://WWW.NCBI.NLM.NIH.GOV/) and SwissProt (http://www.expasy.ch/) sequence databases. Brief 'how to use' information is given later in this chapter. Knockout and mutation analysis in mice is becoming increasingly important in understanding the function of adhesion molecules and there are some useful databases containing this information (see Table 1). Likewise, there are a large and expanding number of web sites that can be bookmarked to cover additional information, for example on ESTs, protein motifs and glycosylation sites. We have included a list of our favourite web sites in Table 1. Their use is largely intuitive and they all have good homepage listings of contents, on-line help sections and e-mail addresses for queries. One surprise, though, is that there were very few really valuable sites specifically orientated towards adhesion molecules; those we have found are listed in Table 1.

NCBI SERVER FOR GENBANK AND OTHER DATABASES

A homepage for GenBank (http://WWW.NCBI.NLM.NIH.GOV/) enables a variety of searches to be formulated. The most commonly used is likely to be the Entrez browser, which then links to search engines for DNA (GenBank), protein and other databases. Entering the nucleotide or protein databases enables the simple keyword search function to be operated which then links to the citation(s) for the sequence requested. For 'nucleotide' sequences this will produce a GenBank report, which contains information on key references, accession numbers, the nucleotide sequence and its derived amino acid sequence with some information on structural motifs. Medline references are directly linked to those cited in the report. The report form also enables the information to be displayed in different formats and access to be made to a variety of related databases, such as Medline, GenPept (protein) reports, lists of related sequences in the database, etc. The 'protein' search reveals similar information in an identical format but only contains the amino acid sequence of the molecule.

Table 1 *Useful databases*

Cell adhesion molecule web resources

[a]**CD Molecules:**
 http://www.ncbi.nlm.nih.gov/prow/cd/index_molecule.htm
HLDA Homepage:
 http://mol.genes.nig.ac.jp/hlda/
Cell adhesion Domain:
 http://www.cell-adhesion.net/
Adhesion proteins in drosophila:
 http://flybase.bio.indiana.edu/allied-data/lk/interactive-
 fly/aimain/6transm.htm#dafka2
Adhesion proteins in neurobiology:
 http://www.neuro.wustl.edu/neuromuscular/lab/adhesion.htm
Useful bioresource search engine:
 http://www.expasy.ch/BioHunt

Gene knockout resources

[a]**Mouse Genome Informatics homepage at Jackson Laboratory:**
 http://www.informatics.jax.org/
[a]**Searchable knockout and transgenic mouse database (TBASE):**
 http://tbase.jax.org/
Mouse gene knockout listing:
 http://www.bioscience.org/knockout/alphabet.htm
BioMedNet (Mouse Knockout and Mutation Database) listing:
 http://research.bmn.com/mkmd
Internet transgenic resource links:
 http://tbase.jax.org/docs/databases.html

Quick sequence searches (GenBank and SwissProt and links)

[a]**NCBI server:**
 http://WWW.NCBI.NLM.NIH.GOV/
[a]**Swiss Institute of Bioinformatics server:**
 http://www.expasy.ch/

Human genetic diseases

[a]**OMIM (Online Inheritance in Man):**
 http://www.ncbi.nlm.nih.gov/Omim/
Human Gene Mutation Database (HGMD):
 http://www.uwcm.ac.uk/uwcm/mg/hgmd0.html

EST databases

NCBI EST search:
http://WWW.NCBI.NLM.NIH.GOV/dbEST/index.html

Human EST project:
http://genome.wustl.edu/est/esthmpg.html

EST projects at Washington University:
http://genome.wustl.edu/gsc/est/navest.pl

Molecular biology, protein modelling and other links

Databases and search engines

Resources at the Sanger Centre:
http://www.sanger.ac.uk/

The Genome Database:
http://gdbwww.gdb.org/

UK Human Genome Mapping Project resources:
http://www.hgmp.mrc.ac.uk/

European Bioinformatics Institute (database access, sequence search tools, etc.):
http://www.ebi.ac.uk/

Sequence searches at EMBL:
http://www.embl-heidelberg.de/Services/index.html

Other useful sites for databases, etc.

DNA and protein analysis tools:
http://www.rockefeller.edu/rucs/toolkit/toolkit.html

Protein and biological databases at Johns Hopkins University:
http://www.bis.med.jhmi.edu/

Molecular biology and protein links:
http://www.cbs.dtu.dk/biolink.html

Bioinformatics at the Weizmann Institute:
http://bioinformatics.weizmann.ac.il/

Harvard links:
http://mcb.harvard.edu/BioLinks.html

TIGR database:
http://www.tigr.org/tdb/

TIGR human gene index:
http://www.tigr.org/tdb/hgi/hgi.html

Nucleotide and protein analysis including structural motifs.

The Protein Data Bank (protein modelling resource):
http://www.rcsb.org/pdb/

Swiss Institute of Bioinformatics server (including 2D gel electrophoresis):
http://www.expasy.ch/

The Protein Domain Database:
http://www.toulouse.inra.fr/prodom.html

Table 1 *Continued*

Protein modelling and motifs listing:
   http://guitar.rockefeller.edu/modbase/databases.html
Nucleotide and protein structure prediction resources (Center for Biological
   Sequence Analysis):
   http://www.cbs.dtu.dk/services/
Current Biology's Macromolecular Structures Database (searchable by keywords):
   http://research.bmn.com/msd
Protein modelling at UCL:
   http://www.biochem.ucl.ac.uk/bsm/biocomp/index.html

DNA clone access

I.M.A.G.E. DNA clones: http://www-bio.llnl.gov/bbrp/image/image.html

Glycosylation analysis

[a]Sugars, proteoglycans and GAGs:
   http://bssv01.lancs.ac.uk/gig/pages/toppage.htm
O-glycosylation sites:
   http://www.cbs.dtu.dk/services/NetOGlyc/

---

[a]www sites found to be particularly useful in the preparation of this *FactsBook*.

SWISSPROT PROTEIN SEQUENCE DATABASE

Entry is made through the ExPASy molecular biology server (http://www.expasy.ch/),
which provides lists of databases and tools on its homepage. Click SwissProt and
TrEMBL under 'databases' to enter the SwissProt database. This is accessed by
clicking on SRS ('sequence retrieval system') and then the 'start button' on the
Sequence Retrieval System homepage, which then allows you to select individual
sequence databases to search. Press the 'query form' button and enter the keyword
to be searched. This will result in the downloading of individual sequence-related
information in a standardized form (NiceProt view). This contains key references
(which are directly linked to Medline abstracts), a description of the molecule and
its function, its amino acid sequence and accession number, including those for
the GenBank DNA sequence, and descriptions of the location of basic motifs (e.g.
signal sequence, transmembrane domain, glycosylation sites, cleavage sites) and
specific motifs (e.g. Ig domains, repetitive amino acid sequences, etc.) within the
molecule. Further information on domain structure and homologous sequences
can be downloaded via two interactive 'domain viewers', 'FT table viewer' and
'Domain structure' with simple online instructions.

OMIM

Enter the database (http://www.ncbi.nlm.nih.gov/Omim/) via the use of the
'search the OMIM database' prompt, and enter keywords or OMIM accession
number into the search section. The search will yield a page where individual

entries are given by their OMIM number, which can be accessed directly to yield text and references concerning the item of interest. The text also contains linked PubMed and OMIM accession numbers, which can be directly sourced from the body of the database entry. Furthermore, a variety of related links are provided to other databases as listed:

Medline	Linked references on PubMed.
Protein	Access to GenPept protein sequence database.
DNA	Link to GenBank DNA information on molecule.
HGMD	Link to Human Gene Mutation Database entry.
LOCUS	Link to chromosomal information and derived resources.
GenMap	Link to genetic locus, which can be further accessed to the MIM GenMap giving chromosomal locations and adjacent genes.
GDB	A series of linked genetic and sequence resources and their accession numbers.
Nomenclature	A searchable database to definitions of correct usage of gene and nomenclature symbols.

MOUSE KNOCKOUT DATABASES

A resource that can be utilized to access information is held by the Mouse Genome Informatics (MGI) database at the Jackson Laboratory (http://www.informatics.jax.org/). Here, entry is via the MGI homepage and entering a keyword term either in the 'quick gene search' box or the MGI number (in correct format, for example enter MGI:88354 for E-cadherin) and subsequent links to enter the relevant results page. For more refined searches enter dctails via the 'Genetic and Phenotypic Data' section. Clicking on the highlighted gene symbol in the results page enters a subfile that contains details of the mouse chromosomal location of the gene, links to accession numbers in GenBank, SwissProt and other databases. Clicking on the 'phenotype' gives details of the knockout mouse; general information on gene structure and protein function are also included. These are all linked by a unique accession number, the MGI number (quoted in the individual molecule entries under 'Knockout'). Again, unfortunately, the MGI number cannot be used as a unique identifier for a gene on the MGI homepage search engine.

Another way to access knockout information is to enter the TBASE database (Transgenic/Targeted Mutation Database) on mouse knockout phenotypes is to access the 'citation database' (http://tbase.jax.org/), enter a keyword and search. The search reveals related links to the published literature, which can then be further investigated by a series of clickable links, which finally produces the full TBASE printout containing information on the mouse knockout mice, details of the DNA constructs used, information on the host mouse strain, linked references and, of course, its phenotype. This is given in further detail via a link within the 'phenotype results' section, which can be clicked to provide full details, including unpublished data. Files are identified by a unique TBASE accession number.

CELL ADHESION MOLECULES

Cadherins

E-Cadherin

Family

Cadherin (type I subfamily)

Structure

Molecular weights

Amino acids	882
Polypeptide	97 456
SDS-PAGE reduced	120 kDa

Carbohydrate

N-linked sites	2
O-linked sites	

Gene location   16q22.1

Gene structure   ~100 kb, 16 exons

Alternative forms

Alternatively spliced isoforms have been detected in some tumours.

Structure

Like all classical type I cadherins, E-cadherin is expressed initially as a precursor with an N-terminal propeptide domain, which is cleaved off inside the cell during protein maturation. The mature form of the protein forms a homodimer with adjacent molecules at the cell surface. The extracellular domain contains five cadherin repeats of approximately 110 amino acids separated by four calcium binding sites. The N-terminal domain contains the conserved HAV motif required for homophilic binding to E-cadherin expressed on neighbouring cells[1]. The cytoplasmic domain associates with β-catenin, which in turn binds to α-catenin or plakoglobin, thereby linking E-cadherin to the actin cytoskeleton[1]. The solution structure of the N-terminal domain[2] and crystal structure of the two N-terminal domains[3] have been solved. The structure of the cadherins is discussed further in Section I.

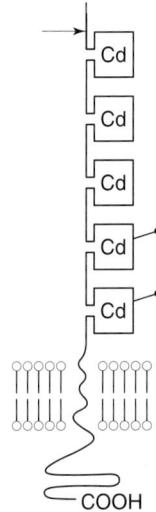

Ligands

E-cadherin mediates Ca^{2+}-dependent homotypic adhesion. In addition, E-cadherin is a ligand for the $\alpha_E\beta_7$ and $\alpha_2\beta_1$ integrins and this binding is to a site in the E-cadherin N-terminal domain which is distinct from the homotypic binding site[4,5].

Function

E-cadherin is a key mediator of epithelial cell:cell adhesion in embryonic and adult tissues. It functions both in the formation of organized cell:cell contacts[6] and in tissue morphogenesis and the maintenance of the epithelium[7]. Antibody blocking or dominant negative experiments demonstrate that loss of functional E-cadherin results in the loss of a polarized epithelial phenotype. These experiments together with *in vivo* data (see below) demonstrate that E-cadherin can act as a tumour suppressor. In addition, the binding of E-cadherin to leucocyte integrins mediates the interaction between T cells and mucosal epithelium.

Distribution

E-cadherin is widely expressed in non-neural epithelial cells, where it is localized to cell:cell contact areas and particularly to the adherens junctions. During development, E-cadherin is expressed very early on embryonic cells, is switched off in the mesoderm forming at gastrulation and is then re-expressed during mesenchymal to epithelial transition[8]. Loss of E-cadherin expression is found in many tumours (see below).

Disease association

OMIM 192090

Using *in vitro* and *in vivo* systems it has been demonstrated that loss of E-cadherin expression in tumour cells is associated with a more invasive phenotype and that re-expression can result in reversion to a more benign phenotype[9]. Consequently, E-cadherin has been implicated to function as a tumour suppressor. Using a mouse model of pancreatic β-cell carcinogenesis, Perl et al. have directly demonstrated that loss of E-cadherin has a causal role in the transition from adenoma to carcinoma[10]. Loss or reduction in E-cadherin expression in tumours can result from: (a) allelic deletion or loss or heterozygosity on chromosome 16; (b) missense or nonsense mutations; (c) alternative splicing resulting in loss of the extracellular or intracellular domains; (d) proteolysis of the extracellular domain; or (e) suppression of expression[9,11–14].

Knockout

MGI:88354
The Gene Knockout FactsBook[15], p101
E-cadherin −/− embryos develop normally to blastula stage and compact due to the presence of maternal E-cadherin. Subsequently cells do not properly polarize resulting in the absence of a blastocyst cavity and trophectodermal epithelium, and the inability to implant[16,17].

Amino acid sequence of human E-cadherin

```
  1 MGPWSRSLSA LLLLLQVSSW LCQEPEPCHP GFDAESYTFT VPRRHLERGR VLGRVNFEDC
 61 TGRQRTAYFS LDTRFKVGTD GVITVKRPLR FHNPQIHFLV YAWDSTYRKF STKVTLNTVG
121 HHHRPPPHQA SVSGIQAELL TFPNSSPGLR RQKRDWVIPP ISCPENEKGP FPKNLVQIKS
181 NKDKEGKVFY SITGQGADTP PVGVFIIERE TGWLKVTEPL DRERIATYTL FSHAVSSNGN
241 AVEDPMEILI TVTDQNDNKP EFTQEVFKGS VMEGALPGTS VMEVTATDAD DDVNTYNAAI
301 AYTILSQDPE LPDKNMFTIN RNTGVISVVT TGLDRESFPT YTLVVQAADL QGEGLSTTAT
```

```
361 AVITVTDTND NPPIFNPTTY KGQVPENEAN VVITTLKVTD ADAPNTPAWE AVYTILNDDG
421 GQFVVTTNPV NNDGILKTAK GLDFEAKQQY ILHVAVTNVV PFEVSLTTST ATVTVDVLDV
481 NEAPIFVPPE KRVEVSEDFG VGGQEITSYTA QEPDTFMEQK ITYRIWRDTA NWLEINPDTG
541 AISTRAELDR EDFEHVKNST YTALIIATDN GSPVATGTGT LLLILSDVND NAPIPEPRTI
601 FFCERNPKPQ VINIIDADLP PNTSPFTAEL THGASANWTI QYNDPTQESI ILKPKMALEV
661 GDYKINLKLM DNQNKDQVTT LEVSVCDCEG AAGVCRKAQP VEAGLQIP_AI_ _LGILGGILAL_
721 _LILILLLLF_ _L_RRRAVVKEP LLPPEDDTRD NVYYYDEEGG GEEDQDFDLS QLHRGLDARP
781 EVTRNDVAPT LMSVPRYLPR PANPDEIGNF IDENLKAADT DPTAPPYDSL LVFDYEGSGS
841 EAASLSSLNS SESDKDQDYD YLNEWGNRFK KLADMYGGGE DD
```

The propeptide sequence underlined and in italics is cleaved to form the mature protein.

Database accession

EMBL/GenBank   Z13009
SwissProt   P12830

References

[1] Yap, A.S. et al. (1997) Annu. Rev. Cell Dev. Biol. 13, 119–146.
[2] Overduin, M. et al. (1995) Science 267, 386–389.
[3] Nagar, B. et al. (1996) Nature 380, 360–364.
[4] Cepek, K.L. et al. (1994) Nature 372, 190–193.
[5] Karecla, P.I. et al. (1996) J. Biol. Chem. 271, 30909–30915.
[6] Adams, C.L. and Nelson, W.J. (1998) Curr. Opin. Cell Biol. 10, 572–577.
[7] Takeichi, M. (1995) Curr. Opin. Cell Biol. 7, 619–627.
[8] Huber, O. et al. (1996) Curr. Opin. Cell Biol. 8, 685–691.
[9] Takeichi, M. (1993) Curr. Opin. Cell Biol. 5, 806–811.
[10] Perl, A.K. et al. (1998) Nature 392, 190–193.
[11] Birchmeier, W. and Behrens, J. (1994) Biochem. Biophys. Acta 1198, 11–26.
[12] Berx, G. et al. (1998) Hum. Mutat. 12, 226–237.
[13] Guilford, P. et al. (1998) Nature 392, 402–405.
[14] Richards, F.M. et al. (1999) Hum. Mol. Genet. 8, 607–610.
[15] Mak, T.W. (1998) The Gene Knockout FactsBook. Academic Press, London.
[16] Riethmacher, D. et al. (1995) Proc. Natl Acad. Sci. USA 92, 855–859.
[17] Larue, L. et al. (1994) Proc. Natl Acad. Sci. USA 91, 8263–8267.

N-Cadherin

Cadherin-2, neural cadherin

Family

Cadherin (type I subfamily)

Structure

Molecular weights

Amino acids	906
Polypeptide	99 851

SDS-PAGE reduced	135–140 kDa

Carbohydrate

N-linked sites	7
O-linked sites	

Gene location	18q11.2

Gene structure	>200 kb, 16 exons

Alternative forms

Structure

Like all classical type I cadherins, N-cadherin is expressed initially as a precursor with an N-terminal propeptide domain, which is cleaved off inside the cell during protein maturation. The mature form of the protein forms a homodimer with adjacent molecules at the cell surface. The extracellular domain contains five cadherin repeats of approximately 110 amino acids separated by four calcium binding sites. The N-terminal domain contains the conserved HAV motif required for homophilic binding to N-cadherin expressed on neighbouring cells[1]. The cytoplasmic domain associates with catenins, thereby linking N-cadherin to the actin cytoskeleton[2]. This interaction is promoted by the non-receptor protein tyrosine phosphatase PTP-1B, which when associated with N-cadherin can dephosphorylate β-catenin[3]. Crystal structure from the N-terminal domain of N-cadherin[4] led to the proposal that classic cadherins form strand dimers whereby cadherin dimers on apposing cells interact to form a linear 'zipper'[4,5]. The structure of the cadherins is discussed further in Section I.

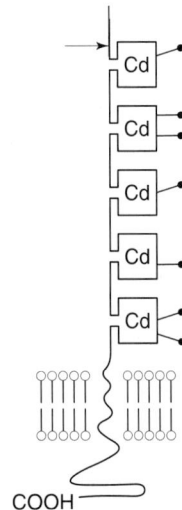

Ligands

N-cadherin mediates Ca^{2+}-dependent homotypic adhesion. In addition, N-cadherin-positive retinal axons can use R-cadherin as a substrate for axon elongation[6]. In at least some cells types, N-cadherin interacts in *cis* with the fibroblast growth factor receptor and activation of the receptor tyrosine kinase is a necessary component of neurite outgrowth[7,8].

Function

N-cadherin functions to promote neurite outgrowth on astrocytes, muscle cells and Schwann cells. It is also a necessary component of cardiac cells compartmentalization during early heart development[9], myoblast fusion[10], skeletal muscle differentiation[11,12] and somite formation[13].

Distribution

N-cadherin is highly expressed in adult neural tissue, cardiac and skeletal muscle, lens and endothelial cells. In the embryo, it is initially expressed in mesoderm migrating from the primitive streak and then later in a number of embryonic tissues including somites, neural tube and heart.

Disease association

OMIM 114020

There is increasing evidence for a role of N-cadherin in promoting the migration, invasion and metastasis of epithelial tumour cells[14,15].

Knockout

MGI:88355

The Gene Knockout FactsBook[16], p103

N-cadherin −/− mice die by day 10 of gestation due to failure of the heart tube to develop normally. In addition, epithelial organization of the somites is disrupted, the mesodermal and endodermal layers of the yolk sac separate, and the neural tube is malformed[17].

Amino acid sequence of human N-cadherin

```
  1 MCRIAGALRT LLPLLLALLQ ASVEASGEIA LCKTGFPEDV YSAVLSKDVH EGQPLLNVKF
 61 SNCNGKRKVQ YESSEPADFK VDEDGMVYAV RSFPLSSEHA KFLIYAQDKE TQEKWQVAVK
121 LSLKPTLTEE SVKESAEVEE IVFPRQFSKH SGHLQRQKRD WVIPPINLPE NSRGPFPQEL
181 VRIRSDRDKN LSLRYSVTGP GADQPPTGIF IINPISGQLS VTKPLDREQI ARFHLRAHAV
241 DINGNQVENP IDIVINVIDM NDNRPEFLHQ VWNGTVPEGS KPGTYVMTVT AIDADDPNAL
301 NGMLRYRIVS QAPSTPSPNM FTINNETGDI ITVAAGLDRE KVQQYTLIIQ ATDMEGNPTY
361 GLSNTATAVI TVTDVNDNPP EFTAMTFYGE VPENRVDIIV ANLTVTDKDQ PHTPAWNAVY
421 RISGGDPTGR FAIQTDPNSN DGLVTVVKPI DFETNRMFVL TVAAENQVPL AKGIQHPPQS
481 TATVSVTVID VNENPYFAPN PKIIRQEEGL HAGTMLTTFT AQDPDRYMQQ NIRYTKLSDP
541 ANWLKIDPVN GQITTIAVLD RESPNVKNNI YNATFLASDN GIPPMSGTGT LQIYLLDIND
601 NAPQVLPQEA ETCETPDPNS INITALDYDI DPNAGPFAFD LPLSPVTIKR NWTITRLNGD
661 FAQLNLKIKF LEAGIYEVPI IITDSGNPPK SNISILRVKV CQCDSNGDCT DVDRIVGAGL
721 GTGAIIAILL CIIILLILVL MFVVWMKRRD KERQAKQLLI DPEDDVRDNI LKYDEEGGGE
781 EDQDYDLSQL QQPDTVEPDA IKPVGIRRMD ERPIHAEPQY PVRSAAPHPG DIGDFINEGL
841 KAADNDPTAP PYDSLLVFDY EGSGSTAGSL SSLNSSSSGG EQDYDYLNDW GPRFKKLADM
901 YGGGDD
```

The propeptide sequence underlined and in italics is cleaved to form the mature protein.

Database accession

EMBL/GenBank   S42303
SwissProt   P19022

References

1. Yap, A.S. et al. (1997) Annu. Rev. Cell Dev. Biol. 13, 119–146.
2. Sacco, P.A. et al. (1995) J. Biol. Chem. 270, 20201–20206.
3. Balsamo, J. et al. (1998) J. Cell Biol. 143, 523–532.
4. Shapiro, L. et al. (1995) Nature 374, 327–337.
5. Tamura, K. et al. (1998) Neuron 20, 1153–1163.
6. Redies, C. and Takeichi, M. (1992) Glia 8, 161–171.
7. Williams, E.J. et al. (1994) Neuron 13, 583–594.
8. Saffell, J.L. et al. (1997) Neuron 18, 231–242.
9. Linask, K.K. et al. (1997) Dev. Biol. 185, 148–164.
10. Charlton, C.A. et al. (1997) J. Cell Biol. 138, 331–336.
11. George-Weinstein, M. et al. (1997) Dev. Biol. 185, 14–24.
12. Goichberg, P. and Geiger, B. (1998) Mol. Biol. Cell 9, 3119–3131.
13. Linask, K.K. et al. (1998) Dev. Biol. 202, 85–102.
14. Nieman, M.T. et al. (1999) J. Cell Biol. 147, 631–644.
15. Hazan, R.B. et al. (2000) J. Cell Biol. 148, 779–790.
16. Mak, T.W. (1998) The Gene Knockout FactsBook. Academic Press, London.
17. Radice, G.L. et al. (1997) Dev. Biol. 181, 64–78.

P-Cadherin
Cadherin-3, placental cadherin

Family

Cadherin (type I subfamily)

Structure

Molecular weights

Amino acids	829
Polypeptide	91 427

SDS-PAGE reduced	115–120 kDa

Carbohydrate

N-linked sites	2
O-linked sites	

Gene location	16q22.1
Gene structure	45 kb, 15 exons (mouse)

Alternative forms

Structure

Like all classical type I cadherins, P-cadherin is expressed initially as a precursor with an N-terminal propeptide domain, which is cleaved off inside the cell during protein maturation. The extracellular domain contains five cadherin repeats of approximately 110 amino acids separated by four calcium binding sites. The N-terminal domain contains the conserved HAV motif required for homophilic binding to P-cadherin expressed on neighbouring cells[1,2]. The structure of the cadherins is discussed further in Section I.

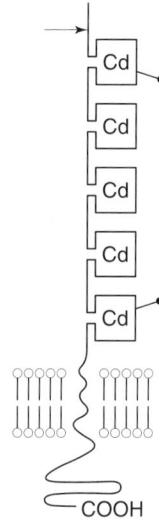

Ligands

P-cadherin mediates Ca^{2+}-dependent homotypic adhesion[3].

Function

As demonstrated by the knockout mice (see below), the precocious mammary development in the absence of P-cadherin suggests a role in maintaining the undifferentiated state of the myoepithleium.

Distribution

P-cadherin is predominantly expressed in the placenta, epidermis and mammary gland.

Disease association

OMIM 114021

Knockout

MGI:88356

The Gene Knockout FactsBook[4], p104

P-cadherin −/− mice are viable and fertile but with virgin females exhibiting precocious differentation of the mammary gland and mammary hyperplasia with age[5].

Amino acid sequence of human P-cadherin

```
  1 MGLPRGPLAS LLLLQVCWLQ CAASEPCRAV FREAEVTLEA GGAEQEPGQA LGKVFMGCPG
 61 QEPALFSTDN DDFTVRNGET VQERRSLKER NPLKIFPSKR ILRRHKRDWV VAPISVPENG
121 KGPFPQRLNQ LKSNKDRDTK IFYSITGPGA DSPPEGVFAV EKETGWLLLN KPLDREEIAK
181 YELFGHAVSE NGASVEDPMN ISIIVTDQND HKPKFTQDTF RGSVLEGVLP GTSVMQVTAT
241 DEDDAIYTYN GVVAYSIHSQ EPKDPHDLMF TIHRSTGTIS VISSGLDREK VPEYTLTIQA
301 TDMDGDGSTT TAVAVVEILD ANDNAPMFDP QKYEAHVPEN AVGHEVQRLT VTDLDAPNSP
361 AWRATYLIMG GDDGDHFTIT THPESNQGIL TTRKGLDFEA KNQHTLYVEV TNEAPFVLKL
421 PTSTATIVVH VEDVNEAPVF VPPSKVVEVQ EGIPTGEPVC VYTAEDPDKE NQKISYRILR
481 DPAGWLAMDP DSGQVTAVGT LDREDEQFVR NNIYEVMVLA MDNGSPPTTG TGTLLLTLID
541 VNDHGPVPEP RQITICNQSP VRHVLNITDK DLSPHTSPFQ AQLTDDSDIY WTAEVNEEGD
601 TVVLSLKKFL KQDTYDVHLS LSDHGNKEQL TVIRATVCDC HGHVETCPGP WKGGFILPVL
661 GAVLALLFLL LVLLLLVRKK RKIKEPLLLP EDDTRDNVFY YGEEGGGEED QDYDITQLHR
721 GLEARPEVVL RNDVAPTIIP TPMYRPRPAN PDEIGNFIIE NLKAANTDPT APPYDTLLVF
781 DYEGSGSDAA SLSSLTSSAS DQDQDYDYLN EWGSRFKKLA DMYGGGEDD
```

The propeptide sequence underlined and in italics is cleaved to form the mature protein.

Database accession

EMBL/GenBank   X63629
SwissProt   P22223

References

[1] Nose, A. and Takeichi, M. (1986) J. Cell Biol. 103, 2649–2658.

[2] Yap, A.S. et al. (1997) Annu. Rev. Cell Dev. Biol. 13, 119–146.

[3] Takeichi, M. (1995) Curr. Opin. Cell Biol. 7, 619–627.

[4] Mak, T.W. (1998) The Gene Knockout FactsBook. Academic Press, London.

[5] Radice, G.L. et al. (1997) J. Cell Biol. 139, 1025–1032.

R-Cadherin

Cadherin-4, retinal cadherin

Family

Cadherin (type I subfamily)

Structure

Molecular weights

Amino acids	916
Polypeptide	100 446

SDS-PAGE reduced	120–130 kDa

Carbohydrate

N-linked sites	6
O-linked sites	

Gene location 20q13.3

Gene structure

Alternative forms

Structure

Like all classical type I cadherins, R-cadherin is expressed initially as a precursor with an N-terminal propeptide domain, which is cleaved off inside the cell during protein maturation. The extracellular domain contains five cadherin repeats of approximately 110 amino acids separated by four calcium binding sites. The N-terminal domain contains the conserved HAV motif required for homophilic binding to R-cadherin expressed on neighbouring cells[1-4]. The structure of the cadherins is discussed further in Section I.

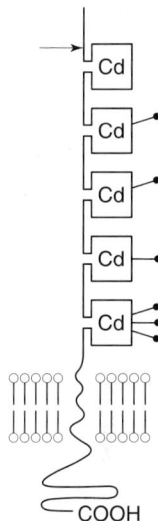

Ligands

R-cadherin mediates Ca^{2+}-dependent homotypic adhesion. In addition, N-cadherin-positive retinal axons can use R-cadherin as a substrate for axon elongation[5,6].

Function

It has been proposed that the cell:cell adhesion mediated by R-cadherin may be important in the formation of striated muscle and epithelial tissues[7], the segregation of exocrine and endocrine cells during pancreatic development[2], as a substrate for the outgrowth of N-cadherin retinal axons[6] and in retinal morphogenesis[1].

Distribution

In the embryo, R-cadherin is expressed in developing muscle in the somitic mytome, early skeletal muscle and smooth muscle, but not cardiac muscle.

In addition, expression is seen in developing epithelia in the kidney, lung, thymus[7] and in the early optic nerve glia[6].

Disease association

OMIM 603006

Knockout

MGI:99218

Amino acid sequence of human R-cadherin

```
  1 MTAGAGVLLL LLSLSGALRA HNEDLTTRET CKAGFSEDDY TALISQNILE GEKLLQVKFS
 61 SCVGTKGTQY ETNSMDFKVG ADGTVFATRE LQVPSEQVAF TVTAWDSQTA EKWDAVVRLL
121 VAQTSSPHSG HKPQKGKKVV ALDPSPPPKD TLLPWPQHQN ANGLRRRKRD WVIPPINVPE
181 NSRGPFPQQL VRIRSDKDND IPIRYSITGV GADQPPMEVF SINSMSGRMY VTRPMDREEH
241 ASYHLRAHAV DMNGNKVENP IDLYIYVIDM NDNHPEFINQ VYNCSVDEGS KPGTYVMTIT
301 ANDADDSTTA NGMVRYRIVT QTPQSPSQNM FTINSETGDI VTVAAGWDRE KVQQYTVIVQ
361 ATDMEGNLNY GLSNTATAII TVTDVNDNPS EFTASTFAGE VPENSVETVV ANLTVMDRDQ
421 PHSPNWNAVY RIISGDPSGH FSVRTDPVTN EGMVTVVKAV DYELNRAFML TVMVSNQAPL
481 ASGIQMSFQS TAGVTISIMD INEAPYFPSN HKLIRLEEGV PPGTVLTTFS AVDPDRFMQQ
541 AVRYSKLSDP ASWLHINATN GQITTVAVLD RESLYTKNNV YEATFLAADN GIPPASGTGT
601 LQIYLIDIND NAPELLPKEA QICERPNLNA INITAADADV HPNIGPYVFE LPFVPAAVRK
661 NWTITRLNGD YAQLSLRILY LEAGMYDVPI IVTDSGNPPL SNTSIIKVKV CPCDDNGDCT
721 TIGAVAAAGL GTGAIVAILI CILILLTMVL LFVMWMKRRE KERHTKQLLI DPEDDVREKI
781 LKYDEEGGGE EDQDYDLSQL QQPEAMGHVP SKAPGVRRVD ERPVGPEPQY PIRPMVPHPG
841 DIGDFINEGL RAADNDPTAP PYDSLLVFDY EGSGSTAGSV SSLNSSSSGD QDYDYLNDWG
901 PRFKKLADMY GGGEED
```

The propeptide sequence underlined and in italics is cleaved to form the mature protein.

Database accession

EMBL/Genbank   L34059
SwissProt   P55283

References

[1] Inuzuka, H. et al. (1991) Neuron 7, 69–79.
[2] Hutton, J.C. et al. (1993) Mol. Endocrinol. 7, 1151–1160.
[3] Tanihara, H. et al. (1994) Cell Adhes. Commun. 2, 15–26.
[4] Yap, A.S. et al. (1997) Annu. Rev. Cell Dev. Biol. 13, 119–146.
[5] Matsunami, H. et al. (1993) J. Cell Sci. 106, 401–409.
[6] Redies, C. and Takeichi, M. (1992) Glia 8, 161–171.
[7] Rosenberg, P. et al. (1997) Dev. Biol. 187, 55–70.

VE-Cadherin
Cadherin-5, vascular endothelial cadherin, CD144

Family
Cadherin (type II subfamily)

Structure

Molecular weights
Amino acids	784
Polypeptide	87 528
SDS-PAGE reduced	130 kDa

Carbohydrate
N-linked sites	7
O-linked sites	

Gene location
16q22.1

Gene structure
36 kb, 12 exons (mouse)

Alternative forms

Structure
Like the classical type I cadherins, VE-cadherin is expressed initially as a precursor with an N-terminal propeptide domain, which is cleaved off inside the cell during protein maturation. The extracellular domain contains five cadherin repeats of approximately 110 amino acids separated by four calcium binding sites. In contast to the classical type I cadherins, VE-cadherin has a shorter propeptide sequence and a VIV rather than HAV motif in the homotypic binding site in the N-terminal domain[1,2]. The cytoplasmic domain associates with catenins thereby linking VE-cadherin to the actin cytoskeleton[3,4]. The structure of the cadherins is discussed further in Section I.

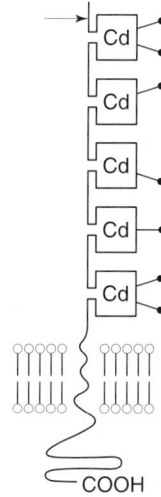

Ligands
VE-cadherin mediates Ca^{2+}-dependent homotypic adhesion.

Function
The restriction of VE-cadherin to the endothelial junctions suggests a role in promoting homotypic interactions. Generation of VE-cadherin $-/-$ embryonic stem (ES) cells and antibody blocking experiments *in vitro* were used to demonstrate a requirement for this cadherin in vascular development[5,6]. In addition, VE-cadherin is necessary for maintaining endothelial integrity as introduction of anti-VE-cadherin antibodies *in vivo* results in increased vascular permeability and accelerated neutrophil recruitment into the peritoneum[7]. Recent data from VE-cadherin deficient embryos indicate that this cadherin is required *in vivo* for endothelial cell survival. Loss of VE-cadherin, or deletion of the cytoplasmic domain, induces endothelial apoptosis. This is due to an inability of the cells to respond to vascular endothelial growth factor (VEGF)-A which in wild type cells signals via the VEGF receptor-2 to activate PI3-kinase and Akt to increase levels of the anti-apoptotic mediator Bcl2[8].

Distribution

VE-cadherin expression is restricted to all types of endothelium. In these cells VE-cadherin is localized to the intercellular junctions whereas the other endothelial cadherin, N-cadherin, is dispersed over the plasma membrane[3,4].

Disease association

OMIM 601120
VE-cadherin expression is reduced in human angiosarcomas

Knockout

MGI:105057
VE-cadherin –/– embryos retain the ability to assemble endothelial cells into vessels but die at E9.5 due to a failure in remodelling and maturation of the vascular network[8].

Amino acid sequence of human VE-cadherin

```
  1 MQRLMMLLAT SGACLGLLAV AAVAAAGANP AQRDTHSLLP THRRQKRDWI WNQMHIDEEK
 61 NTSLPHHVGK IKSSVSRKNA KYLLKGEYVG KVFRVDAETG DVFAIERLDR ENISEYHLTA
121 VIVDKDTGEN LETPSSFTIK VHDVNDNWPV FTHRLFNASV PESSAVGTSV ISVTAVDADD
181 PTVGDHASVM YQILKGKEYF AIDNSGRIIT ITKSLDREKQ ARYEIVVEAR DAQGLRGDSG
241 TATVLVTLQD INDNFPFFTQ TKYTFVVPED TRVGTSVGSL FVEDPDEPQN RMTKYSILRG
301 DYQDAFTIET NPAHNEGIIK PMKPLDYEYI QQYSFIVEAT DPTIDLRYMS PPAGNRAQVI
361 INITDVDEPP IFQQPFYHFQ LKENQKKPLI GTVLAMDPDA ARHSIGYSIR RTSDKGQFFR
421 VTKKGDIYNE KELDREVYPW YNLTVEAKEL DSTGTPTGKE SIVQVHIEVL DENDNAPEFA
481 KPYQPKVCEN AVHGQLVLQI SAIDKDITPR NVKFKFILNT ENNFTLTDNH DNTANITVKY
541 GQFDREHTKV HFLPVVISDN GMPSRTGTST LTVAVCKCNE QGEFTFCEDM AAQVGVSIQA
601 VVAILLCILT ITVITLLIFL RRRLRKQARA HGKSVPEIHE QLVTYDEEGG GEMDTTSYDV
661 SVLNSVRRGG AKPPRPALDA RPSLYAQVQK PPRHAPGAHG GPGEMAAMIE VKKDEADHDG
721 DGPPYDTLHI YGYEGSESIA ESLSSLGTDS SDSDVDYDFL NDWGPRFKML AELYGSDPRE
781 ELLY
```

The propeptide sequence underlined and in italics is cleaved to form the mature protein.

Database accession

EMBL/GenBank   X79981
SwissProt   P33151

References

[1] Lampugnani, M.G. et al. (1992) J. Cell Biol. 118, 1511–1522.

[2] Yap, A.S. et al. (1997) Annu. Rev. Cell Dev. Biol. 13, 119–146.

[3] Lampugnani, M.G. et al. (1995) J. Cell Biol. 129, 203–217.

[4] Navarro, P. et al. (1998) J. Cell Biol. 140, 1475–1484.

[5] Vittet, D. et al. (1997) Proc. Natl Acad. Sci. USA 94, 6273–6278.

[6] Bach, T.L. et al. (1998) Exp. Cell Res. 238, 324–334.

[7] Gotsch, U. et al. (1997) J. Cell Sci. 110, 583–588.

[8] Carmeliet, P. et al. (1999) Cell 98, 147–157.

K-Cadherin | Cadherin-6, kidney cadherin

Family
Cadherin (type II subfamily)

Structure

Molecular weights
Amino acids	790
Polypeptide	88 308

SDS-PAGE reduced   120–130 kDa

Carbohydrate
N-linked sites	6 (1 removed with propeptide)
O-linked sites	

Gene location

Gene structure   5p14-p15.1

Alternative forms

Structure
Like the classical type I cadherins, K-cadherin is predicted to be expressed initially as a precursor with an N-terminal propeptide domain, which is cleaved off inside the cell during protein maturation. The extracellular domain contains five cadherin repeats of approximately 110 amino acids separated by four calcium binding sites. In contast to the classical type I cadherins, K-cadherin has a shorter propeptide sequence and a QAI rather than HAV motif in the N-terminal domain, which does not appear to be involved in the adhesive function[1,2]. In addition, binding of Ca^{2+} does not protect K-cadherin from proteolysis[3]. The cytoplasmic domain associates with catenins thereby linking K-cadherin to the actin cytoskeleton[4]. The structure of the cadherins is discussed further in Section I.

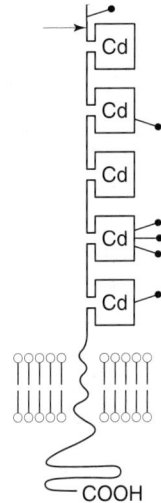

Ligands
K-cadherin mediates Ca^{2+}-dependent homotypic adhesion. In addition, in mixed aggregation assays, K-cadherin partially interacts with cadherin-14[3].

Function
Predicted to function as a homotypic adhesion receptor in K-cadherin-positive tissues.

Distribution
K-cadherin is highly expressed in kidney, brain and cerebellum. Weaker expression is seen in lung, pancreas and gastric mucosa.

Disease association

OMIM 603007

Expression is upregulated in some liver and kidney carcinomas[1,4].

Knockout

MGI:107435

Amino acid sequence of human K-cadherin

```
  1 MRTYRYFLLL FWVGQPYPTL STPLSKRTSG FPAKKRALEL SGNSKNELNR SKRSWMWNQF
 61 FLLEEYTGSD YQYVGKLHSD QDRGDGSLKY ILSGDGAGDL FIINENTGDI QATKRLDREE
121 KPVYILRAQA INRRTGRPVE PESEFIIKIH DINDNEPIFT KEVYTATVPE MSDVGTFVVQ
181 VTATADDDPT YGNSAKVVYS ILQGQPYFSV ESETGIIKTA LLNMDRENRE QYQVVIQAKD
241 MGGQMGGLSG TTTVNITLTD VNDNPPRFPQ STYQFKTPES SPPGTPIGRI KASDADVGEN
301 AEIEYSITDG EGLDMFDVIT DQETQEGIIT VKKLLDFEKK KVYTLKVEAS NPYVEPRFLY
361 LGPFKDSATV RIVVEDVDEP PVFSKLAYIL QIREDAQINT TIGSVTAQDP DAARNPVKYS
421 VDRHTDMDRI FNIDSGNGSI FTSKLLDRET LLWHNITVIA TEINNPKQSS RVPLYIKVLD
481 VNDNAPEFAE FYETFVCEKA KADQLIQTLH AVDKDDPYSG HQFSFSLAPE AASGSNFTIQ
541 DNKDNTAGIL TRKNGYNRHE MSTYLLPVVI SDNDYPVQSS TGTVTVRVCA CDHHGNMQSC
601 HAEALIHPTG LSTGALVAIL LCIVILLVTV VLFAALRRQR KKEPLIISKE DIRDNIVSYN
661 DEGGGEEDTQ AFDIGTLRNP EAIEDNKLRR DIVPEALFLP RRTPTARDNT DVRDFINQRL
721 KENDTDPTAP PYDSLATYAY EGTGSVADSL SSLESVTTDA DQDYDYLSDW GPRFKKLADM
781 YGGVDSDKDS
```

The propeptide sequence underlined and in italics is cleaved to form the mature protein.

Database accession

EMBL/GenBank   D31784

SwissProt   P55285

References

[1] Xiang, Y.Y. et al. (1994) Cancer Res. 54, 3034–3041.

[2] Shimoyama, Y. et al. (1995) Cancer Res. 55, 2206–2211.

[3] Shimoyama, Y. et al. (1999) J. Biol. Chem. 274, 11987–11994.

[4] Paul, R. et al. (1997) Cancer Res. 57, 2741–2748.

Cadherin-8

Family

Cadherin (type II subfamily)

Structure

Molecular weights

Amino acids	793
Polypeptide	87 570

SDS-PAGE reduced   130 kDa

Carbohydrate

N-linked sites	6 (2 removed with propeptide)
O-linked sites	

Gene location   16q21–16q22.1

Gene structure

Alternative forms

In rat and human, cDNAs encoding an altered form of cadherin-8 truncated in the fifth cadherin repeat have been isolated. A 95-kDa form of cadherin-8 has been detected in rat brain[1].

Structure

Like the classical type I cadherins, cadherin-8 is predicted to be expressed initially as a precursor with an N-terminal propeptide domain cleaved off inside the cell during protein maturation. The extracellular domain contains five cadherin repeats of approximately 110 amino acids separated by four calcium binding sites. In contrast to the classical type I cadherins, cadherin-8 has a shorter propeptide sequence, a QAV rather than an HAV motif in the N-terminal domain homotypic binding site, and binding of Ca^{2+} does not protect cadherin-8 from proteolysis[1-3]. The cytoplasmic domain associates with β-catenin[1]. The structure of the cadherins is discussed further in Section I.

Ligands

Cadherin-8 transfected into L cells mediates weak homotypic adhesion[1].

Function

The distribution of cadherin-8 suggests a function in the formation of particular neural circuits[4] and/or interactions of thymocytes with stroma[5].

Distribution

In the adult, expression is mainly detected in the brainstem, cerebellum[4] and CD4+ CD8+ thymocytes[5]. During development, cadherin-8 is

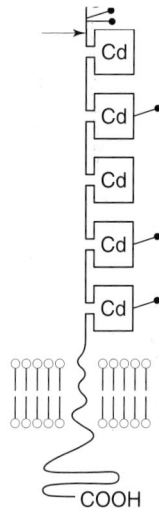

expressed in restricted areas of the central nervous system and the thymus[6,7]. In transfected L cells, cadherin-8 was found localized to the cell periphery mainly at cell:cell contact sites[1].

Disease association

OMIM 603008

Knockout

MGI:107434

Amino acid sequence of human cadherin-8

```
  1 MLLDLWTPLI ILWITLPPCI YMAPMNQSQV LMSGSPLELN SLGEEQRILN RSKRGWVWNQ
 61 MFVLEEFSGP EPILVGRLHT DLDPGSKKIK YILSGDGAGT IFQINDVTGD IHAIKRLDRE
121 EKAEYTLTAQ AVDWETSKPL EPPSEFIIKV QDINDNAPEF LNGPYHATVP EMSILGTSVT
181 NVTATDADDP VYGNSAKLVY SILEGQPYFS IEPETAIIKT ALPNMDREAK EEYLVVIQAK
241 DMGGHSGGLS GTTTLTVTLT DVNDNPPKFA QSLYHFSVPE DVVLGTAIGR VKANDQDIGE
301 NAQSSYDIID GDGTALFEIT SDAQAQDGII RLRKPLDFET KKSYTLKDEA ANVHIDPRFS
361 GRGPFKDTAT VKIVVEDADE PPVFSSPTYL LEVHENAALN SVIGQVTARD PDITSSPIRF
421 SIDRHTDLER QFNINADDGK ITLATPLDRE LSVWHNITII ATEIRNHSQI SRVPVAIKVL
481 DVNDNAPEFA SEYEAFLCEN GKPGQVIQTV SAMDKDDPKN GHYFLYSLLP EMVNNPNFTI
541 KKNEDNSLSI LAKHNGFNRQ KQEVYLLPII ISDSGNPPLS STSTLTIRVC GCSNDGVVQS
601 CNVEAYVLPI GLSMGALIAI LACIILLLVI VVLFVTLRRH QKNEPLIIKD DEDVRENIIR
661 YDDEGGGEED TEAFDIATLQ NPDGINGFLP RKDIKPDLQF MPRQGLAPVP NGVDVDEFIN
721 VRLHEADNDP TAPPYDSIQI YGYEGRGSVA GSLSSLESTT SDSDQNFDYL SDWGPRFKRL
781 GELYSVGESD KET
```

The propeptide sequence underlined and in italics is cleaved to form the mature protein.

Database accession

EMBL/GenBank   L34060
SwissProt   P55286

References

[1] Kido, M. et al. (1998) Genomics 28, 186–194.
[2] Tanihara, H. et al. (1994) Cell Adhes. Commun. 2, 15–26.
[3] Yap, A.S. et al. (1997) Annu. Rev. Cell Dev. Biol. 13, 119–146.
[4] Korematsu, K. et al. (1998) Neuroscience 87, 303–315.
[5] Munro, S.B. et al. (1996) Cell. Immunol. 169, 309–312.
[6] Korematsu, K. and Redies, C. (1997) Dev. Dyn. 208, 178–189.
[7] Korematsu, K. and Redies, C. (1997) J. Comp. Neurol. 387, 291–306.

OB-Cadherin

Family

Cadherin (type II subfamily)

Structure

Molecular weights

Amino acids	796
Polypeptide	88 049
SDS-PAGE reduced	115 kDa

Carbohydrate

N-linked sites	2
O-linked sites	

Gene location   16q21–16q22.1

Gene structure

Alternative forms

An alternatively spliced truncated form of OB-cadherin with a different cytoplasmic domain results from an insertion of 179 bp in an intron[1].

Structure

Like the classical type I cadherins, OB-cadherin is predicted to be expressed initially as a precursor with an N-terminal propeptide domain cleaved off inside the cell during protein maturation. The extracellular domain contains five cadherin repeats of approximately 110 amino acids separated by four calcium binding sites. In contrast to the classical type I cadherins, OB-cadherin has a shorter propeptide sequence and a QAV rather than HAV motif in the homotypic binding site in the N-terminal domain[2-4]. The alternatively spliced form of OB-cadherin is expressed on the cell surface with intact OB-cadherin[1]. The cytoplasmic domain of the intact form associates with β-catenin[1]. The structure of the cadherins is discussed further in Section I.

Ligands

OB-cadherin expressed in L-cells mediates Ca^{2+}-dependent cell–cell adhesion[1,2,5] and this interaction is strengthened by co-expression of the truncated OB-cadherin isoform[1].

Function

The expression pattern of OB-cadherin suggests that it may be involved in mesenchymal differentiation[1], segmentation and compartmentalization of the developing brain[6], mediating trophoblast–endometrium interactions[7] and bone formation[2].

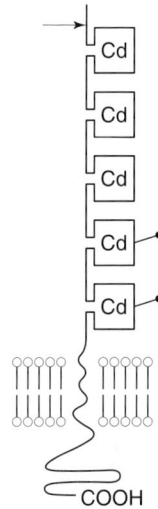

Distribution

OB-cadherin was initially identified in osteoblastic cell lines, precursor osteoblasts and primary osteoblastic cells from calvaria. Low level expression is detected *in vivo* in a variety of tissues including bone[2]. During development OB-cadherin is a marker of mesenchymal cells and restricted areas of the brain[5,6,8,9].

Disease association

OMIM 600023

Knockout

MGI:99217

Amino acid sequence of human OB-cadherin

```
  1 MKENYCLQAA LVCLGMLCHS HAFAPERRGH LRPSFHGHHE KGKEGQVLQR SKRGWVWNQF
 61 FVIEEYTGPD PVLVGRLHSD IDSGDGNIKY ILSGEGAGTI FVIDDKSGNI HATKTLDREE
121 RAQYTLMAQA VDRDTNRPLE PPSEFIVKVQ DINDNPPEFL HETYHANVPE RSNVGTSVIQ
181 VTASDADDPT YGNSAKLVYS ILEGQPYFSV EAQTGIIRTA LPNMDREAKE EYHVVIQAKD
241 MGGHMGGLSG TTKVTITLTD VNDNPPKFPQ RLYQMSVSEA AVPGEEVGRV KAKDPDIGEN
301 GLVTYNIVDG DGMESFEITT DYETQEGVIK LKKPVDFETE RAYSLKVEAA NVHIDPKFIS
361 NGPFKDTVTV KISVEDADEP PMFLAPSYIH EVQENAAAGT VVGRVHAKDP DAANSPIRYS
421 IDRHTDLDRF FTINPEDGFI KTTKPLDREE TAWLNITVFA AEIHNRHQEA QVPVAIRVLD
481 VNDNAPKFAA PYEGFICESD QTKPLSNQPI VTISADDKDD TANGPRFIFS LPPEIIHNPN
541 FTVRDNRDNT AGVYARRGGF SRQKQDLYLL PIVISDGGIP PMSSTNTLTI KVCGCDVNGA
601 LLSCNAEAYI LNAGLSTGAL IAILACIVIL LVIVVLFVTL RRQKKEPLIV FEEEDVRENI
661 ITYDDEGGGE EDTEAFDIAT LQNPDGINGF IPRKDIKPEY QYMPRPGLRP APNSVDVDDF
721 INTRIQEADN DPTAPPYDSI QIYGYEGRGS VAGSLSSLES ATTDSDLDYD YLQNWGPRFK
781 KLADLYGSKD TFDDDS
```

The propeptide sequence underlined and in italics is cleaved to form the mature protein.

Data base accession

EMBL/GenBank   L34056
SwissProt   P55287

References

[1] Kawaguchi, J. et al. (1999) J. Bone Miner. Res. 14, 764–775.
[2] Okazaki, M. et al. (1994) J. Biol. Chem. 269, 12092–12098.
[3] Tanihara, H. et al. (1994) Cell Adhes. Commun. 2, 15–26.
[4] Yap, A.S. et al. (1997) Annu. Rev. Cell Dev. Biol. 13, 119–146.
[5] Kimura, Y. et al. (1995) Dev. Biol. 169, 347–358.
[6] Kimura, Y. et al. (1996) Dev. Dyn. 206, 455–462.
[7] MacCalman, C.D. et al. (1996) Dev. Dyn. 206, 201–211.
[8] Simonneau, L. et al. (1995) Cell Adhes. Commun. 3, 115–130.
[9] Simonneau, L. and Thiery, J.P. (1998) Cell Adhes. Commun. 6, 431–450.

Br-Cadherin

Cadherin-12, brain cadherin, N-cadherin 2

Family

Cadherin (type II subfamily)

Structure

Molecular weights

Amino acids	794
Polypeptide	88 274

SDS-PAGE reduced   120–125 kDa

Carbohydrate

N-linked sites	4
O-linked sites	unknown

Gene location   5p14–5p13

Gene structure

Alternative forms

Structure

Like the classical type I cadherins, Br-cadherin is predicted to be expressed initially as a precursor with an N-terminal propeptide domain cleaved off inside the cell during protein maturation. The extracellular domain contains five cadherin repeats of approximately 110 amino acids separated by four calcium binding sites. In contrast to the classical type I cadherins, Br-cadherin has a shorter propeptide sequence and a QAV rather than HAV motif in the homotypic binding site in the N-terminal domain[1-3]. The structure of the cadherins is discussed further in Section I.

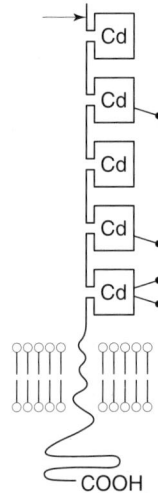

Ligands

Predicted to mediate Ca^{2+}-dependent homotypic adhesion.

Function

The expression pattern of Br-cadherin is consistent with a role in neuronal development.

Distribution

Br-Cadherin is expressed exclusively in neurons in the central nervous system. During development, expression is detectable with the onset of synaptogenesis and dendrite outgrowth[2].

Disease association

OMIM 600562

Knockout

MGI:109503

Amino acid sequence of human Br-cadherin

```
  1 MLTRNCLSLL LWVLFDGGLL TPLQPQPQQT LATEPRENVI HLPGQRSHFQ RVKRGWVWNQ
 61 FFVLEEYVGS EPQYVGKLHS DLDKGEGTVK YTLSGDGAGT VFTIDETTGD IHAIRSLDRE
121 EKPFYTLRAQ AVDIETRKPL EPESEFIIKV QDINDNEPKF LDGPYVATVP EMSPVGAYVL
181 QVKATDADDP TYGNSARVVY SILQGQPYFS IDPKTGVIRT ALPNMDREVK EQYQVLIQAK
241 DMGGQLGGLA GTTIVNITLT DVNDNPPRFP KSIFHLKVPE SSPIGSAIGR IRAVDPDFGQ
301 NAEIEYNIVP GDGGNLFDIV TDEDTQEGVI KLKKPLDFET KKAYTFKVEA SNLHLDHRFH
361 SAGPFKDTAT VKISVLDVDE PPVFSKPLYT MEVYEDTPVG TIIGAVTAQD LDVGSGAVRY
421 FIDWKSDGDS YFTIDGNEGT IATNELLDRE STAQYNFSII ASKVSNPLLT SKVNILINVL
481 DVNEFPPEIS VPYETAVCEN AKPGQIIQIV SAADRDLSPA GQQFSFRLSP EAAIKPNFTV
541 RDFRNNTAGI ETRRNGYSRR QQELYFLPVV IEDSSYPVQS STNTMTIRVC RCDSDGTILS
601 CNVEAIFLPV GLSTGALIAI LLCIVILLAI VVLYVALRRQ KKKHTLMTSK EDIRDNVIHY
661 DDEGGGEEDT QAFDIGALRN PKVIEENKIR RDIKPDSLCL PRQRPPMEDN TDIRDFIHQR
721 LQENDVDPTA PPIDSLATYA YEGSGSVAES LSSIDSLTTE ADQDYDYLTD WGPRFKVLAD
781 MFGEEESYNP DKVT
```

The propeptide sequence underlined and in italics is cleaved to form the mature protein.

Database accession

EMBL/GenBank   L34057
SwissProt   P55289

References

[1] Tanihara, H. et al. (1994) Cell Adhes. Commun. 2, 15–26.

[2] Selig, S. et al. (1997) Proc. Natl Acad. Sci. USA 94, 2398–2403.

[3] Yap, A.S. et al. (1997) Annu. Rev. Cell Dev. Biol. 13, 119–146.

H-Cadherin

Cadherin-13, heart cadherin, T-cadherin, truncated cadherin

Family

Cadherin (modified cadherin)

Structure

Molecular weights

Amino acids	713
Polypeptide	78 286
SDS-PAGE reduced	105 kDa

Carbohydrate

N-linked sites	8 (2 in propeptide sequence)
O-linked sites	unknown

Gene location   16q24

Gene structure

Alternative forms

Structure

Like the classical type I cadherins, H-cadherin is predicted to be expressed initially as a precursor with an N-terminal propeptide domain cleaved off inside the cell during protein maturation. The extracellular domain contains five cadherin repeats of approximately 110 amino acids separated by four calcium binding sites. In contrast to the classical type I cadherins, H-cadherin has a EV/TT rather than HAV motif in the homotypic binding site in the N-terminal domain and is attached to the cell membrane by a GPI anchor[1,2]. In addition, 130 kDa forms of H-cadherin representing molecules retaining the propeptide can also be detected at the cell surface[3]. The structure of the cadherins is discussed further in Section I.

Ligands

Not known.

Function

The ability of H-cadherin to reverse the invasive phenotype in H-cadherin negative breast cancer cell lines[4,5] suggests a role in the maintenance of the normal epithelial phenotype. Without a cytoplasmic domain it is predicted that H-cadherin may not function in mediating direct cell–cell interactions, rather it may initiate signalling pathways due to its localization with signalling complexes in caveoli[6].

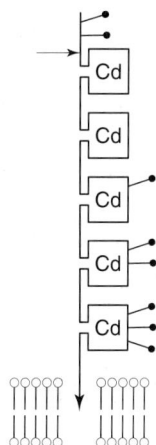

Distribution

H-cadherin is highly expressed in heart and breast epithelium. In vascular smooth muscle cells, H-cadherin co-localizes with caveolin rather than to cell–cell or cell–matrix contact areas[6].

Disease association

OMIM 601364

H-cadherin expression is lost in many breast and lung cancers[4,7]. Monitoring H-cadherin levels may have prognostic value as a marker for breast cancer development[4].

Knockout

MGI:99551

Amino acid sequence of human H-cadherin

```
  1 MQPRTPLVLC VLLSQVLLLT SAEDLDCTPG FQQKVFHINQ PAEFIEDQSI LNLTFSDCKG
 61 NDKLRYEVSS PYFKVNSDGG LVALRNITAV GKTLFVHART PHAEDMAELV IVGGKDIQGS
121 LQDIFKFART SPVPRQKRSI VVSPILIPEN QRQPFPRDVG KVVDSDRPER SKFRLTGKGV
181 DQEPKGIFRI NENTGSVSVT RTLDREVIAV YQLFVETTDV NGKTLEGPVP LEVIVIDQND
241 NRPIFREGPY IGHVMEGSPT GTTVMRMTAF DADDPATDNA LLRYNIRQQT PDKPSPNMFY
301 IDPEKGDIVT VVSPALLDRE TLENPKYELI IEAQDMAGLD VGLTGTATAT IMIDDKNDHS
361 PKFTKKEFQA TVEEGAVGVI VNLTVEDKDD PTTGAWRAAY TIINGNPGQS FEIHTNPQTN
421 EGMLSVVKPL DYEISAFHTL LIKVENEDPL VPDVSYGPSS TATVHITVLD VNEGPVFYPD
481 PMMVTRQEDL SVGSVLLTVN ATDPDSLQHQ TIRYSVYKDP AGWLNINPIN GTVDTTAVLD
541 RESPFVDNSV YTALFLAIDS GNPPATGTGT LLITLEDVND NAPFIYPTVA EVCDDAKNLS
601 VVILGASDKD LHPNTDPFKF EIHKQAVPDK VWKISKINNT HALVSLLQNL NKANYNLPIM
661 VTDSGKPPMT NITDLRVQVC SCRNSKVDCN AAGALRFSLP SVLLLSLFSL ACL
```

Notes
1 G693 represents the site of lipid attachment.
2 The propeptide sequence underlined and in italics is cleaved to form the mature protein.

Database accession

EMBL/GenBank   L34058
SwissProt   P55290

References

[1] Tanihara, H. et al. (1994) Cell Adhes. Commun. 2, 15–26.
[2] Yap, A.S. et al. (1997) Annu. Rev. Cell Dev. Biol. 13, 119–146.
[3] Stambolsky, D.V. et al. (1999) Biochem. Biophys. Acta 1416, 155–160.
[4] Lee, S.W. (1996) Nature Med. 2, 776–782.
[5] Lee, S.W. et al. (1998) Carinogenesis 19, 1157–1159.
[6] Philippova, M.P. et al. (1998) FEBS Lett. 429, 207–210.
[7] Sato, M. et al. (1998) Hum. Genet. 103, 96–101.

Cadherin-14

Family

Cadherin family (type II subfamily)

Structure

Molecular weights
amino acids	790
polypeptide	88 017

SDS-PAGE reduced   105 kDa

Carbohydrate
N-linked sites	3
O-linked sites	

Gene location

Gene structure

Alternative forms

Structure

Like the classical type I cadherins, cadherin-14 is predicted to be expressed initially as a precursor with an N-terminal propeptide domain cleaved off inside the cell during protein maturation. The extracellular domain contains 5 cadherin repeats of approximately 110 amino acids separated by four calcium binding sites. In contrast to the classical type I cadherins, cadherin-14 has a shorter propeptide sequence and a QAI rather than HAV motif in the N-terminal domain[1-3]. The cytoplasmic domain associates with catenin complexes[3]. The structure of the cadherins is further discussed in Section I.

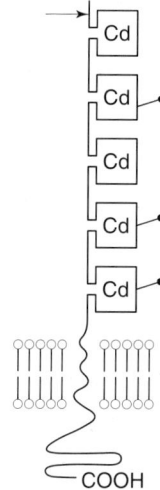

Ligands

Cadherin-14 mediates Ca^{2+}-dependent homotypic adhesion. In addition, in mixed aggregation assays, cadherin-14 partially interacts with cadherin-6[3].

Function

The restricted expression of cadherin-14 to the central nervous system suggests a role in regulating neural morphogenesis.

Distribution

Cadherin-14 exhibits neural-specific expression and by Northern blot analysis is found broadly expressed in the adult central nervous system. In addition, cadherin-14 is detected in small-cell lung carcinoma cell lines which have a neuroectodermal phenotype[1].

Disease association

Knock out
MGI:1344366

Amino acid sequence of human cadherin-14

```
  1 MKITSTSCIC PVLVCLCFVQ RCYGTAHHSS IKVMRNQTKH IEGETEVHHR PKRGWVWNQF
 61 FVLEEHMGPD PQYVGKLHSN SDKGDGSVKY ILTGEGAGTI FIIDDTTGDI HSTKSLDREQ
121 KTHYVLHAQA IDRRTNKPLE PESEFIIKVQ DINDNAPKFT DGPYIVTVPE MSDMGTSVLQ
181 VTATDADDPT YGNSARVVYS ILQGQPYFSV DPKTGVIRTA LHNMDREARE HYSVVIQAKD
241 MAGQVGGLSG STTVNITLTD VNDNPPRFPQ KHYQLYVPES AQVGSAVGKI KANDADTGSN
301 ADMTYSIING DGMGIFSIST DKETREGILS LKKPLNYEKK KSYTLNIEGA NTHLDFRFSH
361 LGPFKDATML KIIVGDVDEP PLFSMPSYLM EVYENAKIGT VVGTVLAQDP DSTNSLVRYF
421 INYNVEDDRF FNIDANTGTI RTTKVLDREE TPWYNITVTA SEIDNPDLLS HVTVGIRVLD
481 VNDNPPELAR EYDIIVCENS KPGQVIHTIS ATDKDDFANG PRFNFFLDER LPVNPNFTLK
541 DNEDNTASIL TRRRRFSRTV QDVYYLPIMI SDGGIPSLSS SSTLTIRVCA CERDGRVRTC
601 HAEAFLSSAG LSTGALIAIL LCVLILLAIV VLFITLRRSK KEPLIISEED VRENVVTYDD
661 EGGGEEDTEA FDITALRNPS AAEELKYRRD IRPEVKLTPR HQTSSTLESI DVQEFIKQRL
721 AEADLDPSVP PYDSLQTYAY EGQRSEAGSI SSLDSATTQS DQDYHYLGDW GPEFKKLAEL
781 YGEIESERTT
```

The propeptide sequence underlined and in italics is cleaved to form the mature protein.

Data base accession
EMBL/Genbank   U59325
SwissProt   Q13634

References
[1] Shibata, T. et al. (1997) J. Biol. Chem. 272, 5236–5240.
[2] Yap, A.S. et al. (1997) Ann. Rev. Cell Dev. Biol. 13, 119–146.
[3] Shimoyama, Y. et al. (1999) J. Biol. Chem. 274, 11987–11994.

M-Cadherin

Cadherin-15, muscle cadherin

Family

Cadherin (type II subfamily)

Structure

Molecular weights

Amino acids	814
Polypeptide	88 915
SDS-PAGE reduced	125 kDa

Carbohydrate

N-linked sites	4
O-linked sites	unknown

Gene location   16q24.3

Gene structure

Alternative forms

Structure

Like the classical type I cadherins, M-cadherin is predicted to be expressed initially as a precursor with an N-terminal propeptide domain cleaved off inside the cell during protein maturation. The extracellular domain contains five cadherin repeats of approximately 110 amino acids separated by four calcium binding sites. In contrast to the classical type I cadherins, M-cadherin has a shorter propeptide sequence and lacks the HAV motif in the N-terminal domain. In addition, M-cadherin exhibits a higher sensitivity to proteolysis in the presence of Ca^{2+}[1,2]. The cytoplasmic domain associates with catenin complexes thereby linking of M-cadherin with the actin cytoskeleton[2]. The structure of the cadherins is discussed further in Section I.

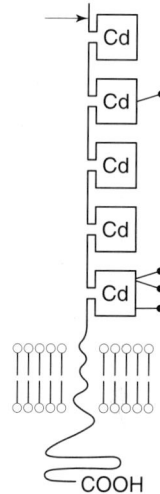

Ligands

M-cadherin mediates Ca^{2+}-dependent homotypic adhesion[2].

Function

The expression of M-cadherin in myogenic cells during embryogenesis and in regenerating adult muscle suggests that M-cadherin in involved in morphogenesis of skeletal muscle cells. In particular, blocking studies have implicated M-cadherin in their terminal differentiation[3].

Distribution

During development, M-cadherin is present at low levels in myoblasts and is upregulated in myotube forming cells[4,5]. Shortly before birth, M-cadherin expression is downregulated but is re-expressed in regenerating skeletal muscle myoblasts[5-7]. In addition M-cadherin is expressed in specialized adherens junctions in the granule cell layer of the cerebellar glomerulus[8].

Disease association

OMIM 114019

Knockout

Amino acid sequence of human M-cadherin

```
  1 MDAAFLLVLG LLAQSLCLSL GVPGWRRPTT LYPWRRAPAL SRVRRAWVIP PISVSENHKR
 61 LPYPLVQIKS DKQQLGSVIY SIQGPGVDEE PRGVFSIDKF TGKVFLNAML DREKTDRFRL
121 RAFALDLGGS TLEDPTDLEI VVVDQNDNRP AFLQEAFTGR VLEGAVPGTY VTRAEATDAD
181 DPETDNAALR FSILQQGSPE LFSIDELTGE IRTVQVGLDR EVVAVYNLTL QVADMSGDGL
241 TATASAIITL DDINDNAPEF TRDEFFMEAI EAVSGVDVGR LEVEDRDLPG SPNWVARFTI
301 LEGDPDGQFT IRTDPKTNEG VLSIVKALDY ESCEHYELKV SVQNEAPLQA AALRAERGQA
361 KVRVHVQDTN EPPVFQENPL RTSLAEGAPP GTLVATFSAR DPDTEQLQRL SYSKDYDPED
421 WLQVDAATGR IQTQHVLSPA SPFLKGGWYR AIVLAQDDAS QPRTATGTLS IEILEVNDHA
481 PVLAPPPPGS LCSEPHQGPG LLLGATDEDL PPHGAPFHFQ LSPRLPELGR NWSLSQVNVS
541 HARLRPRHQV PEGLHRLSLL LRDSGQPPQQ REQPLNVTVC RCGKDGVCLP GAAALLAGGT
601 GLSLGALVIV LASALLLLVL VLLVALRARF WKQSRGKGLL HGPQDDLRDN VLNYDEQGGG
661 EEDQDAYDIS QLRHPTALSL PLGPPPLRRD APQGRLHPQP PRVLPTSPLD IADFINDGLE
721 AADSDPSVPP YDTALIYDYE GDGSVAGTLS SILSSQGDED QDYDYLRDWG PRFARLADMY
781 GHPCGLEYGA RWDHQAREGL SPGALLPRHR GRTA
```

The propeptide sequence underlined and in italics is cleaved to form the mature protein.

Database accession

EMBL/GenBank   D83542
SwissProt   P55291

References

[1] Yap, A.S. et al. (1997) Annu. Rev. Cell Dev. Biol. 13, 119–146.
[2] Kuch, C. et al. (1997) Exp. Cell Res. 232, 331–338.
[3] Zeschnigk, M. et al. (1995) J. Cell Sci. 108, 2973–2981.
[4] Donalies, M. et al. (1991) Proc. Natl Acad. Sci. USA 88, 8024–8028.
[5] Moore, R. and Walsh, F.S. (1993) Development 117, 1409–1420.
[6] Irintchev, A. et al. (1994) Dev. Dyn. 199, 326–337.
[7] Pouliot, Y. et al. (1994) Dev. Dyn. 200, 305–312.
[8] Rose, O. et al. (1995) Proc. Natl Acad. Sci. USA 92, 6022–6026.

Ksp-Cadherin

Cadherin-16, kidney-specific cadherin

Family

Cadherin (modified cadherin)

Structure

Molecular weights

Amino acids	829
Polypeptide	89 866

SDS-PAGE reduced	130 kDa

Carbohydrate

N-linked sites	4
O-linked sites	unknown

Gene location   16q21–16q22

Gene structure

Alternative forms

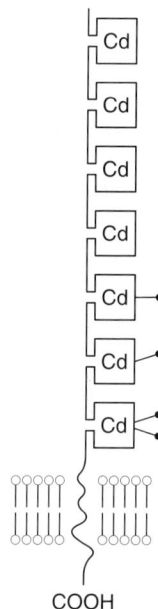

COOH

Structure

Unlike the classical type I cadherins[1], the extracellular domain of Ksp-cadherin contains seven rather than five cadherin repeats of approximately 110 amino acids, it lacks the HAV adhesion recognition sequence in the N-terminal domain, there is no evidence to support the existence of a propeptide sequence and the cytoplasmic domain is relatively short[2,3]. The structure of the cadherins is discussed further in Section I.

Ligands

Not known but based on homology to LI-cadherin, Ksp-cadherin is predicted to mediate Ca^{2+}-dependent homotypic adhesion.

Function

Predicted to mediate cell–cell adhesion in the kidney.

Distribution

Ksp-cadherin expression is restricted to the kidney where it is found on the basolateral membranes of the proximal tubule, the thick and thin limbs of the loop of Henle, the distal convoluted tubule and a subpopulation of cells in the connecting tubule and collecting duct[2,4].

Disease association

OMIM 603118

Knockout

MGI:106671

Amino acid sequence of human Ksp-cadherin

```
  1 MVPAWLWLLC VSVPQALPKA QPAELSVEVP ENYGGNFPLY LTKLPLPREG AEGQIVLSGD
 61 SGKATEGPFA MDPDSGFLLV TRALDREEQA EYQLQVTLEM QDGHVLWGPQ PVLVHVKDEN
121 DQVPHFSQAI YRARLSRGTR PGIPFLFLEA SDRDEPGTAN SDLRFHILSQ APAQPSPDMF
181 QLEPRLGALA LSPKGSTSLD HALERTYQLL VQVKDMGDQA SGHQATATVE VSIIESTWVS
241 LEPIHLAENL KVLYPHHMAQ VHWSGGDVHY HLESHPPGPF EVNAEGNLYV TRELDREAQA
301 EYLLQVRAQN SHGEDYAAPL ELHVLVMDEN DNVPICPPRD PTVSIPELSP PGTEVTRLSA
361 EDADAPGSPN SHVVYQLLSP EPEDGVEGRA FQVDPTSGSV TLGVLPLRAG QNILLLVLAM
421 DLAGAEGGFS STCEVEVAVT DINDHAPEFI TSQIGPISLP EDVEPGTLVA MLTAIDADLE
481 PAFRLMDFAI ERGDTEGTFG LDWEPDSGHV RLRLCKNLSY EAAPSHEVVV VVQSVAKLVG
541 PGPGPGATAT VTVLVERVMP PPKLDQESYE ASVPISAPAG SFLLTIQPSD PISRTLRFSL
601 VNDSEGWLCI EKFSGEVHTA QSLQGAQPGD TYTVLVEAQD TDEPRLSASA PLVIHFLKAP
661 PAPALTLAPV PSQYLCTPRQ DHGLIVSGPS KDPDLASGHG PYSFTLGPNP TVQRDWRLQT
721 LNGSHAYLTL ALHWVEPREH IIPVVVSHNA QMWQLLVRVI VCRCNVEGQC MRKVGRMKGM
781 PTKLSAVGIL VGTLVAIGIF LILIFTHWTM SRKKDPDQPA DSVPLKATV
```

Database accession

EMBL/GenBank   AF016272
TrEMBL   O75309

References

1 Yap, A.S. et al. (1997) Annu. Rev. Cell Dev. Biol. 13, 119–146.
2 Thomson, R.B. et al. (1995) J. Biol. Chem. 270, 17594–17601.
3 Thomson, R.B. et al. (1998) Genomics 51, 445–451.
4 Thomson, R.B. and Aronson, P.S. (1999) Am. J. Physiol. 277, F146–156

LI-Cadherin

Family

Cadherin (modified cadherin)

Structure

Molecular weights

Amino acids	832
Polypeptide	92 149

SDS-PAGE reduced	120 kDa

Carbohydrate

N-linked sites	7
O-linked sites	

Gene location   8q22.1

Gene structure

Alternative forms

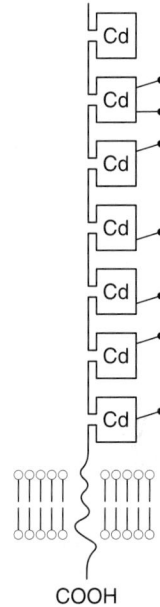

COOH

Structure

Unlike the classical type I cadherins[1], the extracellular domain of LI-cadherin contains seven rather than five cadherin repeats of approximately 110 amino acids, it lacks the HAV adhesion recognition sequence in the N-terminal domain, there is no evidence to support the existence of a propeptide sequence and the cytoplasmic domain is relatively short[2]. The structure of the cadherins is discussed further in Section I.

Ligands

LI-cadherin can mediate Ca^{2+}-dependent cell:cell adhesion in transfected cells[2]. Unlike the classical cadherins, this adhesion is independent of the cytoplasmic domain as generation of a GPI-anchored LI-cadherin resulted in cell–cell adhesion in transfected cells equivalent to that mediated by intact LI-cadherin[3].

Function

Predicted to mediate cell–cell adhesion in liver and intestine with an adhesive function complementary to that of the co-expressed classical cadherins[3].

Distribution

LI-cadherin has a restricted expression in liver and intestine, where it is found on the basolateral surface of hepatocytes and the lateral surface of intestinal epithelial cells[2].

Disease association

OMIM 603017

Knockout

MGI:1095414

Amino acid sequence of human LI-cadherin

```
  1 MILQAHLHSL CLLMLYLATG YGQEGKFSGP LKPMTFSIYE GQEPSQIIFQ FKANPPAVTF
 61 ELTGETDNIF VIEREGLLYY NRALDRETRS THNLQVAALD ANGIIVEGPV PITIKVKDIN
121 DNRPTFLQSK YEGSVRQNSR PGKPFLYVNA TDLDDPATPN GQLYYQIVIQ LPMINNVMYF
181 QINNKTGAIS LTREGSQELN PAKNPSYNLV ISVKDMGGQS ENSFSDTTSV DIIVTENIWK
241 APKPVEMVEN STDPHPIKIT QVRWNDPGAQ YSLVDKEKLP RFPFSIDQEG DIYVTQPLDR
301 EEKDAYVFYA VAKDEYGKPL SYPLEIHVKV KDINDNPPTC PSPVTVFEVQ ENERLGNSIG
361 TLTAHDRDEE NTANSFLNYR IVEQTPKLPM DGLFLIQTYA GMLQLAKQSL KKQDTPQYNL
421 TIEVSDKDFK TLCFVQINVI DINDQTPIFE KSDYGNLTLA EDTNIGSTIL TIQATDADEP
481 FTGSSKILYH IIKGDSEGRL GVDTDPHTNT GYVIIKKPLD FETAAVSNIV FKAENPEPLV
541 FGVKYNASSF AKFTLIVTDV NEAPQFSQHV FQAKVSEDVA IGTKVGNVTA KDPEGLDISY
601 SLRGDTRGWL KIDHVTGEIF SVAPLDREAG SPYRVQVVAT EVGGSSLSSV SEFHLILMDV
661 NDNPPRLAKD YTGLFFCHPL SAPGSLIFEA TDDDQHLFRG PHFTFSLGSG SLQNDWEVSK
721 INGTHARLST RHTEFEEREY VVLIRINDGG RPPLEGIVSL PVTFCSCVEG SCFRPAGHQT
781 GIPTVGMAVG ILLTTLLVIG IILAVVFIRI KKDKGKDNVE SAQASEVKPL RS
```

Database accession

EMBL/GenBank U07969 (human)
SwissProt   P55281 (rat)

References

[1] Yap, A.S. et al. (1997) Annu. Rev. Cell Dev. Biol. 13, 119–146.
[2] Berndorff, D. et al. (1994) J. Cell Biol. 125, 1353–1369.
[3] Kreft, B. et al. (1997) J. Cell Biol. 136, 1109–1121.

Desmocollin 1

Family

Cadherin (desmosomal subfamily)

Structure

Molecular weights

Amino acids   894
Polypeptide   100 044

SDS-PAGE reduced   120 kDa

Carbohydrate

N-linked sites   2
O-linked sites

Gene location   18 (closely linked to desmocollin 2)

Gene structure

Alternative forms

Types IA (desmosomal glycoprotein 2) and IB (desmosomal glycoprotein 3) are produced by alternate splicing leading to different cytoplasmic tails.

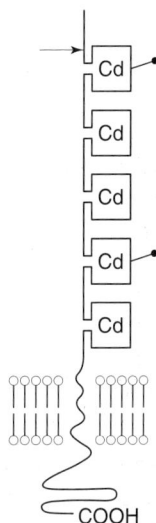

Structure

Desmocollin 1[1-3] is a type I membrane glycoprotein that shows overall homology to the classic cadherins, and similarly has a cleaved N-terminal propeptide and five extracellular cadherin repeats of approximately 110 acids length. The mature form of the protein forms a homodimer with adjacent molecules at the cell surface, utilizing conserved membrane proximal cysteine residues. An adhesion recognition sequence that is involved in homophilic interactions with cadherins on neighbouring cells is F/YAT/S, rather than HAV as seen in the classical cadherins. A short conserved sequence (C domain) in the cytoplasmic tail of desmocollin 1 interacts with intermediate filaments (keratin in skin and desmin in heart) via plakoglobin (γ-catenin) and other desmosome-associated proteins, such as desmoplakins and plectin (reviewed in refs 4–9).

Ligands

Calcium-dependent homophilic adhesion occurs between cadherin repeats of molecules in adjacent cells.

Function

Desmocollin 1 is a component of the intercellular desmosomal junction[4-9], maintaining the strength and integrity of the epithelium[10-12]. It mediates calcium-dependent cell–cell adhesion and the interaction of membrane plaque proteins with intracellular intermediate filaments. It may also be involved in keratinization (epithelial maturation) and differential cellular adhesion via desmocollin variants interacting with different cytoplasmic components[4,13,14].

Distribution

There is strong expression in the epidermis but not in other, non-keratinizing epithelia; there is lower expression in a restricted number of other tissues such as lymph node.

Disease association

OMIM 125643

Knockout

MGI:109173

Amino acid sequence of human desmocollin 1

```
  1 MALASAAPGS IFCKQLLFSL LVLTLLCDAC QKVYLRVPSH LQAETLVGKV NLEECLKSAS
 61 LIRSSDPAFR ILEDGSIYTT HDLILSSERK SFSIFLSDGQ RREQQEIKVV LSARENKSPK
121 KRHTKDTALK RTKRRWAPIP ASLMENSLGP FPQHVQQIQS DAAQNYTIFY SISGPGVDKE
181 PFNLFYIEKD TGDIFCTRSI DREKYEQFAL YGYATTADGY APEYPLPLII KIEDDNDNAP
241 YFEHRVTIFT VPENCRSGTS VGKVTATDLD EPDTLHTRLK YKILQQIPDH PKHFSIHPDT
301 GVITTTTPFL DREKCDTYQL IMEVRDMGGQ PFGLFNTGTI TISLEDENDN PPSFTETSYV
361 TEVEENRIDV EILRMKVQDQ DLPNTPHSKA VYKILQGNEN GNFIISTDPN TNEGVLCVVK
421 PLNYEVNRQV ILQVGVINEA QFSKAASSQT PTMCTTTVTV KIIDSDEGPE CHPPVKVIQS
481 QDGFPAGQEL LGYKALDPEI SSGEGLRYQK LGDEDNWFEI NQHTGDLRTL KVLDRESKFV
541 KNNQYNISVV AVDAVGRSCT GTLVVHLDDY NDHAPQIDKE VTICQNNEDF AVLKPVDPDG
601 PENGPPFQFF LDNSASKNWN IEEKDGKTAI LRQRQNLDYN YYSVPIQIKD RHGLVATHML
661 TVRVCDCSTP SECRMKDKST RDVRPNVILG RWAILAMVLG SVLLLCILFT CFCVTAKRTV
721 KKCFPEDIAQ QNLIVSNTEG PGEEVTEANI RLPMQTSNIC DTSMSVGTVG GQGIKTQQSF
781 EMVKGGYTLD SNKGGGHQTL ESVKGVGQGD TGRYAYTDWQ SFTQPRLGEK VYLCGQDEEH
841 KHCEDYVFSY NYEGKGSLAG SVGCCSDRQE EEGLEFLDHL EPKFRTLAKT CIKK
```

Notes

1 The propeptide sequence underlined and in italics is cleaved to form the mature protein.

2 In form 3b amino acids 830–840 are replaced by ESIRGHTLIKN, and 841 to 894 are deleted.

Database accession

EMBL/GenBank   Z34522
SwissProt   Q08554

References

1 Troyanovsky, S.M. et al. (1993) Cell 72, 561–574.
2 Theis, D.G. et al. (1993) Int. J. Dev. Biol. 37, 101–110.
3 King, I.A. et al. (1993) Genomics 18, 185–194.
4 Buxton, R.S. and Magee, A.I. (1992) Semin. Cell Biol. 3, 157–167.
5 Koch, P.J. and Franke, W.W. (1994) Curr. Opin. Cell Biol. 6, 682–687.
6 Green, K.J. and Jones, J.C. (1996) FASEB J. 10, 871–881.

7 Garrod, D. et al. (1996) Curr. Opin. Cell Biol. 8, 670–678
8 Troyanovsky, S.M. and Leube, R.E. (1998) Subcell. Biochem. 31, 263–289.
9 Kowalczyk, A.P. et al. (1999) Int. Rev. Cytol. 185, 237–302.
10 King, I.A. et al. (1991) FEBS Lett. 286, 9–12.
11 King, I.A. et al. (1996) J. Invest. Dermatol. 107, 531–538.
12 Burdett, I.D. (1998) Micron 29, 309–328.
13 Buxton, R.S. et al. (1993) J. Cell Biol. 121, 481–483.
14 Schmidt, A. et al. (1994) Eur. J. Cell Biol. 65, 229–245.

Desmocollin 2

Family

Cadherin (desmosomal subfamily)

Structure

Molecular weights

Amino acids	901
Polypeptide	99 961
SDS-PAGE reduced	115 kDa

Carbohydrate

N-linked sites	5
O-linked sites	

Gene location   18q12.1

Gene structure

Alternative forms

Types 2A (desmosomal glycoprotein 2) and 2B (desmosomal glycoprotein 3) are produced by alternate splicing leading to different cytoplasmic tails[1].

Structure

Desmocollin 2[1-3] is a type I glycoprotein of the cadherin family composed of five cadherin repeats with a cleaved N-terminal propeptide as in the classical cadherins (see desmocollin 1 for general structural details, which are well reviewed)[4-9].

Ligands

Calcium-dependent homophilic adhesion occurs between cadherin repeats of molecules in adjacent cells.

Function

Desmocollin 2 is a component of the intercellular desmosomal junction[4-9], maintaining the strength and integrity of the epithelium and heart muscle[10-12]. It mediates calcium-dependent cell:cell adhesion and the interaction of membrane plaque proteins with intracellular intermediate filaments. It may also be involved in keratinization (epithelial maturation) and differential cellular adhesion via desmocollin variants interacting with different cytoplasmic components.

Distribution

Strong expression in all epithelia, lymph nodes and heart occurs.

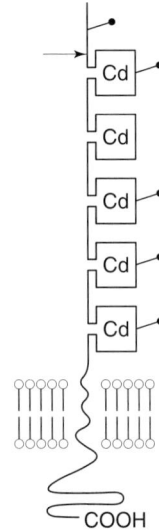

Disease association

OMIM 125645

Knockout

MGI:103221

Amino acid sequence of human desmocollin 2

```
  1 MAAVGSMRSG SPAFGLGHLL TLAILAL*ASD ACKEVVLQVP SELPAEKFVG RVNLMDCLKS*
 61 *ADIVHLSDPD FQVLEDGSVY TTSSVVLSSG QRSFTIWLFS TDSQEEREIS VHLEGPVEVL*
121 *NKRPHTEKVL SRAKR*RWAPI PCSMLENSLG PFPLFLQQIQ SDTAQNYTIY YSIRGPGVDK
181 EPLNLFYVER DTGNLYCTGR VDREQYESFE LTAFATTPDG YTPEYPLPLL IKIEDENDNY
241 PIFTQKLYSF TVQENSRIGS IVGEVCATDL DEPDTMHTRL RYSILEQSPS PPMLFTMHPS
301 TGVITTTSAQ LDRELIDKYQ LLIKVQDMDG QYFGLHTTAK CIITIEDVND NLPTFTRTTY
361 VTSVEENTVN VEILRLTVQD KDLVNSPNWR ANYTILKGNE NGNFKIVTDP KTNEGILCVI
421 KPLDYEERQQ VTLQIGVVNE APYTREASSK SPMSTATVTV TVTNQDEGPE CIPPMQTVRI
481 QENVPVGTRN DGYKAYDPET RSSSGIRYRK LSDPRGWVTV NEDSGSITIF RALDREAETV
541 RNGIYNITVL ALDADGRSCT GTLGIILEDV NDNGPFIPKQ TVVICKATMS SAEIVAVDLD
601 DPVNGPPFDF SLESSDSEVQ RMWRLTRIND TAARLSYQND PSFGSYAVPI RVTDRLGLSS
661 VTTLNVLVCD CITESDCTLR SGERTGYADV RLGPWAILAI LLGIALLFCI LFTLVCSVSR
721 ASKQQKILPD DLAQQNLIVS NTEAPGDDKV YSTNGLTTQT MGASGQTAFT TMGTGVKSGG
781 QETIEMVKGG QQTLDSRRGA GYHHHTLDPC RGGHVEVDNY RHTYSEWYNF IQPRLGDKVQ
841 FCHTDDNQKL AQDYVLTYNY EGKGSAAGSV GCCSDLQEED GLEFLDHLEP KFRTLAEVCA
901 KR
```

Notes

1 The propeptide sequence underlined and in italics is cleaved to form the mature protein.
2 In form 2b amino acids 837–847 are replaced by ESIRGHTLIKN, and 848 to 901 are deleted.

Database accession

EMBL/GenBank   X56807
SwissProt   Q02487

References

1 Parker, A.E. et al (1991) J. Biol. Chem. 266, 10438–10445.
2 Buxton, R.S. et al. (1994) Genomics 21, 510–516.
3 Lorimer, J.E. et al. (1994) Mol. Membrane Biol. 11, 229–236.
4 Buxton, R.S. and Magee, A.I. (1992) Semin. Cell Biol. 3, 157–167.
5 Koch, P.J. and Franke, W.W. (1994) Curr. Opin. Cell Biol. 6, 682–687.
6 Green, K.J. and Jones, J.C. (1996) FASEB J. 10, 871–881.
7 Garrod, D. et al. (1996) Curr. Opin. Cell Biol. 8, 670–678.
8 Troyanovsky, S.M. and Leube, R.E. (1998) Subcell. Biochem. 31, 263–289.
9 Kowalczyk, A.P. et al. (1999) Int. Rev. Cytol. 185, 237–302.
10 King, I.A. et al. (1991) FEBS Lett. 286, 9–12.
11 King, I.A. et al. (1996) J. Invest. Dermatol. 107, 531–538.
12 Burdett, I.D. (1998) Micron 29, 309–328.

Desmocollin 3 — Formerly desmocollin 4

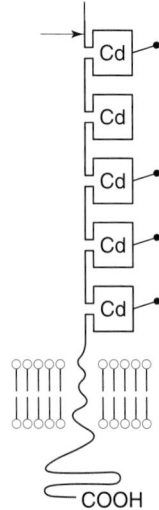

Family

Cadherin (desmosomal subfamily)

Structure

Molecular weights

Amino acids	896
Polypeptide	99 968

SDS-PAGE reduced   115 kDa

Carbohydrate

N-linked sites	4
O-linked sites	

Gene location   18q12.1

Gene structure

Alternative forms

Types 3A and 3B are produced by alternate splicing leading to different cytoplasmic tails.

Structure

Desmocollin 3[1,2] is a type I glycoprotein of the cadherin family composed of five cadherin repeats with a cleaved N-terminal propeptide as in the classical cadherins (see desmocollin 1 for general structural details, which are well reviewed)[3-8].

Ligands

Calcium-dependent homophilic adhesion occurs between cadherin repeats of molecules in adjacent cells.

Function

Its location in epithelial desmosomes and cadherin structure[3-8] implies it has a role in epithelial cell–cell adhesion[9-11].

Distribution

Keratinocytes and all other epithelia.

Disease association

OMIM 600271

Knockout

MGI:1194993

Amino acid sequence of human desmocollin 3

```
  1 MAAAGPRRSV RGAVCLHLLL TLVIFSRAGE ACKKVILNVP SKLEADKIIG RVNLEECFRS
 61 ADLIRSSDPD FRVLNDGSVY TARAVALSDK KRSFTIWLSD KRKQTQKEVT VLLEHQKKVS
121 KTRHTRETVL RRAKRRWAPI PCSMQENSLG PFPLFLQQVE SDAAQNYTVF YSISGRGVDK
181 EPLNLFYIER DTGNLFCTRP VDREEYDVFD LIAYASTADG YSADLPLPLP IRVEDENDNH
241 PVFTEAIYNF EVLESSRPGT TVGVVCATDR DEPDTMHTRL KYSILQQTPR SPGLFSVHPS
301 TGVITTVSHY LDREVVDKYS LIMKVQDMDG QFFGLIGTST CIITVTDSND NAPTFRQNAY
361 EAFVEENAFN VEILRIPIED KDLINTANWR VNFTILKGNE NGHFKISTDK ETNEGVLSVV
421 KPLNYEENRQ VNLEIGVNNE APFARDIPRV TALNRALVTV HVRDLDEGPE CTPAAQYVRI
481 KENLAVGSKI NGYKAYDPEN RNGNGLRYKK LHDPKGWITI DEISGSIITS KILDREVETP
541 KNELYNITVL AIDKDDRSCT GTLAVNIEDV NDNPPEILQE YVVICKPKMG YTDILAVDPD
601 EPVHGAPFYF SLPNTSPEIS RLWSLTKVND TAARLSYQKN AGFQEYTIPI TVKDRAGQAA
661 TKLLRVNLCE CTHPTQCRAT SRSTGVILGK WAILAILLGI ALLFSVLLTL VCGVFGATKG
721 KRFPEDLAQQ NLIISNTEAP GDDRVCSANG FMTQTTNNSS QGFCGTMGSG MKNGGQETIE
781 MMKGGNQTLE SCRGAGHHHT LDSCRGGHTE VDNCRYTYSE WHSFTQPRLG EKLHRCNQNE
841 DRMPSQDYVL TYNYEGRGSP AGSVGCCSEK QEEDGLDFLN NLEPKFITLA EACTKR
```

Notes

1 The propeptide sequence underlined and in italics is cleaved to form the mature protein.
2 In form 3b amino acids 832–839 are replaced by ESIRGHTG, and 840 to 896 are deleted.

Database accession

EMBL/GenBank   D17427
SwissProt   Q14574

References

[1] Kawamura, K. et al. (1994) J. Biol. Chem. 269, 26295–26302.
[2] King, I.A. et al. (1995) J. Invest. Dermatol. 105, 314–321.
[3] Buxton, R.S. and Magee, A.I. (1992) Semin. Cell Biol. 3, 157–167.
[4] Koch, P.J. and Franke, W.W. (1994) Curr. Opin. Cell Biol. 6, 682–687.
[5] Green, K.J. and Jones, J.C. (1996) FASEB J. 10, 871–881.
[6] Garrod, D. et al. (1996) Curr. Opin. Cell Biol. 8, 670–678.
[7] Troyanovsky, S.M. and Leube, R.E. (1998) Subcell. Biochem. 31, 263–289.
[8] Kowalczyk, A.P. et al. (1999) Int. Rev. Cytol. 185, 237–302.
[9] King, I.A. et al. (1991) FEBS Lett. 286, 9–12.
[10] King, I.A. et al. (1996) J. Invest. Dermatol. 107, 531–538.
[11] Burdett, I.D. (1998) Micron 29, 309–328.

Desmoglein 1

Family

Cadherin (desmosomal subfamily)

Structure

Molecular weights

Amino acids	1049
Polypeptide	113 715
SDS-PAGE reduced	165 kDa

Carbohydrate

N-linked sites	3
O-linked sites	

Gene location   18q12.1–12.2

Gene structure

Alternative forms

Structure

Desmoglien 1[1-3] is a type I membrane glycoprotein of the cadherin family. The extracellular N-terminus of desmoglein 1 differs from classical cadherins in having a short, 29 amino acid long propeptide and only four cadherin repeat domains. The mature form of the protein forms a homodimer with adjacent molecules at the cell surface, utilizing conserved membrane proximal cysteine residues. The adhesion motif, R/YAL, utilized in homophilic adhesion with cadherins on adjacent cells also differs from that utilized in classical cadherins (HAV). The intracellular part of desmoglein 1 diverges substantially from other cadherins in containing five, 28–30 amino acid long, repeat sequences (desmoglein repeats). Whilst smaller, these show some homology to cadherin repeats and are predicted to have an antiparallel beta-sheet structure. A short conserved sequence (C domain) in the cytoplasmic tail of desmoglein 1 interacts with intermediate filaments (keratin in skin and desmin in heart) via plakoglobin (γ-catenin) and other desmosome-associated proteins, such as desmoplakins and plectin (reviewed in refs 4–9).

Ligands

Calcium-dependent homophilic adhesion occurs between cadherin repeats of molecules in adjacent cells.

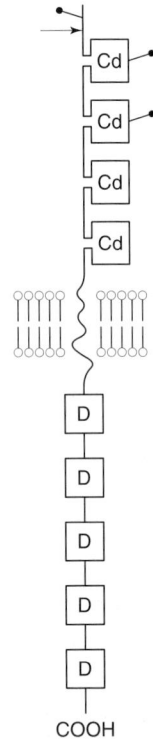

Function

Desmoglein 1 is a component of intercellular desmosomal junction[4-9], maintaining the strength and integrity of the epithelium. It mediates calcium-dependent cell–cell adhesion and the interaction of membrane plaque proteins with intracellular intermediate filaments[10-12]. It may also be involved in epithelial maturation and keratinization.

Distribution

Suprabasal expression in keratinizing epidermis, including tongue, tonsil and oesophagus.

Disease association

OMIM 125670
Desmoglein 1 is the target antigen for the development of autoantibodies in the blistering skin disease, pemphigus foliaceus[13-16].

Knockout

MGI:94930

Amino acid sequence of human desmoglein 1

```
   1 MDWSFFRVVA VLFIFLVVVE VNSEFRIQVR DYNTKNGTIK WHSIRRQKRE WIKFAAACRE
  61 GEDNSKRNPI AKIHSDCAAN QQVTYRISGV GIDQPPYGIF VINQKTGEIN ITSIVDREVT
 121 PFFIIYCRAL NSMGQDLERP LELRVRVLDI NDNPPVFSMA TFAGQIEENS NANTLVMILN
 181 ATDADEPNNL NSKIAFKIIR QEPSDSPMFI INRNTGEIRT MNNFLDREQY GQYALAVRGS
 241 DRDGGADGMS AECECNIKIL DVNDNIPYME QSSYTIEIQE NTLNSNLLEI RVIDLDEEFS
 301 ANWMAVIFFI SGNEGNWFEI EMNERTNVGI LKVVKPLDYE AMQSLQLSIG VRNKAEFHHS
 361 IMSQYKLKAS AISVTVLNVI EGPVFRPGSK TYVVTGNMGS NDKVGDFVAT DLDTGRPSTT
 421 VRYVMGNNPA DLLAVDSRTG KLTLKNKVTK EQYNMLGGKY QGTILSIDDN LQRTCTGTIN
 481 INIQSFGNDD RTNTEPNTKI TTNTGRQEST SSTNYDTSTT STDSSQVYSS EPGNGAKDLL
 541 SDNVHFGPAG IGLLIMGFLV LGLVPFLMIC CDCGGAPRSA AGFEPVPECS DGAIHSWAVE
 601 GPQPEPRDIT TVIPQIPPDN ANIIECIDNS GVYTNEYGGR EMQDLGGGER MTGFELTEGV
 661 KTSGMPEICQ EYSGTLRRNS MRECREGGLN MNFMESYFCQ KAYAYADEDE GRPSNDCLLI
 721 YDIEGVGSPA GSVGCCSFIG EDLDDSFLDT LGPKFKKLAD ISLGKESYPD LDPSWPPQST
 781 EPVCLPQETE PVVSGHPPIS PHFGTTTVIS ESTYPSGPGV LHPKPILDPL GYGNVTVTES
 841 YTTSDTLKPS VHVHDNRPAS NVVVTERVVG PISGADLHGM LEMPDLRDGS NVIVTERVIA
 901 PSSSLPTSLT IHHPRESSNV VVTERVIQPT SGMIGSLSMH PELANAHNVI VTERVVSGAG
 961 VTGISGTTGI SGGIGSSGLV GTSMGAGSGA LSGAGISGGG IGLSSLGGTA SIGHMRSSSD
1021 HHFNQTIGSA SPSTARSRIT KYSTVQYSK
```

The propeptide sequence underlined and in italics is cleaved to form the mature protein.

Database accession

EMBL/GenBank X56654
SwissProt Q02413

References

1 Nilles, L.A. et al. (1991) J. Cell Sci. 99, 809–821.
2 Wheeler, G.N. et al. (1991) Proc. Natl Acad. Sci. 88, 4796–4800.
3 Buxton, R.S. et al. (1994) Genomics 21, 510–516.

4 Buxton, R.S. and Magee, A.I. (1992) Semin. Cell Biol. 3, 157–167.
5 Koch, P. J. and Franke, W.W. (1994) Curr. Opin. Cell Biol. 6, 682–687.
6 Green, K.J. and Jones, J.C. (1996) FASEB J. 10, 871–881.
7 Garrod, D. et al. (1996) Curr. Opin. Cell Biol. 8, 670–678.
8 Troyanovsky, S.M. and Leube, R.E. (1998) Subcell. Biochem. 31, 263–289.
9 Kowalczyk, A.P. et al. (1999) Int. Rev. Cytol. 185, 237–302.
10 King, I.A. et al. (1991) FEBS Lett. 286, 9–12.
11 King, I. A. et al. (1996) J. Invest. Dermatol. 107, 531–538.
12 Burdett, I.D. (1998) Micron 29, 309–328.
13 Amagai, M. et al. (1991) Cell 67, 869–877.
14 Chidgey, M.A. (1997) Histol. Histopathol. 12, 1159–1168.
15 Lin, M.S. et al. (1997) Clin. Exp. Immunol. 107 (Suppl. 1), 9–12.
16 Moll, R. and Moll, I. (1998) Virchows Arch. 432, 487–504.

Desmoglein 2 HDGC

Family

Cadherin (desmosomal subfamily)

Structure

Molecular weights

Amino acids	1117
Polypeptide	122 385
SDS-PAGE reduced	165 kDa

Carbohydrate

N-linked sites	5
O-linked sites	

Gene location 18q12.1–12.2

Gene structure

Alternative forms

Structure

Desmoglein 2 is a type I membrane glycoprotein of the cadherin family[1,2]. Like desmoglein 1 (see entry for further details), it is has a short propeptide (25 amino acids) and is composed of four extracellular cadherin repeats (N-terminal) and six intracellular cadherin-like (desmoglein) repeats (C-terminal) (reviewed in refs[3-8]).

Ligands

Homophilic adhesion occurs between cadherin repeats of molecules in adjacent cells.

Function

It is a component of the intercellular desmosomal junction[3-8], maintaining the strength and integrity of the epithelium. It mediates calcium-dependent cell–cell adhesion and the interaction of membrane plaque proteins with intracellular intermediate filaments[9-11]. It may also be involved in epithelial maturation and keratinization. It was originally found in colon cancer and designated 'human-desmoglein-colon' (hence, HDGC)[12].

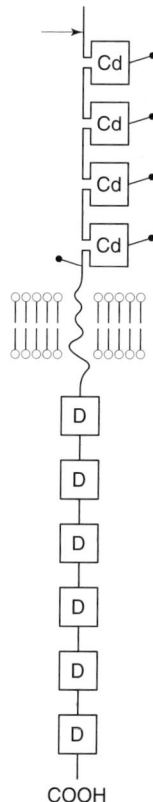

Distribution

It is expressed in all simple and stratified epithelia.

Disease association

OMIM 125671

Knockout

MGI:1196466

Amino acid sequence of human desmoglein 2

```
   1 MARTRDRVRL LLLLICFNVG SGLHLQVLST RNENKLLPKH PHLVRQKRAW ITAPVALREG
  61 EDLSKKNPIA KIHSDLAEER GLKITYKYTG KGITEPPFGI FVFNKDTGEL NVTSILDREE
 121 TPFFLLTGYA LDARGNNVEK PLELRIKVLD INDNEPVFTQ DVFVGSVEEL SAAHTLVMKI
 181 NATDADEPNT LNSKISYRIV SLEPAYPPVF YLNKDTGEIY TTSVTLDREE HSSYTLTVEA
 241 RDGNGEVTDK PVKQAQVQIR ILDVNDNIPV VENKVLEGMV EENQVNVEVT RIKVFDADEI
 301 GSDNWLANFT FASGNEGGYF HIETDAQTNE GIVTLIKEVD YEEMKNLDFS VIVANKAAFH
 361 KSIRSKYKPT PIPIKVKVKN VKEGIHFKSS VISIYVSESM DRSSKGQIIG NFQAFDEDTG
 421 LPAHARYVKL EDRDNWISVD SVTSEIKLAK LPDFESRYVQ NGTYTVKIVA ISEDYPRKTI
 481 TGTVLINVED INDNCPTLIE PVQTICHDAE YVNVTAEDLD GHPNSGPFSF SVIDKPPGMA
 541 EKWKIARQES TSVLLQQSEK KLGRSEIQFL ISDNQGFSCP EKQVLTLTVC EVLHGSGCRE
 601 AQHDSYVGLG PAAIALMILA FLLLLLVPLL LLMCHCGKGA KAFTPIPGTI EMLHPWNNEG
 661 APPEDKVVPS FLPVDQGGSL VGRNGVGGMA KEATMKGSSS ASIVKGQHEM SEMDGRWEEH
 721 RSLLSGRATQ FTGATGAIMT TETTKTARAT GASRDMAGAQ AAAVALNEEF LRNYFTDKAA
 781 SYTEEDENHT AKDCLLVYSQ EETESLNASI GCCSFIEGEL DDRFLDDLGL KFKTLAEVCL
 841 GQKIDINKEI EQRQKPATET SMNTASHSLC EQTMVNSENT YSSGSSFPVP KSLQEANAEK
 901 VTQEIVTERS VSSRQAQKVA TPLPDPMASR NVIATETSYV TGSTMPPTTV ILGPSQPQSL
 961 IVTERVYAPA STLVDQPYAN EGTVVVTERV IQPHGGGSNP LEGTQHLQDV PYVMVRERES
1021 FLAPSSGVQP TLAMPNIAVG QNVTVTERVL APASTLQSSY QIPTENSMTA RNTTVSGAGV
1081 PGPLPDFGLE ESGHSNSTIT TSSTRVTKHS TVQHSYS
```

The propeptide sequence underlined and in italics is cleaved to form the mature protein.

Database accession

EMBL/GenBank   Z26317
SwissProt   Q14126

References

[1] Koch, P.J. et al. (1991) Eur. J. Cell Biol. 55, 200–208.

[2] Schaefer, S. et al. (1994) Exp. Cell Res. 211, 391–399.

[3] Buxton, R.S. and Magee, A.I. (1992) Semin. Cell Biol. 3, 157–167.

[4] Koch, P.J. and Franke, W.W. (1994) Curr. Opin. Cell Biol. 6, 682–687.

[5] Green, K.J. and Jones, J.C. (1996) FASEB J. 10, 871–881.

[6] Garrod, D. et al. (1996) Curr. Opin. Cell Biol. 8, 670–678.

[7] Troyanovsky, S.M. and Leube, R.E. (1998) Subcell. Biochem. 31, 263–289.

[8] Kowalczyk, A.P. et al. (1999) Int. Rev. Cytol. 185, 237–302.

[9] King, I.A. et al. (1991) FEBS Lett. 286, 9–12.

[10] King, I.A. et al. (1996) J. Invest. Dermatol. 107, 531–538.

[11] Burdett, I.D. (1998) Micron 29, 309–328.

[12] Arnemann, J. et al. (1992) Genomics 13, 484–486.

Desmoglein 3 — Pemphigus vulgaris antigen, PVA

Family
Cadherin (desmosomal subfamily)

Structure

Molecular weights
Amino acids	999
Polypeptide	107 503
SDS-PAGE reduced	130 kDa

Carbohydrate
N-linked sites	4
O-linked sites	

Gene location   18q12.1–12.2

Gene structure   23 kb, 15 exons

Alternative forms

Structure
Desmoglein 3 is a type I membrane glycoprotein of the cadherin family[1,2]. Like desmoglein 1 (see entry for further details) it is has a short propeptide (26 amino acids) and is composed of four extracellular cadherin repeats (N-terminal) but only two intracellular cadherin-like (desmoglein) repeats (C-terminal) (reviewed in refs 4–9).

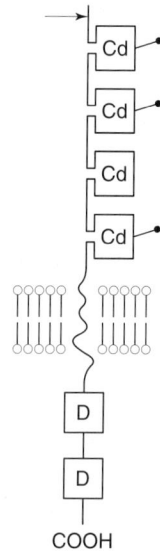

Ligands
Calcium-dependent homophilic adhesion occurs between cadherin repeats of molecules in adjacent cells.

Function
It is a component of the intercellular desmosomal junction[4-9], maintaining the strength and integrity of the epithelium[10-12]. It mediates calcium-dependent cell–cell adhesion and the interaction of membrane plaque proteins with intracellular intermediate filaments.

Distribution
Epidermis, tongue, tonsil, oesophagus, carcinomas.

Disease association
OMIM   169615

Autoantibodies to desmoglein 3 lead to the severe blistering skin disease, pemphigus vulgaris[1,13-16], due to loss of epithelial cell–cell adhesion, and to pemphigus associated with lymphoid neoplasms.

Knockout

MGI:99499

The Gene Knockout FactsBook[17], p262

A mutation in mouse desmoglein 3 results in the phenotype 'balding'[18].

Amino acid sequence of human desmoglein 3

```
  1 MMGLFPRTTG ALAIFVVVIL VHGELRIETK GQYDEEEMTM QQAKRRQKRE WVKFAKPCRE
 61 GEDNSKRNPI AKITSDYQAT QKITYRISGV GIDQPPFGIF VVDKNTGDIN ITAIVDREET
121 PSFLITCRAL NAQGLDVEKP LILTVKILDI NDNPPVFSQQ IFMGEIEENS ASNSLVMILN
181 ATDADEPNHL NSKIAFKIVS QEPAGTPMFL LSRNTGEVRT LTNSLDREQA SSYRLVVSGA
241 DKDGEGLSTQ CECNIKVKDV NDNFPMFRDS QYSARIEENI LSSELLRFQV TDLDEEYTDN
301 WLAVYFFTSG NEGNWFEIQT DPRTNEGILK VVKALDYEQL QSVKLSIAVK NKAEFHQSVI
361 SRYRVQSTPV TIQVINVREG IAFRPASKTF TVQKGISSKK LVDYILGTYQ AIDEDTNKAA
421 SNVKYVMGRN DGGYLMIDSK TAEIKFVKNM NRDSTFIVNK TITAEVLAID EYTGKTSTGT
481 VYVRVPDFND NCPTAVLEKD AVCSSSPSVV VSARTLNNRY TGPYTFALED QPVKLPAVWS
541 ITTLNATSAL LRAQEQIPPG VYHISLVLTD SQNNRCEMPR SLTLEVCQCD NRGICGTSYP
601 TTSPGTRYGR PHSGRLGPAA IGLLLLGLLL LLLAPLLLLT CDCGAGSTGG VTGGFIPVPD
661 GSEGTIHQWG IEGAHPEDKE ITNICVPPVT ANGADFMESS EVCTNTYARG TAVEGTSGME
721 MTTKLGAATE SGGAAGFATG TVSGAASGFG AATGVGICSS GQSGTMRTRH STGGTNKDYA
781 DGAISMNFLD SYFSQKAFAC AEEDDGQEAN DCLLIYDNEG ADATGSPVGS VGCCSFIADD
841 LDDSFLDSLG PKFKKLAEIS LGVDGEGKEV QPPSKDSGYG IESCGHPIEV QQTGFVKCQT
901 LSGSQGASAL SASGSVQPAV SIPDPLQHGN YLVTETYSAS GSLVQPSTAG FDPLLTQNVI
961 VTERVICPIS SVPGNLAGPT QLRGSHTMLC TEDPCSRLI
```

The propeptide sequence underlined and in italics is cleaved to form the mature protein.

Database accession

EMBL/GenBank   M76482

SwissProt   P32926

References

1. Amagai, M. et al. (1991) Cell 67, 869–877.
2. Silos, S.A. et al. (1996) J. Biol. Chem. 271, 17504–17511.
4. Buxton, R.S. and Magee, A.I. (1992) Semin. Cell Biol. 3, 157–167.
5. Koch, P.J. and Franke, W.W. (1994) Curr. Opin. Cell Biol. 6, 682–687.
6. Green, K.J. and Jones, J.C. (1996) FASEB J. 10, 871–881.
7. Garrod, D. et al. (1996) Curr. Opin. Cell Biol. 8, 670–678.
8. Troyanovsky, S.M. and Leube, R.E. (1998) Subcell. Biochem. 31, 263–289.
9. Kowalczyk, A.P. et al. (1999) Int. Rev. Cytol. 185, 237–302.
10. King, I.A. et al. (1991) FEBS Lett. 286, 9–12.
11. King, I.A. et al. (1996) J. Invest. Dermatol. 107, 531–538.
12. Burdett, I.D. (1998) Micron 29, 309–328.
13. Amagai, M. et al. (1998) J. Clin. Invest. 102, 775–782.
14. Chidgey, M.A. (1997) Histol. Histopathol. 12, 1159–1168.
15. Lin, M.S. et al. (1997) Clin. Exp. Immunol. 107 (Suppl. 1), 9–15.
16. Moll, R, and Moll, I. (1998) Virchows Arch. 432, 487–504.
17. Mak, T.W. (1998) The Gene Knockout FactsBook. Academic Press, London.
18. Koch, P.J. et al. (1997) J. Cell Biol. 137, 1091–1102.

Immunoglobulin Superfamily (IgSF)

ALCAM

Family

Immunoglobulin superfamily

Structure

Molecular weights

Amino acids	583
Polypeptide	65 132

SDS-PAGE reduced   100–105 kDa

Carbohydrate

N-linked sites	9
O-linked sites	

Gene location   3q13.1–3q13.2

Gene structure

Alternative forms

COOH

Structure

ALCAM is a type I transmembrane protein containing five Ig-like domains and a short 32 amino acid cytoplasmic domain. The CD6 binding domain has been mapped to the N-terminal Ig domain[1,2].

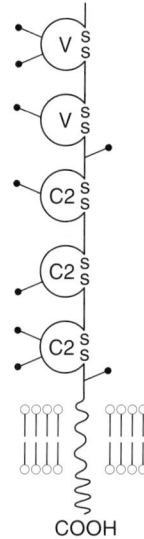

Ligands

ALCAM is a counter-receptor for CD6, but has also been reported to act as a homophilic adhesion molecule in the nervous system[3] and bind to the chicken L1 family member, NgCAM[4].

Function

The ALCAM–CD6 interaction mediates thymocyte–thymic epithelial cell adhesion and may play a role in T cell development[5]. In the nervous system, a role in axon guidance has been postulated[6,7].

Distribution

ALCAM is found expressed on neurons, epithelial cells, fibroblasts, activated T cells and activated monocytes[8–10].

Disease association

OMIM 601662
It is speculated that an interaction between ALCAM and CD6+ T cells could be involved in certain neurodegenerative disease pathologies. In addition, ALCAM expression correlates with the neoplastic progression of melanomas[11].

Knockout

MGI:131266

Amino acid sequence of human ALCAM

```
  1 MESKGASSCR LLFCLLISAT VFRPGLGWYT VNSAYGDTII IPCRLDVPQN LMFGKWKYEK
 61 PDGSPVFIAF RSSTKKSVQY DDVPEYKDRL NLSENYTLSI SNARISDEKR FVCMLVTEDN
121 VFEAPTIVKV FKQPSKPEIV SKALFLETEQ LKKLGDCISE DSYPDGNITW YRNGKVLHPL
181 EGAVVIIFKK EMDPVTQLYT MTSTLEYKTT KADIQMPFTC SVTYYGPSGQ KTIHSEQAVF
241 DIYYPTEQVT IQVLPPKNAI KEGDNITLKC LGNGNPPPEE FLFYLPGQPE GIRSSNTYTL
301 MDVRRNATGD YKCSLIDKKS MIASTAITVH YLDLSLNPSG EVTRQIGDAL PVSCTISASR
361 NATVVWMKDN IRLRSSPSFS SLHYQDAGNY VCETALQEVE GLKKRESLTL IVEGKPQIKM
421 TKKTDPSGLS KTIICHVEGF PKPAIQWTIT GSGSVINQTE ESPYINGRYY SKIIISPEEN
481 VTLTCTAENQ LERTVNSLNV SAISIPEHDE ADEISDENRE KVNDQAKLIV GIVVGLLLAA
541 LVAGVVYWLY MKKSKTASKH VNKDLGNMEE NKKLEENNHK TEA
```

Database accession

EMBL/GenBank   L38608
SwissPort   Q13740

References

[1] Bowen, M.A. et al. (1996) J. Biol. Chem. 271, 17390–17396.
[2] Skonier, J.E. et al. (1996) Biochemistry 35, 12287–12291.
[3] Corbel, C. et al. (1996) Proc. Natl Acad. Sci. USA 93, 2844–2849.
[4] DeBernardo, A.P. et al. (1996) J. Cell Biol. 133, 657–666.
[5] Osario, L.M. et al. (1998) Immunology 93, 358–365.
[6] Burns, F.R. et al. (1991) Neuron 7, 209–220.
[7] Kanki, J.P. et al. (1994) J. Neurobiol. 25, 831–845.
[8] Bowen, M.A. et al. (1995) J. Exp. Med. 181, 2213–2220.
[9] Patel, D.D. et al. (1995) J. Exp. Med. 181, 1563–1568.
[10] Pourquie, O. et al. (1992) Proc. Natl Acad. Sci. USA 89, 5261–5265.
[11] van Kempen, L.C. et al. (2000) Am. J. Pathol. 156, 769–774.

CD22

Siglec-2, BL-CAM, B lymphocyte cell adhesion molecule, Leu-14

Family

Immunoglobulin superfamily
(siglec subfamily)

Structure

Molecular weights

Amino acids	α 647 amino acids
	β 847 amino acids
Polypeptide	α 73 202
	β 95 347
SDS-PAGE reduced	α 130 kDa
	β 140 kDa

Carbohydrate

N-linked sites	α 10
	β 11
O-linked sites	

Gene location 19q13.1

Gene structure 22 kb; 15 exons

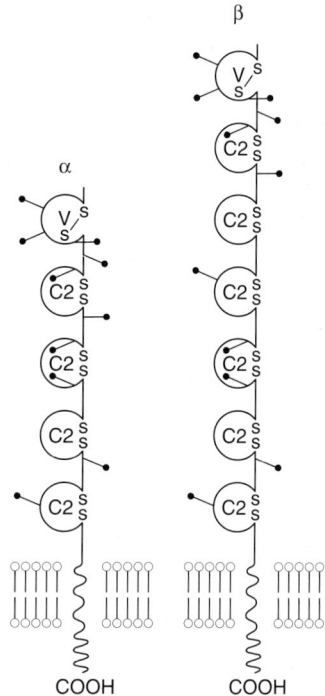

Alternative forms

Alternative splicing of the CD22 gene results in the CD22α isoform lacking Ig domains 3 and 4, and with a truncated cytoplasmic domain[1].

Structure

CD22 is a member of the siglec family of sialic acid binding lectins (sialoadhesin, CD33 and MAG) and, like other family members, is predicted to have an unusual disulphide bond between β strands B and E in domain 1, and a disulphide bond between domains 1 and 2[2]. The predominant form in humans is CD22β. The full cytoplasmic domain contains six tyrosine residues, four of which are in ITIM motifs. Tyrosine phosphorylation results in association with the tyrosine phosphatase SHP-1, the tyrosine kinase Syk, and phospholipase C-γ[3,4]. Other intracellular signalling molecules including Lck, PI3-kinase and Grb2 have also been reported to associate with the CD22 cytoplasmic domain. Some studies suggest that CD22 is a heterodimer of the α-and β-isoforms.

Ligands

CD22 binds to the sialoglycoconjugate NeuAcα2–6Galβ1–4GlcNAc, which is widely present on N-linked carbohydrates[5]. Binding to sialoglycoconjugates requires that CD22 expressing cells do not themselves carry these ligands[6]. CD22 forms a loose complex with the B cell antigen receptor (BCR).

Function

Engagement of the BCR results in an increased tyrosine phosphorylation of CD22, and the recruitment of SHP-1 and other signalling molecules[7,8]. This in turn results in the negative regulation of B cell signalling. In addition, CD22 functions as an adhesion receptor mediating the interaction of B cells with α2–6-linked sialic acid glycoconjugates. A role for CD22 as a B cell homing receptor mediating the interaction with bone marrow sinusoidal endothelium has also been proposed[9].

Distribution

CD22 is present in the cytoplasm of late pro-B cells and appears on the cell surface simultaneously with surface IgD, and is found on most mature peripheral B cells. Expression is lost with terminal B cell differentiation[10].

Disease association

OMIM 107266
The presence of cytoplasmic CD22 is a marker for B cell precursor acute lymphocytic leukaemia.

Knockout

MGI:88322
The Gene Knockout FactsBook[11], p151
CD22 −/− B cells are hyper-responsive to receptor signalling, and display a reduced level of surface IgM, an enhanced Ca^{2+} flux and cell proliferation in response to Ig signalling and an increased rate of apoptosis. Mice have an augmented immune response, contain increased serum titres of autoantibody and an absence of recirculating B cells in the bone marrow. CD22 is not required for embryonic development, B cell development or isotype switching[12–16].

Amino acid sequence of human CD22α

```
  1 MHLLGPWLLL LVLEYLAFSD SSKWVFEHPE TLYAWEGACV WIPCTYRALD GDLESFILFH
 61 NPEYNKNTSK FDGTRLYEST KDGKVPSEQK RVQFLGDKNK NCTLSIHPVH LNDSGQLGLR
121 MESKTEKWME RIHLNVSERP FPPHIQLPPE IQESQEVTLT CLLNFSCYGY PIQLQWLLEG
181 VPMRQAAVTS TSLTIKSVFT RSELKFSPQW SHHGKIVTCQ LQDADGKFLS NDTVQLNVKH
241 PPKKVTTVIQ NPMPIREGDT VTLSCNYNSS NPSVTRYEWK PHGAWEEPSL GVLKIQNVGW
301 DNTTIACAAC NSWCSWASPV ALNVQYAPRD VRVRKIKPLS EIHSGNSVSL QCDFSSSHPK
361 EVQFFWEKNG RLLGKESQLN FDSISPEDAG SYSCWVNNSI GQTASKAWTL EVLYAPRRLR
421 VSMSPGDQVM EGKSATLTCE SDANPPVSHY TWFDWNNQSL PYHSQKLRLE PVKVQHSGAY
481 WCQGTNSVGK GRSPLSTLTV YYSPETIGRR VAVGLGSCLA ILILAICGLK LQRRWKRTQS
541 QQGLQENSSG QSFFVRNKKV RRAPLSEGPH SLGCYNPMME DGISYTTLRF PEMNIPRTGD
601 AESSEMQRPP PDCDDTVTYS ALHKRQVGTM RTSFQIFQKM RGFITQS
```

Amino acid sequence of human CD22β

```
  1 MHLLGPWLLL LVLEYLAFSD SSKWVFEHPE TLYAWEGACV WIPCTYRALD GDLESFILFH
 61 NPEYNKNTSK FDGTRLYEST KDGKVPSEQK RVQFLGDKNK NCTLSIHPVH LNDSGQLGLR
121 MESKTEKWME RIHLNVSERP FPPHIQLPPE IQESQEVTLT CLLNFSCYGY PIQLQWLLEG
181 VPMRQAAVTS TSLTIKSVFT RSELKFSPQW SHHGKIVTCQ LQDADGKFLS NDTVQLNVKH
241 TPKLEIKVTP SDAIVREGDS VTMTCEVSSS NPEYTTVSWL KDGTSLKKQN TFTLNLREVT
301 KDQSGKYCCQ VSNDVGPGRS EEVFLQVQYA PEPSTVQILH SPAVEGSQVE FLCMSLANPL
361 PTNYTWYHNG KEMQGRTEEK VHIPKILPWH AGTYSCVAEN ILGTGQRGPG AELDVQYPPK
421 KVTTVIQNPM PIREGDTVTL SCNYNSSNPS VTRYEWKPHG AWEEPSLGVL KIQNVGWDNT
481 TIACARCNSW CSWASPVALN VQYAPRDVRV RKIKPLSEIH SGNSVSLQCD FSSSHPKEVQ
541 FFWEKNGRLL GKESQLNFDS ISPEDAGSYS CWVNNSIGQT ASKAWTLEVL YAPRRLRVSM
601 SPGDQVMEGK SATLTCESDA NPPVSHYTWF DWNNQSLPHH SQKLRLEPVK VQHSGAYWCQ
661 GTNSVGKGRS PLSTLTVYYS PETIGRRVAV GLGSCLAILI LAICGLKLQR RWKRTQSQQG
721 LQENSSGQSF FVRNKKVRRA PLSEGPHSLG CYNPMMEDGI SYTTLRFPEM NIPRTGDAES
781 SEMQRPPRTC DDTVTYSALH KRQVGDYENV IPDFPEDEGI HYSELIQFGV GERPQAQENV
841 DYVILKH
```

Database accession

	CD22α	CD22β
EMBL/GenBank	X52785	X59350
SwissProt	P20273	P20273

References

[1] Stamenkovic, I. and Seed, B. (1990) Nature 345, 74–77.
[2] Crocker, P.R. et al. (1996) Biochem. Soc. Trans. 24, 150–156.
[3] Doody, G.M. et al. (1995) Science, 269, 242–244.
[4] Law, C.L. et al. (1996) J. Exp. Med. 183, 547–560.
[5] Powell, L.D. and Varki, A. (1995) J. Biol. Chem. 270, 14243–14246.
[6] Razi, N. and Varki, A. (1998) Proc. Natl Acad. Sci. USA 95, 7469–7474.
[7] Doody, G.M. et al. (1996) Curr. Opin. Immunol. 8, 378–382.
[8] Cornall, R.J. et al. (1999) Curr. Top. Microbiol. Immunol. 244, 57–68.
[9] Nitschke, L. et al. (1999) J. Exp. Med. 189, 1513–1518.
[10] Law, C.L. et al. (1994) Immunol. Today 15, 442–449.
[11] Mak, T.W. (1998) The Gene Knockout FactsBook. Academic Press, London.
[12] O'Keefe, T.L. et al. (1996) Science 274, 798–801.
[13] Sato, S. et al. (1996) Immunity 6, 509–517.
[14] Otipoby, K.L. et al. (1996) Nature 384, 634–637.
[15] Nitschke, L. et al. (1997) Curr. Biol. 7, 133–143.
[16] O'Keefe, T.L. et al. (1999) J. Exp. Med. 189, 1307–1313.

Family
Immunoglobulin superfamily

Structure
Molecular weights
Amino acids	738
Polypeptide	82 536

SDS-PAGE reduced	120–140 kDa

Carbohydrate
N-linked sites	9
O-linked sites	

Gene location
17q23

Gene structure
65 kb, 16 exons

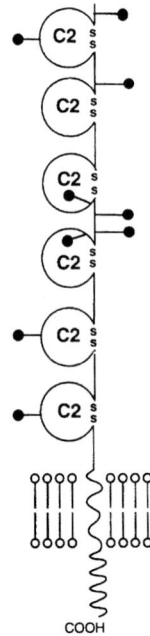

Alternative forms
Alternative splicing in the transmembrane domain (exon 9) can result in soluble CD31 isoforms. In addition the cytoplasmic domain encoded for by exons 10–16 is subject to variable alternative splicing[1]. A Leu-Val polymorphism is found at amino acid 125[2].

Structure
CD31 is a type I transmembrane molecule with six C2 Ig domains. The first two Ig domains, with contributions from domains 3, 5 and 6, mediate the homophilic interaction[3,4], whereas domain 2 is the main domain interacting with integrins[5] and glycosaminoglycans[6]. Alternative splicing and tyrosine phosphorylation within the cytoplasmic domain modulates the configuration of the ligand binding domain, suggesting a mechanism for regulating ligand specificity[1,7,8].

Ligands
CD31 is a homophilic adhesion molecule[3] but can also interact heterotypically with integrin $\alpha_V\beta_3$[5,9] and glycosaminoglycans[6]. Recently malaria (*Plasmodium falciparum*)-infected red blood cells have been shown to adhere to CD31 on vascular endothelium[10].

Function

The majority of CD31 activities result from the homophilic CD31–CD31 interactions. These include leucocyte extravasation, bone marrow haematopoiesis and vascular development[11]. There is increasing evidence that CD31 has signalling function via the ability of tyrosine phosphorylated CD31 to recruit the tyrosine phosphatase SHP-2[12] and other signalling components[13]. Activation of CD31 upregulates β_1 and β_2 integrins resulting in further adhesion interactions, and this receptor crosstalk is mediated by the small GTPase Rap1[14].

Distribution

CD31 is highly expressed by endothelial cells, where it localizes to cell–cell contact areas. In addition, expression is found on platelets, monocytes, granulocytes and other leucocytes.

Disease association

OMIM 173445
The polymorphism at residue 125 can result in acute graft-versus-host disease in mismatched bone marrow transplants.

Knockout

MGI:97537
The Gene Knockout FactsBook[15], p166
CD31 −/− mice develop normally with no obvious vascular or platelet aggregation defects. Leucocytes are able to extravasate into inflammatory sites, although there are subtle defects in this process as evidenced by an increased number of arrested neutrophils between the vascular endothelium and basement membrane[16].

Amino acid sequence of human CD31

```
  1 MQPRWAQGAT MWLGVLLTLL LCSSLEGQEN SFTINSVDMK SLPDWTVQNG KNLTLQCFAD
 61 VSTTSHVKPQ HQMLFYKDDV LFYNISSMKS TESYFIPEVR IYDSGTYKCT VIVNNKEKTT
121 AEYQLLVEGV PSPRVTLDKK EAIQGGIVRV NCSVPEEKAP IHFTIEKLEL NEKMVKLKRE
181 KNSRDQNFVI LEFPVEEQDR VLSFRCQARI ISGIHMQTSE STKSELVTVT ESFSTPKFHI
241 SPTGMIMEGA QLHIKCTIQV THLAQEFPEI IIQKDKAIVA HNRHGNKAVY SVMAMVEHSG
301 NYTCKVESSR ISKVSSIVVN ITELFSKPEL ESSFTHLDQG ERLNLSCSIP GAPPANFTIQ
361 KEDTIVSQTQ DFTKIASKSD SGTYICTAGI DKVVKKSNTV QIVVCEMLSQ PRISYDAQFE
421 VIKGQTIEVR CESISGTLPI SYQLLKTSKV LENSTKNSND PAVFKDNPTE DVEYQCVADN
481 CHSHAKMLSE VLRVKVIAPV DEVQISILSS KVVESGEDIV LQCAVNEGSG PITYKFYREK
541 EGKPFYQMTS NATQAFWTKQ KASKEQEGEY YCTAFNRANH ASSVPRSKIL TVRVILAPWK
601 KGLIAVVIIG VIIALLIIAA KCYFLRKAKA KQMPVEMSRP AVPLLNSNNE KMSDPNMEAN
661 SHYGHNDDVR NHAMKPINDN KEPLNSDVQY TEVQVSSAES HKDLGKKDTE TVYSEVRKAV
721 PDAVESRYSR TEGSLDGT
```

Database accession

EMBL/GenBank   M37780/ M28526
SwissProt   P16284

References
[1] Yan, H.C. et al. (1995) J. Exp. Med. 184, 229–239.
[2] Behar, E. et al. (1996) N. Engl. J. Med. 334, 286–291.
[3] Fawcett, J. et al. (1995) J. Cell Biol. 130, 451–460.
[4] Newton, J.P. et al. (1997) J. Biol. Chem. 272, 20555–20563.
[5] Buckley, C.D. et al. (1996) J. Cell Sci. 109, 437–445.
[6] DeLisser, H.M. et al. (1993) J. Biol. Chem. 268, 16037–16046.
[7] Baldwin, H.S. et al. (1994) Development 120, 2539–2553.
[8] Famiglietti, J. et al. (1997) J. Cell Biol. 138, 1425–1435.
[9] Piali, L. et al. (1995) J. Cell Biol. 130, 451–460.
[10] Treutiger, C.J. et al. (1997) Nature Med. 3, 1405–1408.
[11] Newman, P.J. (1997) J. Clin. Invest. 99, 3–8.
[12] Sagawa, K. et al. (1997) J. Biol. Chem. 272, 31086–31091.
[13] Pumphrey, N.J. et al. (1999) FEBS Lett. 450, 77–83.
[14] Reedquist, K.A. et al. (2000) J. Cell Biol. 148, 1151–1158.
[15] Mak, T.W. (1998) The Gene Knockout FactsBook. Academic Press, London.
[16] Duncan, G.S. et al. (1999) J. Immunol. 162, 3022–3030.

Family

Immunoglobulin superfamily (siglec subfamily)

Structure

Molecular weights
Amino acids	364
Polypeptide	39 726
SDS-PAGE reduced	67 kDa

Carbohydrate
N-linked sites	5
O-linked sites	0

Gene location   19q13.3–q13.4

Gene structure   >35 kb

Alternative forms

There is evidence for alternative splicing in the mouse CD33 cytoplasmic domain[1].

Structure

CD33 is the smallest member of the siglec family (sialoadhesin, CD22 and MAG), and like other family members is predicted to have an unusual disulphide bond between β strands B and E in domain 1 and a disulphide bond between domains 1 and 2[2]. The cytoplasmic domain has two ITIM motifs which when tyrosine phosphorylated can bind SH2-containing phosphatases.

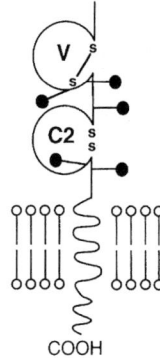

Ligands

CD33 binds to the sialoglycoconjugates NeuAcα2–3Galβ1–3(4)GlcNAc and NeuAcα2–3Galβ1–3GalNAc on glycoproteins. CD33 ligand binding is down-modulated by N-linked glycosylation in the first Ig domain[4].

Function

CD33 mediates cell adhesion *in vitro* but an *in vivo* function in myeloid cells has yet to be determined. The demonstration that tyrosine phosphorylation in the cytoplasmic domain can result in the recruitment of the tyrosine phosphatases SHP-1 and SHP-2 suggests a method for regulation of ligand binding and/or a role as a signalling receptor[5-6].

Distribution

CD33 is absent from pluripotential stem cells, but then appears on myelomonocytic precursors and then the myeloid and monocyte lineages, although it is absent from granulocytes[3,7].

Disease association

OMIM 159590

CD33 is expressed on some acute undifferentiated leukaemias and occasionally by acute lymphoblastic leukaemias.

Knockout

MGI:99440

Amino acid sequence of human CD33

```
  1 MPLLLLLPLL WAGALAMDPN FWLQVQESVT VQEGLCVLVP CTFFHPIPYY DKNSPVHGYW
 61 FREGAIISGD SPVATNKLDQ EVQEETQGRF RLLGDPSRNN CSLSIVDARR RDNGSYFFRM
121 ERGSTKYSYK SPQLSVHVTD LTHRPKILIP GTLEPGHSKN LTCSVSWACE QGTPPIFSWL
181 SAAPTSLGPR TTHSSVLIIT PRPQDHGTNL TCQVKFAGAG VTTERTIQLN VTYVPQNPTT
241 GIFPGDGSGK QETRAGVVHG AIGGAGVTAL LALCLCLIFF IVKTHRRKAA RTAVGRNDTH
301 PTTGSASPKH QKKSKLHGPT ETSSCSGAAP TVEMDEELHY ASLNFHGMNP SKDTSTEYSE
361 VRTQ
```

Database accession

EMBL/GenBank   M23197

SwissProt   P20138

References

1 Simmons, D. and Seed, B. (1988) J. Immunol. 141, 2797–2800.
2 Tchilian, E.Z. et al. (1994) Blood 83, 3188–3198.
3 Freeman, S.D. et al. (1995) Blood 85, 2005–2012.
4 Sgroi, D. et al. (1996) J. Biol. Chem. 271, 18803–18809.
5 Taylor, V.C. et al. (1999) J. Biol. Chem. 274, 11505–11512.
6 Ulyanova, T. et al. (1999) Eur. J. Immunol. 29, 3440–3449.
7 Andrews, R.G. et al. (1989) J. Exp. Med. 169, 1721–1731.

CD96

Family

Immunoglobulin superfamily

Structure

Molecular weights

Amino acids	569
Polypeptide	63 887

SDS-PAGE reduced   160 kDa (160, 180, 240 kDa unreduced)

Carbohydrate

N-linked sites	15
O-linked sites	+++

Gene location

Gene structure

Alternative forms

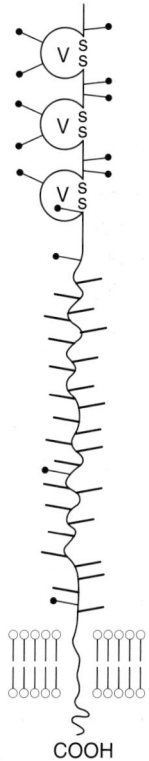

COOH

Structure

CD96 is a type I transmembrane glycoprotein of the Ig superfamily[1] and is composed N-terminally of two V-like and one C2–like Ig domains. The membrane-proximal part of CD96 is formed by a serine/threonine/proline-rich region, which is probably O-glycosylated. It may consist of a disulphide bonded homodimer[1].

Ligands

Not known.

Function

Probably involved in adhesive interactions of activated peripheral T cells and NK cells at late stages of immune response, such as following leucocyte emigration to sites of inflammation.

Distribution

Low levels in peripheral T and NK, but not expressed in B cells; up-regulated following T cell activation[1] and expressed by most T cell lines.

◻ **Disease association**

◻ **Knockout**

Amino acid sequence of human CD96

```
  1 MEKKWKYCAV YYIIQIHFVK GVWEKTVNTE ENVYATLGSD VNLTCQTQTV GFFVQMQWSK
 61 VTNKIDLIAV YHPQYGFYCA YGRPCESLVT FTETPENGSK WTLHLRNMSC SVSGRYECML
121 VLYPEGIQTK IYNLLIQTHV TADEWNSNHT IEIEINQTLE IPCFQNSSSK ISSEFTYAWS
181 VEDNGTQETL ISQNHLISNS TLLKDRVKLG TDYRLHLSPV QIFDDGRKFS CHIRVGPNKI
241 LRSSTTVKVF AKPEIPVIVE NNSTDVLVER RFTCLLKNVF PKANITWFID GSFLHDEKEG
301 IYITNEERKG KDGFLELKSV LTRVHSNKPA QSDNLTIWCM ALSPVPGNKV WNISSEKITF
361 LLGSEISSTD PPLSVTESTL DTQPSPASSV SPARYPATSS VTLVDVSALR PNTTPQPSNS
421 SMTTRGFNYP WTSSGTDTKK SVSRIPSETY SSSPSGAGST LHDNVFTSTA RAFSEVPTTA
481 NGSTKTNHVH ITGIVVNKPK DGMSWPVIVA ALLFCCMILF GLGVRKWCQY QKEIMERPPP
541 FKPPPPPIKY TCIQEPNESD LPYHEMETL
```

◻ **Database accession**
EMBL/GenBank   M88282
Swissprot   P40200

◻ *Reference*
[1] Wang, P.L. et al. (1992) J. Immunol. 148, 2600–2608.

CD147

Family

Immunoglobulin superfamily

Structure

Molecular weights

Amino acids	269
Polypeptide	29 221

SDS-PAGE reduced   55–65 kDa (50–60, unreduced)

Carbohydrate

N-linked sites	3
O-linked sites	

Gene location   19p13.3

Gene structure

Alternative forms

Some evidence of alternate RNA splicing in mouse yielding different N-termini.

Structure

Type I membrane glycoprotein of Ig superfamily[1,2] containing one C2–like and one membrane-proximal V-like domain, that is in the opposite orientation to most Ig superfamily members. N-linked carbohydrates account for approximately 50% of the mass of CD147, varying between tissues. Homologue of the rat OX-47 antigen[3]. Highly conserved transmembrane region between species[1,4].

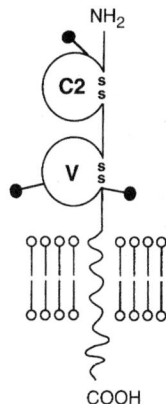

Ligands

Recently it has been shown that CD147 binds interstitial collagenase (MMP-1) at the surface of tumour cells[5].

Function

Intercellular adhesion, such as of retinal cells[6], is inhibited by antibodies to CD147, as is adhesion to matrix proteins (collagen IV, laminin, fibronectin). Recominant CD147 binds to cells, though its receptor is not known. It was first described as a tumour-cell-derived stimulator (EMMPRIN) of collagenase secretion by fibroblasts[7], and is thought via this function to be involved in tumour invasion and metastasis. CD147 carries an oncodevelopmental carbohydrate marker expressed on teratocarcinoma cells.

Distribution

It is widely expressed in embryonic and adult tissues, including the majority of circulating leucocytes (weakly), epithelia and endothelial cells, and is up-regulated upon lymphocyte activation[3]. There are some species differences in distribution[1,2,8,9].

Disease association

OMIM 109480

Knockout

MGI:88208
CD147 knockout mice show abnormal smell sensation and immunological changes[10].

Amino acid sequence of human CD147

```
  1 MAAALFVLLG FALLGTHGAS GAAGTVFTTV EDLGSKILLT CSLNDSATEV TGHRWLKGGV
 61 VLKEDALPGQ KTEFKVDSDD QWGEYSCVFL PEPMGTANIQ LHGPPRVKAV KSSEHINEGE
121 TAMLVCKSES VPPVTDWAWY KITDSEDKAL MNGSESRFFV SSSQGRSELH IENLNMEADP
181 GQYRCNGTSS KGSDQAIITL RVRSHLAALW PFLGIVAEVL VLVTIIFIYE KRRKPEDVLD
241 DDDAGSAPLK SSGQHQNDKG KNVRQRNSS
```

Database accession

EMBL/GenBank   X64364
SwissProt   P35613

References
1 Fossum, S. et al. (1991) Eur. J. Immunol. 21, 671.
2 Seulberger, H. et al. (1990) EMBO J. 9, 2151–2158.
3 Kasinrerk, W. et al. (1992) J. Immunol. 149, 847–854.
4 Miyauchi, T. et al. (1991) J. Biochem. 110, 770–774.
5 Guo, H. et al. (2000) Cancer Res. 60, 888–891.
6 Fadool, J.M. and Linser, P.J. (1993) Dev. Dynamics 196, 252–262.
7 Biswas, C. et al. (1995) Cancer Res. 55, 434–439.
8 Nehme, C.L. et al. (1995) Blood 310, 693–698.
9 Kanekura, T. et al. (1991) Cell Struct. Funct. 16, 23–30.
10 Iagkura, et al. (1996) Biochem. Biophys. Res. Commun. 224, 33–36.

CEACAM family

Carcinoembryonic antigen cell adhesion family, CD66 family, C-CAM family

Members

CEA	CEACAM5, CD66e
CEACAM1	CD66a, Cell-CAM 105, BGP, biliary glycoprotein, NCA-160
CEACAM3	CD66d, CGM1
CEACAM4	CGM7
CEACAM6	CD66c, NCA, NCA-90, CGM6
CEACAM7	CGM2
CEACAM8	CD66b, CGM6, CGM8, NCA-95

Note: The carcinoembryonic antigen (CEA) family nomenclature has recently been redefined[1,2].

Family

Immunoglobulin superfamily

Structure

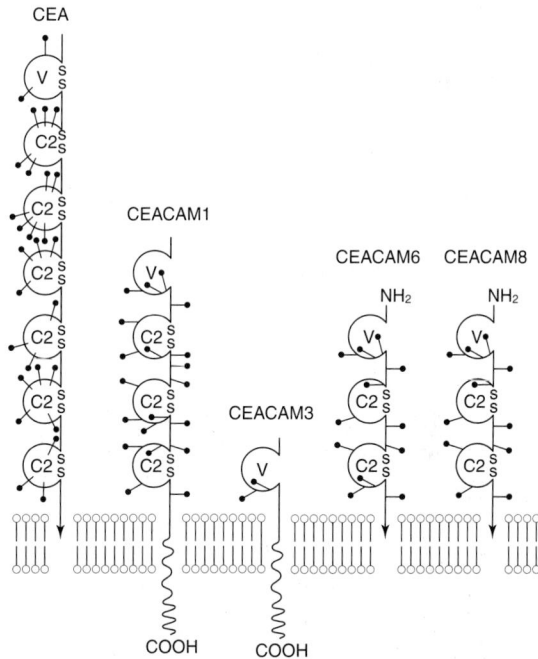

Molecular weights

Amino acids	CEA	702
	CEACAM1	526
	CEACAM3	252
	CEACAM6	344
	CEACAM8	349
Polypeptide	CEA	76 795
	CEACAM1	57 560
	CEACAM3	27 077
	CEACAM6	37 161
	CEACAM8	38 154

SDS-PAGE reduced   CEA   180–200 kDa
                   CEACAM1   140–180 kDa
                   CEACAM3   35 kDa
                   CEACAM6   90–95 kDa
                   CEACAM8   95–100 kDa

Carbohydrate
N-linked sites   CEA   28
                   CEACAM1   20
                   CEACAM3   2
                   CEACAM6   12
                   CEACAM8   11
O-linked sites

Gene location   19q13.1–19q13.2

Gene structure   The CEA family comprises at least 28 separate genes.

Alternative forms
CEACAM1, CEACAM3 and CEACAM7 are alternatively spliced. The largest isoforms are shown here.

Structure
The human CEA family is part of a cluster of at least 28 genes divided into two functional groups. By genomic mapping, the CEACAM subgroup contains seven members and the PSG (pregnancy-specific glycoprotein) subgroup of secreted molecules contains 11 members. The remaining genes are thought to be pseudogenes[1]. Within the best characterized CEACAM members, CEACAM1 and CEACAM3 encode type 1 transmembrane proteins while CEACAM8, CEACAM6 and CEA are GPI anchored in the membrane. All members possess an N-terminal V-type Ig domain followed by between 0 and 6 C2-type Ig domains. Apart from CEACAM3, the extracellular domains are extensively *N*-glycosylated. CEACAM1 and CEACAM3 have putative tyrosine phosphorylation sites in their cytoplasmic domains, which could bind signalling components such as the tyrosine phosphatases SHP-1 and SHP-2; CEACAM1 can associate with the cytoplasmic tyrosine kinases Src, Lyn and Hck[3,4]. Alternative splicing results in CEACAM1, 3 and 7 isoforms with varying numbers of Ig domains and/or shorter cytoplasmic domains[1]. Further structural and sequence information on other CEACAM family members and members of the PSG family can be found in refs 1 and 2.

Ligands
The CEACAM family can mediate homophilic cell–cell adhesion and in certain combinations, heterophilic interactions with other family members[5]. Binding is via the V-type domain[6,7]. In addition, CEACAM1 and CEACAM6 have been reported to bind E-selectin, CEACAM1 and CEACAM3 can act as a receptors for *Neisseria gonorrhoeae* and *Neisseria meningitidis*[8], murine CEACAM1 and CEACAM2 are receptors for murine coronaviruses[9], and CEACAM6 can bind galectins.

Function

Binding assays indicate a role for CEACAM family members in mediating adhesion between granulocytes and/or between granulocytes and epithelial cells, and as microbial receptors. In addition, signalling via CEACAM1 and CEACAM3 cytoplasmic domains[4] may regulate the adhesive activity of the β_2 integrins[10] and the cytolytic function of intraepithelial lymphocytes[11]. Different splice variants of CEACAM1 and 3 display different bacterial tropism and invasion[4]. Importantly, members of the CEACAM family are strongly down-regulated in malignancies, implicating these receptors as putative tumour suppressors[4]. It should be noted that Cell-CAM 105 originally identified in rats and described as a homophilic adhesion molecule involved in the formation and maintenance of hepatocyte polarization and exhibiting ecto-ATPase activity[12,13], is CEACAM1.

Distribution

CEACAM1 and CEACAM6 are abundant on granulocytes and epithelial cells, CEACAM8 and CEACAM3 are restricted to granulocytes, and CEA is mostly found on epithelial cells.

Disease association

OMIM CEA, 114890; CEACAM1, 109770; CEACAM6, 163980.

CEACAM1 and CEA are strongly down-regulated in colon and other carcinomas. Evidence that CEACAM proteins can act as tumour suppressors comes from studies in which transfection of CEACAM1 in carcinoma cells resulted in an inhibition of tumour development in nude mice and conversely down regulation in benign cells resulted in increased tumourigenicity[4]. CEA levels in serum are used routinely as clinical markers in the diagnosis and serial monitoring of cancer patients for recurrent disease or response to therapy.

Knockout

MGI:1347245 (CEACAM1)

Amino acid sequence of human CEA

```
  1 MESPSAPPHR WCIPWQRLLL TASLLTFWNP PTTAKLTIES TPFNVAEGKE VLLLVHNLPQ
 61 HLFGYSWYKG ERVDGNRQII GYVIGTQQAT PGPAYSGREI IYPNASLLIQ NIIQNDTGFY
121 TLHVIKSDLV NEEATGQFRV YPELPKPSIS SNNSKPVEDK DAVAFTCEPE TQDATYLWWV
181 NNQSLPVSPR LQLSNGNRTL TLFNVTRNDT ASYKCETQNP VSARRSDSVI LNVLYGPDAP
241 TISPLNTSYR SGENLNLSCH AASNPPAQYS WFVNGTFQQS TQELFIPNIT VNNSGSYTCQ
301 AHNSDTGLNR TTVTTITVYA EPPKPFITSN NSNPVEDEDA VALTCEPEIQ NTTYLWWVNN
361 QSLPVSPRLQ LSNDNRTLTL LSVTRNDVGP YECGIQNELS VDHSDPVILN VLYGPDDPTI
421 SPSYTYYRPG VNLSLSCHAA SNPPAQYSWL IDGNIQQHTQ ELFISNITEK NSGLYTCQAN
481 NSASGHSRTT VKTITVSAEL PKPSISSNNS KPVEDKDAVA FTCEPEAQNT TYLWWVNGQS
541 LPVSPRLQLS NGNRTLTLFN VTRNDARAYV CGIQNSVSAN RSDPVTLDVL YGPDTPIISP
601 PDSSYLSGAN LNLSCHSASN PSPQYSWRIN GIPQQHTQVL FIAKITPNNN GTYACFVSNL
661 ATGRNNSIVK SITVSASGTS PGLSAGATVG IMIGVLVGVA LI
```

In CEA the C-terminus is proteolytically cleaved and a GPI anchor attached. However, the site of cleavage has not been unambiguously determined.

Amino acid sequence of human CEACAM1

```
  1 MGHLSAPLHR VRVPWQGLLL TASLLTFWNP PTTAQLTTES MPFNVAEGKE VLLLVHNLPQ
 61 QLFGYSWYKG ERVDGNRQIV GYAIGTQQAT PGPANSGRET IYPNASLLIQ NVTQNDTGFY
121 TLQVIKSDLV NEEATGQFHV YPELPKPSIS SNNSNPVEDK DAVAFTCEPE TQDTTYLWWI
181 NNQSLPVSPR LQLSNGNRTL TLLSVTRNDT GPYECEIQNP VSANRSDPVT LNVTYGPDTP
241 TISPSDTYYR PGANLSLSCY AASNPPAQYS WLINGTFQQS TQELFIPNIT VNNSGSYTCH
301 ANNSVTGCNR TTVKTIIVTE LSPVVAKPQI KASKTTVTGD KDSVNLTCST NDTGISIRWF
361 FKNQSLPSSE RMKLSQGNTT LSINPVKRED AGTYWCEVFN PISKNQSDPI MLNVNYNALP
421 QENGLSPGAI AGIVIGVVAL VALIAVALAC FLHFGKTGRA SDQRDLTEHK PSVSNHTQDH
481 SNDPPNKMNE VTYSTLNFEA QQPTQPTSAS PSLTATEIIY SEVKKQ
```

Amino acid sequence of human CEACAM3

```
  1 MGPPSASPHR ECIPWQGLLL TASLLNFWNP PTTAKLTIES MPLSVAEGKE VLLLVHNLPQ
 61 HLFGYSWYKG ERVDGNSLIV GYVIGTQQAT PGAAYSGRET IYTNASLLIQ NVTQNDIGFY
121 TLQVIKSDLV NEEATGQFHV YQENAPGLPV GAVAGIVTGV LVGVALVAAL VCFLLLAKTG
181 RTSIQRDLKE QQPQALAPGR GPSHSSAFSM SPLSSAQAPL PNPRTAASIY EELLKHDTNI
241 YCRMDHKAEV AS
```

Amino acid sequence of human CEACAM6

```
  1 MGPPSAPPCR LHVPWKEVLL TASLLTFWNP PTTAKLTIES TPFNVAEGKE VLLLAHNLPQ
 61 NRIGYSWYKG ERVDGNSLIV GYVIGTQQAT PGPAYSGRET IYPNASLLIQ NVTQNDTGFY
121 TLQVIKSDLV NEEATGQLHV YPELPKPSIS SNNSNPVEDK DAVAFTCEPE VQNTTYLWWV
181 NGQSLPVSPR LQLSNGNMTL TLLSVKRNDA GSYECEIQNP ASANRSDPVT LNVLYGPDGP
241 TISPSKANYR PGENLNLSCH AASNPPAQYS WFINGTFQQS TQELFIPNIT VNNSGSYMCQ
301 AHNSATGLNR TTVTMITVSG SAPVLSAVAT VGITIGVLAR VALI
```

The sequences sequences underlined and in italics are cleaved off to form mature CEACAM6 and a GPI anchor is added.

Amino acid sequence of human CEACAM8

```
  1 MGPISAPSCR WRIPWQGLLL TASLFTFWNP PTTAQLTIEA VPSNAAEGKE VLLLVHNLPQ
 61 DPRGYNWYKG ETVDANRRII GYVISNQQIT PGPAYSNRET IYPNASLLMR NVTRNDTGSY
121 TLQVIKLNLM SEEVTGQFSV HPETPKPSIS SNNSNPVEDK DAVAFTCEPE TQNTTYLWWV
181 NGQSLPVSPR LQLSNGNRTL TLLSVTRNDV GPYECEIQNP ASANFSDPVT LNVLYGPDAP
241 TISPSDTYYH AGVNLNLSCH AASNPPSQYS WSVNGTFQQY TQKLFIPNIT TKNSGSYACH
301 TTNSATGRNR TTVRMITVSD ALVQGSSPGL SARATVSIMI GVLARVALI
```

The sequences sequences underlined and in italics are cleaved off to form mature CEACAM8 and a GPI anchor is added.

Database accession

	EMBL/GenBank	SwissProt
CEA	M17303	P06731
CEACAM1	X16354	P13688
CEACAM3	L00692	P40198
CEACAM6	M29541	P40199
CEACAM8	X52378	P31997

References

[1] Beauchemin, N. et al. (1999) Exp. Cell Res. 252, 243–249.
[2] http://www.uni-freiburg.de/cea
[3] Skubitz, K.M. et al. (1995) J. Immunol. 155, 5382–5390.

[4] Öbrink, B. et al. (1997) Curr. Opin. Cell Biol. 9, 616–626.

[5] Benchimol, S. et al. (1989) Cell 57, 327–334.

[6] Teixeira, A.M. et al. (1994) Blood 84, 211–219.

[7] Yamanaka, T. et al. (1996) Biochem. Biophys. Res. Commun. 219, 842–847.

[8] Virji, M. et al. (1996) Mol. Microbiol. 22, 929–939.

[9] Dveksler, G.S. et al. (1991) J. Virol. 65, 6881–6891.

[10] Skubitz, K.M. et al. (1996) J. Leukocyte Biol. 60, 106–117.

[11] Morales, V.N. et al. (1999) J. Immunol. 163, 1363–1370.

[12] Lin, S. H. (1989) J. Biol. Chem. 264, 14408–14414.

[13] Sippel, C.J. et al. (1996) J. Biol. Chem. 271, 33095–33104.

Family

Immunoglobulin superfamily

Structure

Molecular weights

Amino acids	1018
Polypeptide	113 320
SDS-PAGE reduced	130 kDa

Carbohydrate

N-linked sites	9
O-linked sites	
Other	

Gene location   12q11–q12

Gene structure

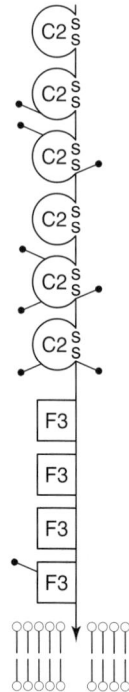

Alternative forms

Two isoforms in adult human brain with differing N-termini are formed by RNA splicing.

Structure

Contactin-1 is a member of the IgSF and contains six N-terminal C2-type Ig domains and four membrane proximal fibronectin type III (Fn3) repeats[1]. It is predicted that contactin-1 is GPI-linked[2] in a similar manner to the murine orthologue F3 and the chicken orthologue F11[3,4]. F3 bears the CD57 (HNK-1) carbohydrate epitope.

Ligands

Contactin binds tenascin[5]. It mediates cell–cell adhesion *in vitro*, possibly via the 190-kDa associated transmembrane glyocoprotein, CASPR (p190, 'contactin-associated protein', neurexin 4; OMIM 602346), which contains multiple protein–protein interaction sequences. In addition, contactin-1 can form *cis* complexes with members of the L1 family[6].

Function

Selective cellular and compartmental distribution during development - suggests a role in axonal patterning[7]. *In vitro* experiments suggest an involvement in intercellular aggregation and neurite outgrowth[8,9] possibly by interacting in *cis* with L1 family members and modulating their adhesive activities[10]. Evidence that contactin-1 can modulate signalling pathways has come from the demonstration that it can form stable interactions in *cis* with protein tyrosine phosphatase-alpha (PTPα) which can dephosphorylate and activate non-receptor tyrosine kinases[11].

Distribution

Restricted distribution to neuronal axonal subpopulations in developing murine cerebellum[8,12], maximally expressed in adult brain.

Disease association

OMIM 600016

Knockout

MGI:105980
Contactin –/– animals display a severe ataxic phenotype and die before postnatal day 18 of cerebellum defects. Analysis of the mutant mice demonstrates a role for contactin in controlling the interactions of cerebellar interneuronbs[13].

Amino acid sequence for human contactin-1

```
  1 MKMWLLVSHL VIISITTCLA EFTWYRRYGH GVSEEDKGFG PIFEEQPINT IYPEESLEGK
 61 VSLNCRARAS PFPVYKWRMN NGDVDLTSDR YSMVGGNLVI NNPDKQKDAG IYYCLASNNY
121 GMVRSTEATL SFGYLDPFPP EERPEVRVKE GKGMVLLCDP PYHFPDDLSY RWLLNEFPVF
181 ITMDKRRFVS QTNGNLYIAN VEASDKGNYS CFVSSPSITK SVFSKFIPLI PIPERTTKPY
241 PADIVVQFKD VYALMGQNVT LECFALGNPV PDIRWRKVLE PMPSTAEIST SGAVLKIFNI
301 QLEDEGIYEC EAENIRGKDK HQARIYVQAF PEWVEHINDT EVDIGSDLYW PCVATGKPIP
361 TIRWLKNGYA YHKGELRLYD VTFENAGMYQ CIAENTYGAI YANAELKILA LAPTFEMNPM
421 KKKILAAKGG RVIIECKPKA APKPKFSWSK GTEWLVNSSR ILIWEDGSLE INNITRNDGG
481 IYTCFAENNR GKANSTGTLV ITDPTRIILA PINADITVGE NATMQCAASF DPALDLTFVW
541 SFNGYVIDFN KENIHYQRNF MLDSNGELLI RNAQLKHAGR YTCTAQTIVD NSSASADLVV
601 RGPGPGPPGL RIEDIRATSV ALTWSRGSDN HSPISKYTIQ TKTILSDDWK DAKTDPPIIE
661 GNMEAARAVD LIPWMEYEFR VVATNTLGRG EPSIPSNRIK TDGAAPNVAP SDVGGGGGRN
721 RELTITWAPL SREYHYGNNF GYIVAFKPFD GEEWKKVTVT NPDTGRYVHK DETMSPSTAF
781 QVKVKAFNNK GDGPYSLVAV INSAQDAPSE APTEVGVKVL SSSEISVHWE HVLEKIVESY
841 QIRYWAAHDK EEAANRVQVT SQEYSARLEN LLPDTQYFIE VGACNSAGCG PPSDMIEAFT
901 KKAPPSQPPR IISSVRSGSR YIITWDHVVA LSNESTVTGY KVLYRPDGQH DGKLYSTHKH
961 SIEVPIPRDG EYVVEVRAHS DGGDGVVSQV KISGAPTLSP SLLGLLLPAF GILVYLEF
```

The sequences underlined and in italics are cleaved off to form mature contactin 1 and a GPI anchor is added.

Database accession

EMBL/GenBank	U07819
SwissProt	Q12860

References

1 Reid, R.A. and Hemperly, J.J. (1994) Brain Res. Mol. Brain Res. 21, 1–8.
2 Berglund, E.O. and Ranscht, B. (1994) Genomics 21, 571–582.
3 Brummendorf, T. et al. (1989) Neuron 2, 1351–1361.
4 Wolff, J.M. et al. (1989) Biochem. Biophys. Res. Commun. 161, 931–938.
5 Zisch, A.H. et al. (1992) J. Cell Biol. 119, 203–213.
6 Sakurai, T. et al. (1997) J Cell Biol. 136, 907–918.
7 Ranscht, B. and Dours, M.T. (1988) J. Cell Biol. 107, 1561–1573.
8 Gennarini, G. et al. (1991) Neuron 6, 595–606.
9 Durbec, P. et al. (1992) J. Cell Biol. 117, 877–889.
10 Brummendorf, T. et al. (1998) Curr. Opin. Neurobiol. 8, 87–97.
11 Zeng, L. et al. (1999) J. Cell Biol. 147, 707–714.
12 Faivre-Sarrailh, C. et al. (1992) J. Neurosci. 12, 257–267.
13 Berglund, E.O. et al. (1999) Neuron 24, 739–750.

Family

Immunoglobulin superfamily

Structure

Molecular weights

Amino acids:	532
Polypeptide:	57 826

SDS-PAGE reduced   90–115 kDa

Carbohydrate

N-linked sites	8
O-linked sites	

Gene location   19p13.3–p13.2

Gene structure   12 kb, 7 exons

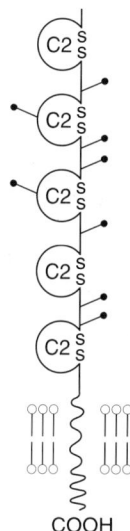

COOH

Alternative forms

A number of alternatively expressed forms lacking combinations of the second, third and fourth Ig domain have been detected in the mouse[1]. Soluble forms have been detected in sera[2].

Structure

ICAM-1 is a type I transmembrane molecule with five C2 Ig domains. The integrin $\alpha_L\beta_2$ binding site is within Ig domain 1 and the top of Ig domain 2[3,4]. The integrin $\alpha_M\beta_2$ binding site is within Ig-domain 3[5]. Rhinoviruses bind an overlapping site with integrin $\alpha_L\beta_2$ on domain 1 (and possibly 2)[6,7] and *Plasmodium falciparum* binds a distinct site in Ig domain 1[2,8]. Crystal structure from the first two Ig domains has been obtained[7,9].

Ligands

ICAM-1 binds to the leucocyte integrins $\alpha_L\beta_2$, $\alpha_M\beta_2$ and $\alpha_X\beta_2$[5,6,10]. In addition it is the receptor for a group of rhinoviruses[11] and *Plasmodium falciparum*-infected erythrocytes[2].

Function

ICAM-1 plays an important role in mediating immune and inflammatory responses. It mediates the adhesion of monocytes, lymphocytes and neutrophils to activated endothelium, and contributes to the extravasation of leucocytes from blood vessels into inflammatory sites. ICAM-1 on antigen presenting cells has a key role in antigen-specific T cell activation[12,13].

Distribution

ICAM-1 is basally expressed in significant amounts on a wide range of both haematopoietic and non-haematopoietic cells[14]. Expression on leucocytes and endothelium can be up-regulated by activation or inflammatory mediators such as IFNγ, IL-1β, TNFα and LPS[12,15].

Disease association

OMIM 147840
Expression of ICAM-1 has been implicated in tumour metastasis[16]. A high-frequency African polymorphism in the N-terminal domain of ICAM-1 is associated with increased susceptibility to cerebral malaria[17].

Knockout

MGI:96392
The Gene Knockout FactsBook[18], p535
ICAM-1 knockout mice show no gross abnormalities in development or fertility. However, ICAM-1 −/− mice have elevated numbers of neutrophils and lymphocytes, and exhibit abnormal inflammatory responses including decreased allogeneic T cell responses, contact hypersensitivity to 2,4–dinitrofluorobenzene and a reduced neutrophil emigration in response to thioglycollate-induced peritonitis and endotoxin challenge. Mutant cells provide no stimulation in a mixed lymphocyte reaction[13,19].

Amino acid sequence of human ICAM-1

```
  1 MAPSSPRPAL PALLVLLGAL FPGPGNAQTS VSPSKVILPR GGSVLVTCST SCDQPKLLGI
 61 ETPLPKKELL LPGNNRKVYE LSNVQEDSQP MCYSNCPDGQ STAKTFLTVY WTPERVELAP
121 LPSWQPVGKN LTLRCQVEGG APRANLTVVL LRGEKELKRE PAVGEPAEVT TTVLVRRDHH
181 GANFSCRTEL DLRPQGLELF ENTSAPYQLQ TFVLPATPPQ LVSPRVLEVD TQGTVVCSLD
241 GLFPVSEAQV HLALGDQRLN PTVTYGNDSF SAKASVSVTA EDEGTQRLTC AVILGNQSQE
301 TLQTVTIYSF PAPNVILTKP EVSEGTEVTV KCEAHPRAKV TLNGVPAQPL GPRAQLLLKA
361 TPEDNGRSFS CSATLEVAGQ LIHKNQTREL RVLYGPRLDE RDCPGNWTWP ENSQQTPMCQ
421 AWGNPLPELK CLKDGTFPLP IGESVTVTRD LEGTYLCRAR STQGEVTREV TVNVLSPRYE
481 IVIITVVAAA VIMGTAGLST YLYNRQRKIK KYRLQQAQKG TPMKPNTQAT PP
```

Database accession

EMBL/GenBank   X06990
SwissProt   P05362

References

[1] King, P.D. et al. (1995) J. Immunol. 154, 6080–6093.
[2] Berendt, A.R. et al. (1991) Cell 68, 71–87.
[3] Rosenstein, Y. et al. (1991) Nature 354, 233–235.
[4] Staunton, D.E. et al. (1988) Cell 52, 925–933.
[5] Diamond, M.S. et al. (1991) Cell 65, 961–971.
[6] Staunton, D.E. et al. (1990) Cell 61, 243–254.
[7] Bella J. et al. (1998) Proc. Natl Acad. Sci. USA 95, 4140–4145.

[8] Ockenhouse, C.F. et al. (1992) Cell 68, 63–69.

[9] Casasnovas, J.M. et al. (1998) Proc. Natl Acad. Sci. USA 95, 4134–4139.

[10] de Fougerolles, A.R. et al. (1995) Eur. J. Immunol. 25, 1008–1012.

[11] Greve, J.M. et al. (1989) Cell 56, 839–847.

[12] Springer, T.A. et al. (1990) Nature 346, 425–434.

[13] Xu, H. et al. (1994) J. Exp. Med. 180, 95–109.

[14] Smith, M.E.F. and Thomas, J.A. (1990) J. Clin. Pathol. 43, 893–900.

[15] Dustin, M.L. and Springer, T.A. (1991) Annu. Rev. Immunol. 9, 27–66.

[16] Johnson, J.P. et al. (1989) Proc. Natl Acad. Sci. USA 86, 641–644.

[17] Fernandez-Reyes, D. et al. (1997) Hum. Molec. Genet. 6, 1357–1360.

[18] Mak, T.W. (1998) The Gene Knockout FactsBook. Academic Press, London

[19] Sligh J.E. Jr. et al. (1993) Proc. Natl Acad. Sci. USA 90, 8529–8533.

ICAM-2

Intercellular adhesion molecule-2, CD102

Family

Immunoglobulin superfamily

Structure

Molecular weights
Amino acids	275
Polypeptide:	30 653

SDS-PAGE reduced	55–65 kDa

Carbohydrate
N-linked sites

Gene location	17q23–q25

Gene structure	4 exons

Alternative forms

Structure
ICAM-2 is a type I transmembrane molecule with two C2 Ig domains. X-ray crystallography reveals a distinctive integrin recognition domain[1]. The cytoplasmic domain of ICAM-2 can bind the ERM (ezrin, radixin, moesin) membrane–cytoskeleton linker molecules[2,3] and α-actinin[4].

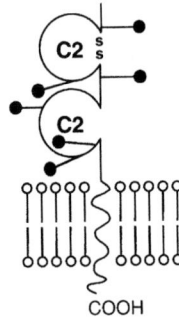

Ligands

ICAM-2 is a counter-receptor for the leucocyte integrins $\alpha_L\beta_2$ and $\alpha_M\beta_2$[5,6].

Function

ICAM-2-mediated adhesion can provide a co-stimulatory signal for T cell activation and may therefore be important where antigen-presenting cells express little or no ICAM-1[7]. Antibody blocking experiments and using ICAM-1 −/− endothelium suggest that ICAM-2 plays a role in leucocyte transendothelial migration;[8,9] however ICAM-2 −/− mice show no differences in lymphocyte recirculation.

Distribution

ICAM-2 is expressed by a subpopulation of lymphocytes, monocytes, splenic sinusoids, dendritic cells, platelets, stromal cells and at high levels on vascular endothelium[10]. In endothelial cells, ICAM-2 is down-regulated by pro-inflammatory mediators[11].

Disease association

OMIM 146630
Increased endothelial ICAM-2 expression has been detected in lymphoid malignancies[12].

Knockout

MGI:96394
The Gene Knockout FactsBook[13], p537
ICAM-2 −/− mice show an impaired accumulation of eosinophils in the lung in response to allergens due to lack of ICAM-2 in the endothelia. No differences in lymphocyte recirculation to the lymph nodes or in NK responses were observed.

Amino acid sequence of human ICAM-2

```
  1 MSSFGYRTLT VALFTLICCP GSDEKVFEVH VRPKKLAVEP KGSLEVNCST TCNQPEVGGL
 61 ETSLNKILLD EQAQWKHYLV SNISHDTVLQ CHFTCSGKQE SMNSNVSVYQ PPRQVILTLQ
121 PTLVAVGKSF TIECRVPTVE PLDSLTLFLF RGNETLHYET FGKAAPAPQE ATATFNSTAD
181 REDGHRNFSC LAVLDLMSRG GNIFHKHSAP KMLEIYEPVS DSQMVIIVTV VSVLLSLFVT
241 SVLLCFIFGQ HLRQQRMGTY GVRAAWRRLP QAFRP
```

Database accession

EMBL/GenBank   X15606
SwissProt   P13598

References

[1] Casasnovas, J.M. et al. (1997) Nature 387, 312–315.
[2] Heiska, L. et al. (1998) J. Biol. Chem. 273, 21893–21900.
[3] Yonemura, S. et al. (1998) J. Cell Biol. 140, 885–895.
[4] Heiska, L. et al. (1996) J. Biol. Chem. 271, 26214–26219.
[5] Staunton, D.E. et al. (1989) Nature 339, 61–64.
[6] Xie, J. et al. (1995) J. Immunol. 155, 3619–3628.
[7] Damle, N.K. et al. (1992) J. Immunol. 148, 665–671.
[8] Reiss, Y. et al. (1998) Eur. J. Immunol. 28, 3086–3099.
[9] Issekutz, A.C. et al. (1999) J. Leukocyte Biol. 65, 117–126.
[10] de Fougerolles, A.R. et al. (1991) J. Exp. Med. 174, 253–267.
[11] McLaughlin, F. et al. (1998) Cell Adhes. Commun. 6, 381–400.
[12] Renkonen, R. et al. (1992) Am. J. Pathol. 140, 763–767.
[13] Mak, T.W. (1998) The Gene Knockout FactsBook. Academic Press, London.

ICAM-3
Intercellular adhesion molecule-3, CD50

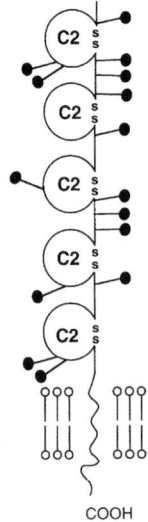

Family
Immunoglobulin superfamily

Structure

Molecular weights
Amino acids	547
Polypeptide:	59 383

SDS-PAGE reduced	110–135 kDa

Carbohydrate
N-linked sites	15
O-linked sites	

Gene location
19p13.3–p13.2

Gene structure

Alternative forms

Structure
ICAM-3 is a heavily glycosylated type I transmembrane molecule with five C2 Ig domains. The cytoplasmic domain is phosphorylated on tyrosine residues following cell activation[1]. Binding to integrin $\alpha_L\beta_2$ is mediated by ICAM-3 Ig domain 1[2] and, unlike ICAM-1, this binding utilizes both faces of the domain[3].

Ligands
ICAM-3 displays calcium-dependent binding to integrin $\alpha_L\beta_2$[4,5] and $\alpha_D\beta_2$[6]. Unlike ICAM-1 and ICAM-2, ICAM-3 does not bind integrins $\alpha_M\beta_2$ or $\alpha_X\beta_2$[7]. Recently, a novel ICAM-3 counter-receptor, DC-SIGN, has been identified on dendritic cells[8].

Function
ICAM-3 mediates intercellular adhesion between leucocytes. It is constitutively expressed on resting antigen-presenting cells, including dendritic cells, and appears to be important in the initial interactions between T cells and dendritic cells leading to T cell activation[9]. Optimal integrin $\alpha_L\beta_2$ binding requires activation of the integrin expressing cells[10] but binding of integrin $\alpha_D\beta_2$ is constitutive. Antibody cross-linking of ICAM-3 on T cells can stimulate T cell adhesion and proliferation[11,12], mobilize intracellular Ca^{2+} [11], and activate tyrosine phosphorylation and the association of ICAM-3 with the tyrosine kinases Fyn and Lck. ICAM-3 can regulate ICAM-1/integrin $\alpha_L\beta_2$-mediated adhesion[13]. Interaction of ICAM-3 on resting T cells with dendritic DC-SIGN enables engagement of the T cell receptor by stabilization of the dendritic-T cell contact zone[8].

Distribution

Constitutively high on leucocytes[9], epidermal Langerhans cells. ICAM-3 is not expressed on endothelia. Soluble forms are detected in sera[14,15].

Disease association

OMIM 146631

Knockout

MGI:101792

Amino acid sequence of human ICAM-3

```
  1 MATMVPSVLW PRACWTLLVC CLLTPGVQGQ EFLLRVEPQN PVLSAGGSLF VNCSTDCPSS
 61 EKIALETSLS KELVASGMGW AAFNLSNVTG NSRILCSVYC NGSQITGSSN ITVYGLPERV
121 ELAPLPPWQP VGQNFTLRCQ VEGGSPRTSL TVVLLRWEEE LSRQPAVEEP AEVTATVLAS
181 RDDHGAPFSC RTELDMQPQG LGLFVNTSAP RQLRTFVLPV TPPRLVAPRF LEVETSWPVD
241 CTLDGLFPAS EAQVYLALGD QMLNATVMNH GDTLTATATA TARADQEGAR EIVCNVTLGG
301 ERREARENLT VFSFLGPIVN LSEPTAHEGS TVTVSCMAGA RVQVTLDGVP AAAPGQPAQL
361 QLNATESDDG RSFFCSATLE VDGEFLHRNS SVQLRVLYGP KIDRATCPQH LKWKDKTRHV
421 LQCQARGNPY PELRCLKEGS SREVPVGIPF FVNVTHNGTY QCQASSSRGK YTLVVVMDIE
481 AGSSHFVPVF VAVLLTLGVV TIVLALMYVF REHQRSGSYH VREESTYLPL TSMQPTEAMG
541 EEPSRAE
```

Database accession

EMBL/GenBank   S50015/X69711
SwissProt   P32942

References

1 Skubitz, K.M. et al. (1995) J. Immunol. 154, 2888–2895.
2 Van Kooyk, Y. et al. (1996) J. Exp. Med. 183, 1247–1252.
3 Bell, E.D. et al. (1998) J. Immunol. 161, 1363–1370.
4 Fawcett, J. et al. (1992) Nature 360, 481–484.
5 Vazeux, R. et al. (1992) Nature 360, 485–488.
6 van der Vieren, M., et al. (1995) Immunity 3, 683–690.
7 de Fougerolles, A.R. et al. (1995) Eur. J. Immunol. 25, 1008–1012.
8 Geijtenbeek, T.B. et al. (2000) Cell 100, 575–585.
9 Starling, G.C. et al. (1995) Eur. J. Immunol. 25, 2528–2532.
10 de Fougerolles, A.R. et al. (1994) J. Exp. Med. 179, 619–629.
11 Juan, M. et al. (1994) J. Exp. Med. 179, 1747–1756.
12 Hernandez-Caselles, T. et al. (1993) Eur. J. Immunol. 23, 2799–2806.
13 Campanero, M.R. et al. (1993) J. Cell Biol. 123, 1007–1016.
14 Del Pozo, M.A. et al. (1994) Eur. J. Immunol. 24, 2586–2594.
15 Pino-Otin, M.R. et al. (1995) J. Immunol. 24, 2586–2594.

ICAM-4

Intercellular adhesion molecule-4, CD242,
Landsteiner–Wiener blood group antigen, LW

Family

Immunoglobulin superfamily

Structure

Molecular weights

Amino acids	271
Polypeptide:	29 265

SDS-PAGE reduced   42–43 kDa

Carbohydrate

N-linked sites	4
O-linked sites	

Gene location   19p13.3

Gene structure   2.65 kb, 3 exons[1]

Alternative forms

A cDNA encoding an alternatively spliced form, which lacks the transmembrane and cytoplasmic domains, has been isolated[2].

Structure

ICAM-4 is a type I transmembrane molecule, which, like ICAM-2, contains two C2 Ig domains with partial conservation of critical residues involved in integrin binding[1]. The molecular basis for the Landsteiner–Wiener (LW)(a)/LW(b) polymorphism is a single basepair mutation changing Gln to Arg at amino acid 100[3].

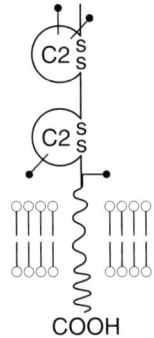

Ligands

ICAM-4 binds to the leucocyte integrins $\alpha_L\beta_2$, $\alpha_M\beta_2$ and $\alpha_X\beta_2$[4].

Function

ICAM-4 is predicted to function as an intercellular adhesion molecule in haematopoietic cells.

Distribution

ICAM-4 is found on erythrocytes and subsets of both B and T cells[5].

Disease association

OMIM 111250
ICAM-4 may be depressed during pregnancy and in some diseases, e.g. Hodgkin's disease, lymphoma, leukaemia and sarcoma.

Knockout

Amino acid sequence of human ICAM-4

```
  1 MGSLFPLSLL FFLAAAYPGV GSALGRRTKR AQSPKGSPLA PSGTSVPFWV RMSPEFVAVQ
 61 PGKSVQLNCS NSCPQPQNSS LRTPLRQGKT LRGPGWVSYQ LLDVRAWSSL AHCLVTCAGK
121 TRWATSRITA YKPPHSVILE PPVLKGRKYT LRCHVTQVFP VGYLVVTLRH GSRVIYSESL
181 ERFTGLDLAN VTLTYEFAAG PRDFWQPVIC HARLNLDGLV VRNSSAPITL MLAWSPAPTA
241 LASGSIAALV GILLTVGAAY LCKCLAMKSQ A
```

Database accession
EMBL/GenBank   L27671
SwissProt   Q14773

References
1 Hermand, P. et al. (1996) Blood 87, 2962–2967.
2 Bailly, P. et al. (1994) Proc. Natl Acad. Sci. USA 91, 5306–5310.
3 Hermand, P. et al. (1995) Blood 86, 1590–1594.
4 Bailly, P. et al. (1995) Eur. J. Immunol. 25, 3316–3320.
5 Oliveira, O.L. et al. (1984) J. Immunogenet. 11, 297–303.

ICAM-5

Intercellular adhesion molecule-5, telencephalin

Family

Immunoglobulin superfamily

Structure

Molecular weights

Amino acids	924
Polypeptide:	97 184
SDS-PAGE reduced	130 kDa

Carbohydrate

N-linked sites	13
O-linked sites	

Gene location 19p13.21[1]

Gene structure

Alternative forms

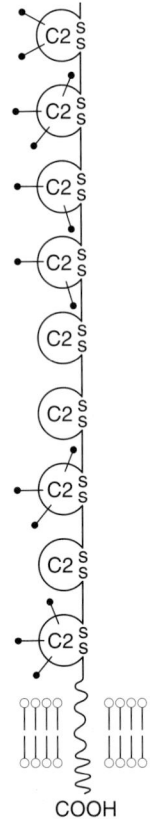

Structure

ICAM-5 is a type I transmembrane protein with nine Ig-like extracellular domains, and is most closely related to ICAM-1 and ICAM-3[2,3].

Ligands

ICAM-5 binds integrin $\alpha_L\beta_2$. Unusually this binding is divalent cation-independent and not activated by protein kinase C[3,4].

Function

ICAM-5 functions in signalling specific subsets of telencephalic neurons to make correct synaptic connections with growing axons[5] and to mediate leukocyte binding to neurons in the central nervous system[6].

Distribution

ICAM-5 is a brain-specific adhesion molecule expressed on the soma-dendritic membrane of subsets of telencephalic neurons. Expression is first detected at birth parallel with dendritic development and synapse formation in the telencephalon[1,5].

Disease association

OMIM 601852

Increased serum concentrations of ICAM-5 are indicative of damage in the mesiotemporal or hippocampal areas of the brain, and may provide a differential diagnosis of seizures[7].

Knockout

MGI:109430

Amino acid sequence of human ICAM-5

```
  1 MPGPSPGLRR ALLGLWAALG LGLFGLSAVS QEPFWADLQP RVAFVERGGS LWLNCSTNCP
 61 RPERGGLETS LRRNGTQRGL RWLARQLVDI REPETQPVCF FRCARRTLQA RGLIRTFQRP
121 DRVELMPLPP WQPVGENFTL SCRVPGAGPR ASLTLTLLRG AQELIRRSFA GEPPRARGAV
181 LTATVLARRE DHGANFSCRA ELDLRPHGLG LFENSSAPRE LRTFSLSPDA PRLAAPRLLE
241 VGSERPVSCT LDGLFPASEA RVYLALGDQN LSPDVTLEGD AFVATATATA SAEQEGARQL
301 ICNVTLGGEN RETRENVTIY SFPAPLLTLS EPSVSEGQMV TVTCAAGTQA LVTLEGVPAA
361 VPGQPAQLQL NATENDDRRS FFCDATLDVD GETLIKNRSA ELRVLYAPRL DDSDCPRSWT
421 WPEGPEQTLR CEARGNPEPS VHCARSDGGA VLALGLLGPV TRALSGTYRC KAANDQGEAV
481 KDVTLTVEYA PALDSVGCPE RITWLEGTEA SLSCVAHGVP PPDVICVRSG ELGAVIEGLL
541 RVAREHAGTY RCEATNPRGS AAKNVAVTVE YGPRFEEPSC PSNWTWVEGS GRLFSCEVDG
601 KPQPSVKCVG SGGATEGVLL PLAPPDPSPR APRIPRVLAP GIYVCNATNR HGSVAKTVVV
661 SAESPPEMDE STCPSHQTWL EGAEASALAC AARGRPSPGV RCSREGIPWP EQQRVSREDA
721 GTYHCVATNA HGTDSRTVTV GVEYRPVVAE LAASPPGGVR PGGNFTLTCR AEAWPPAQIS
781 WRAPPGALNI GLSSNNSTLS VAGAMGSHGG EYECARTNAH GRHARRITVR VAGPWLWVAV
841 GGAAGGAALL AAGAGLAFYV QSTACKKGEY NVQEAESSGE AVCLNGAGGG AGGAAGAEGG
901 PEAAGGAAES PAEGEVFAIQ LTSA
```

Database accession

EMBL/GenBank	U72671
TrEMBL	Q9Y6F3

References
[1] Kilgannon, P. et al. (1998) Genomics 54, 328–330.
[2] Yoshihara, Y. et al. (1994) Neuron 12, 541–553.
[3] Mizuno, T. et al. (1997) J. Biol. Chem. 272, 1156–1163.
[4] Tian, L. et al. (1997) J. Immunol. 158, 928–936.
[5] Yoshihara, Y. and Mori, K. (1994) Neurosci. Res. 21, 119–124.
[6] Tian, L. (2000) Eur. J. Immunol. 30, 810–818.
[7] Rieckmann, P. et al. (1998) Lancet 352, 370–371.

Family

Immunoglobulin superfamily

Structure

Molecular weights

Amino acids	299
Polypeptide	35 216

SDS-PAGE reduced   36–41 kDa

Carbohydrate

N-linked sites	2
O-linked sites	0

Gene location

Gene structure

Alternative forms

Structure

JAM is a type I transmembrane protein with two extracellular V-type Ig domains[1,2].

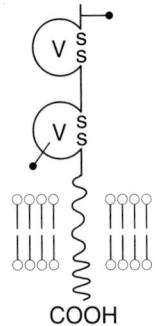

Ligands

Presumed to be homotypic.

Function

In vitro experiments suggest that JAM promotes cell–cell adhesion, possibly by participating in the assembly of tight junctions. Anti-JAM antibodies are effective in blocking leucocyte transmigration *in vivo* and *in vitro*[3] and are able to attenuate cytokine-induced meningitis in mice[4].

Distribution

JAM is found *in vivo* and *in vitro* on endothelial and epithelial cells, where it localizes to sites of cell–cell contact in particular to tight junctions. In addition expression is detected on mesothelial cells and megakaryocytes and, suprisingly, constitutive JAM expression is found on human but not murine peripheral blood leucocytes. Treatment of endothelial cells with TNFα and IFNγ results in a redistribution of JAM from the junctions to the non-junctional areas of the cell membrane[1–3].

Disease association

Knockout

MGI:1321398

Amino acid sequence of human JAM

```
  1 MGTKAQVERK LLCLFILAIL LCSLALGSVT VHSSEPEVRI PENNPVKLSC
 51 AYSGFSSPRV EWKFDQGDTT RLVCYNNKIT ASYEDRVTFL PTGITFKSVT
101 REDTGTYTCM VSEEGGNSYG EVKVKLIVLV PPSKPTVNIP SSATIGNRAV
151 LTCSEQDGSP PSEYTWFKDG IVMPTNPKST RAFSNSSYQL NPTTGELVFD
201 PLSASDTGEY SCEARNGYGT PMTSNAVRME AVERNVGVIV AAVLVTLILL
251 GILVFGIWFA YSRGHFDRTK KGTSSKKVIY SQPSARSEGE FKQTSSFLV
```

Database accession

EMBL/GenBank   AF111713
SwissProt   —

References

[1] Ozaki, H. et al. (1999) J. Immunol. 163, 553–557.
[2] Williams, L.A. (1999) Mol. Immunol. 36, 1175–1188.
[3] Martin-Padura, I. et al. (1998) J. Cell Biol. 142, 117–127.
[4] del Maschio, A. et al. (1999) J. Exp. Med. 190, 1351–1356.

Family

Immunoglobulin superfamily

Structure

Molecular weights

Amino acids	1257
Polypeptide	140 002
SDS-PAGE reduced	200 kDa

Carbohydrate

N-linked sites	20
O-linked sites	

Gene location   Xq28

Gene structure   ~16 kb, 28 exons

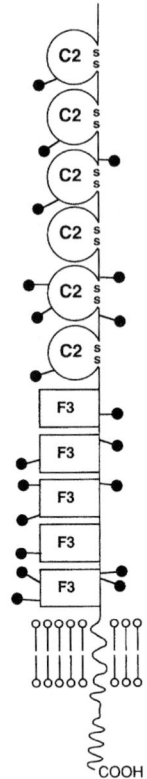

Alternative forms

Alternative splicing can result in a four amino acid insertion in the cytoplasmic domain and five amino acid insertion in the N-terminus of neural but not non-neural L1 (see amino acid sequence).

Structure

L1 is a type I transmembrane protein comprising six C2 type Ig domains followed by five fibronectin type III (F3) domains. Proteolytic processing can result in cleavage after Arg864 within the third fibronectin type III domain, resulting in a extracellular fragment non-covalently associated with a membrane-bound fragment. The cytoplasmic domain associates with the cytoskeleton via its ability to bind ankyrin[1].

Ligands

L1 can mediate homophilic binding. This binding is enhanced by L1 interacting in *cis* with other adhesion molecules such as TAG-1/axonin and NCAM, and with signalling receptors such as fibroblast growth factor receptor[1-4]. L1 also binds integrin $\alpha_v\beta_3$ and $\alpha_5\beta_1$[5-8].

Function

L1 mediates neuron–neuron and neuron–glia interactions. It is implicated in neuronal cell migration, fasciculation and outgrowth[1,9]. In haematopoietic cells, L1 homotypic interactions can mediate the aggregation of activated B cells[10] and the interaction of L1 with integrins may provide an alternative mechanism for the binding of CD31–ve lymphocytes to endothelial cells[5].

Distribution

L1 is widely expressed in the adult and embryonic nervous system. On post-mitotic, premigratory neurons, L1 is localized to the cell bodies. On post-migratory neurons, L1 is localized predominantly to the axons. L1 is also present on Schwann cells and a subpopulation of epithelial cells[9]. In the haematopoietic system, L1 is present on 10–20% of peripheral blood lymphocytes with a mutually exclusive expression to that of CD31[5].

Disease association

OMIM 308840

Mutations in L1 are associated with a number of X-linked neurological deficits including X-lined hydrocephalus, MASA syndrome (adducted thumbs) and spastic paraplegia[11-13].

Knockout

MGI:96721

L1 −/− mice act as an animal model of the L1 mutation X-linked neurological deficits. The size of the corticospinal tract is reduced and non-myelinating Schwann cells show impaired association with axons. The mice are smaller than wild type, less sensitive to touch and pain, and have weakened, uncoordinated hind legs[14,15].

Additional information

There is some controversy as to the relationship of L1 with related molecules from other species, particularly chicken NgCAM and chicken NrCAM. All three receptors have a similar structure in that they contain six C2 type Ig domains and five fibronectin type III domains, and a conserved cytoplasmic domain which can interact with ankyrin. They are capable of homotypic adhesion, can interact with TAG-1/axonin, and function to mediate neuronal adhesion and outgrowth[1,3]. NgCAM and NrCAM are not given separate entries here, but their database accession numbers are given below. In general, chicken NgCAM is regarded as an L1 orthologue, whereas NrCAM, together with murine CHL-1 and neurofascin, are regarded as vertebrate L1 family members. The relationship of L1 with *Drosophila* neuroglian and leech tractin has yet to be established.

Amino acid sequence of human L1

```
   1 MVVALRYVWP LLLCSPCLLI QIPEEYEGHH VMEPPVITEQ SPRRLVVFPT DDISLKCEAS
  61 GKPEVQFRWT RDGVHFKPKE ELGVTVYQSP HSGSFTITGN NSNFAQRFQG IYRCFASNKL
 121 GTAMSHEIRL MAEGAPKWPK ETVKPVEVEE GESVVLPCNP PPSAEPLRIY WMNSKILHIK
 181 QDERVTMGQN GNLYFANVLT SDNHSDYICH AHFPGTRTII QKEPIDLRVK ATNSMIDRKP
 241 RLLFPTNSSS HLVALQGQPL VLECIAEGFP TPTIKWLRPS GPMPADRVTY QNHNKTLQLL
 301 KVGEEDDGEY RCLAENSLGS ARHAYYVTVE AAPYWLHKPQ SHLYGPGETA RLDCQVQGRP
 361 QPEVTWRING IPVEELAKDQ KYRIQRGALI LSNVQPSDTM VTQCEARNRH GLLLANAYIY
 421 VVQLPAKILT ADNQTYMAVQ GSTAYLLCKA FGAPVPSVQW LDEDGTTVLQ DERFFPYANG
 481 TLGIRDLQAN DTGRYFCLAA NDQNNVTIMA NLKVKDATQI TQGPRSTIEK KGSRVTFTCQ
 541 ASFDPSLQPS ITWRGDGRDL QELGDSDKYF IEDGRLVIHS LDYSDQGNYS CVASTELDVV
 601 ESRAQLLVVG SPGPVPRLVL SDLHLLTQSQ VRVSWSPAED HNAPIEKYDI EFEDKEMAPE
 661 KWYSLGKVPG NQTSTTLKLS PYVHYTFRVT AINKYGPGEP SPVSETVVTP EAAPEKNPVD
 721 VKGEGNETTN MVITWKPLRW MDWNAPQVQY RVQWRPQGTR GPWQEQIVSD PFLVVSNTST
 781 FVPYEIKVQA VNSQGKGPEP QVTIGYSGED YPQAIPELEG IEILNSSAVL VKWRPVDLAQ
 841 VKGHLRGYNV TYWREGSQRK HSKRHIHKDH VVVPANTTSV ILSGLRPYSS YHLEVQAFNG
 901 RGSGPASEFT FSTPEGVPGH PEALHLECQS NTSLLLRWQP PLSHNGVLTG YVLSYHPLDE
 961 GGKGQLSFNL RDPELRTHNL TDLSPHLRYR FQLQATTKEG PGEAIVREGG TMALSGISDF
1021 GNISATAGEN YSVVSWVPKE GQCNRFRHIL FKALGEEKGG ASLSPQYVSY NQSSYTQWDL
1081 QPDTDYEIHL FKERMFRHQM AVKTNGTGRV RLPPAGFATE GWFIGFVSAI ILLLLVLLIL
1141 CFIKRSKGGK YSVKDKEDTQ VDSEARPMKD ETFGEYRSLE SDNEEKAFGS SQPSLNGDIK
1201 PLGSDDSLAD YGGSVDVQFN EDGSFIGQYS GKKEKEAAGG NDSSGATSPI NPAVALE
```

The two sequences underlined and in italics are encoded for by exons 2 and 27 and are spliced out in non-neural cells.

Database accession

	EMBL/GenBank	SwissProt
L1	X59847	P32004
NgCAM (chicken)	X56969	Q03696
NrCAM (chicken)	X58482	P35331

References

1 Brummendorf, T. et al. (1998) Curr. Opin. Neurobiol. 8, 87–97.
2 Kadmon, G. et al. (1990) J. Cell Biol. 132, 475–485.
3 Walsh, F.S. and Doherty, P. (1997) Annu. Rev. Cell Dev. Biol. 13, 425–456.
4 Malhotra, J.D. et al. (1998) J. Biol. Chem. 273, 33354–33359.
5 Ebeling, O. et al. (1996) Eur. J. Immunol. 26, 2508–2516.
6 Montgomery, A.M.P et al. (1996) J. Cell Biol. 132, 475–485.
7 Ruppert, M. et al. (1995) J. Cell Biol. 131, 1881–1891.
8 Blaess, S. et al. (1998) J. Neurochem. 71, 2615–2625.
9 Schachner, M. (1990) Ciba Found. Symp. 145, 156–172.
10 Kowitz, A. et al. (1993) Clin. Exp. Metastasis 11, 419–429.
11 Rosenthal, A. et al. (1992) Nature Gen. 2, 107–112.
12 Jouet, M. et al. (1994) Nature Gen. 7, 402–407.
13 Kenwrick, S. and Doherty, P. (1998) Bioessays 20, 668–675.
14 Dahme, M. et al. (1997) Nature Gen. 17, 346–349.
15 Cohen, N.R. et al. (1998) Curr. Biol. 8, 26–33.

MAdCAM-1
Mucosal addressin cell adhesion molecule 1

Family
Immunoglobulin superfamily

Structure

Molecular weights
Amino acids	406
Polypeptide:	42 664

SDS-PAGE reduced	58–66 kDa

Carbohydrate
N-linked sites	2
O-linked sites	+++

Gene location   19q13.3[1]

Gene structure   5 exons

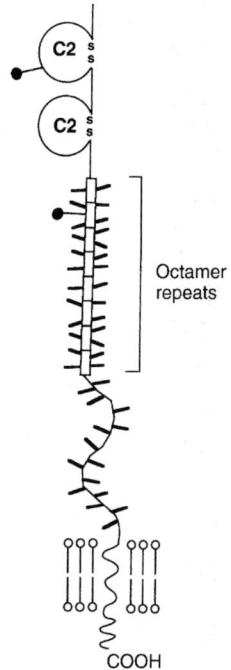

Alternative forms
Alternatively spliced transcripts have been detected in human brain[2].

Structure
The two C2–type Ig at the N-terminus are related to equivalent domains in the other integrin binding proteins: VCAM-1, ICAM-1, ICAM-2 and ICAM-3. The crystal structure of these two domains reveals two separate integrin-recognition motifs[3]. The Ig domains are followed by a mucin-like region rich in Ser, Thr and Pro residues, which includes eight copies of the octameric repeated (P/S)PDTTS(Q/P)E and is modified by O-linked glycosylation. The mucin-like sequences are highly divergent between murine and human MAdCAM-1[4].

Ligands
MAdCAM-1 binds integrin $\alpha_4\beta_7$ via its Ig domains and L-selectin via the sialoglycoconjugates in the mucin-like stalk[5-7].

Function
Endothelial MAdCAM-1 binds to integrin $\alpha_4\beta_7$ and L-selectin on leucocytes thereby contributing to the recirculation of naive lymphocytes to Peyer's patches and mesenteric lymph nodes, and the homing of a subpopulation of

activated or memory lymphocytes to the gut lamina propria[8]. By binding to both integrins and selectins, MAdCAM-1 has a proposed role in mediating both the rolling and the firm adhesion of lymphocytes[8].

Distribution

MAdCAM-1 is expressed at high levels of the high endothelial venules of Peyer's patches and mesenteric lymph nodes and on flat-walled venules within the gut lamina propria. It is also expressed on vascular endothelium in mammary glands, pancreas and the spleen marginal sinus and on blood vessels in areas of chronic inflammation[4,8]. *In vitro*, MAdCAM-1 expression is inducible by TNFα and IL-1.

Disease association

OMIM 102670

Knockout

MGI:103579

Amino acid sequence of human MAdCAM-1

```
  1 MDFGLALLLA GLLGLLLGQS LQVKPLQVEP PEPVVAVALG ASRQLTCRLA CADRGASVQW
 61 RGLDTSLGAV QSDTGRSVLT VRNASLSAAG TRVCVGSCGG RTFQHTVQLL VYAFPDQLTV
121 SPAALVPGDP EVACTAHKVT PVDPNALSFS LLVGGQELEG AQALGPEVQE EEEEPQGDED
181 VLFRVTERWR LPPLGTPVPP ALYCQATMRL PGLELSHRQA IPVLHSPTSP EPPDTTSPEP
241 PNTTSPESPD TTSPESPDTT SQEPPDTTSQ EPPDTTSQEP PDTTSPEPPD KTSPEPAPQQ
301 GSTHTPRSPG STRTRRPEIS QAGPTQGEVI PTGSSKPAGD QLPAALWTSS AVLGLLLLAL
361 PTYHLWKRCR HLAEDDTHPP ASLRLLPQVS AWAGLRGTGQ VGISPS
```

Database accession

EMBL/GenBank   U43628
TrEMBL   Q13477

References

[1] Leung, E. et al. (1997) Immunogenetics 46, 111–119.

[2] Leung, E. et al. (1996) Immun. Cell Biol. 74, 490–496.

[3] Tan, K. et al. (1998) Structure 6, 793–801.

[4] Shyjan, A.M. et al. (1996) J. Immunol. 156, 2851–2857.

[5] Berg, E.L. et al. (1993) Nature 366, 695–698.

[6] Berlin, C. et al. (1993) Cell 74, 185–195.

[7] Briskin, M.J. et al. (1996) J. Immunol. 156, 719–726.

[8] Butcher, E.C. and Picker, L.J. (1996) Science 272, 60–66.

Family

Immunoglobulin superfamily (siglec subfamily)

Structure

Molecular weights

Amino acids	626
Polypeptide:	69 068
SDS-PAGE reduced	100 kDa

Carbohydrate

N-linked sites	8
O-linked sites	

Gene location   19q12–19q13.2

Gene structure   16 kb, 13 exons

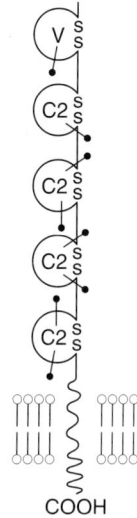

Alternative forms

Alternative use of exons 12 and 13 results in long (L)- and short (S)-cytoplasmic domain MAG isoforms.

Structure

MAG is a member of the siglec family of sialic acid binding lectins (CD33, CD22 and sialoadhesin). Structurally MAG contains one V type and four C2 type Ig-like domains[1]. A related protein Scwann cell myelin protein (SMP; siglec-4b) has been identified in quail (GenBank accession number S83711).

Ligands

MAG binds to the sialoglycoconjugate NeuAcα2–3Galβ1–3GalNAc on N- and O-linked oligosaccharides and gangliosides[2,3]. In a similar manner to CD22 and CD33, MAG ligand binding can be modulated by its own sialylation[4]. In addition, MAG has been reported to bind to collagens I–VI[5].

Function

In vitro MAG-expressing cells bind specifically to neuronal glycoconjugates[2] and blocking experiments originally implicated a role for MAG in nerve myelination. Analysis of MAG −/− mice (see below) suggests MAG function in the integrity and long-term maintenance of both myelin and axons rather than the process of myelination itself.

Distribution

MAG is expressed by oligodendrocytes and Schwann cells.

Disease association

OMIM 159460
Mapping of the mouse gene led to the suggestion that mutations in MAG underlie the neurological 'quivering' (qv) disorder.

Knockout

MGI:96912
In MAG −/− mice there is no gross abnormality in the degree of myelination or its compaction. Mutant animals appear normal in motor coordination and spatial learning, but show a subtle intention tremor. However, the organization of the periaxonal region is partially impaired and ultrastructural studies show that there is a decrease in the proportion of myelinated axons in the optic nerve and a variation in the amount of myelin laid down[6-9]. Moreover, after 8 months of age mice show both axon and myelin degeneration.

Amino acid sequenc of human MAG

```
  1 MIFLTALPLF WIMISASRGG HWGAWMPSSI SAFEGTCVSI PCRFDFPDEL RPAVVHGVWY
 61 FNSPYPKNYP PVVFKSRTQV VHESFQGRSR LLGDLGLRNC TLLLSNVSPE LGGKYYFRGD
121 LGGYNQYTFS EHSVLDIVNT PNIVVPPEVV AGTEVEVSCM VPDNCPELRP ELSWLGHEGL
181 GEPAVLGRLR EDEGTWVQVS LLHFVPTREA NGHRLGCQAS FPNTTLQFEG YASMDVKYPP
241 VIVEMNSSVE AIEGSHVSLL CGADSNPPPL LTWMRDGTVL REAVAESLLL ELEEVTPAED
301 GVYACLAENA YGQDNRTVGL SVMYAPWKPT VNGTMVAVEG ETVSILCSTQ SNPDPILTIF
361 KEKQILSTVI YESELQLELP AVSPEDDGEY WCVAENQYGQ RATAFNLSVE FAPVLLLESH
421 CAAARDTVQC LCVVKSNPEP SVAFELPSRN VTVNESEREF VYSERSGLVL TSILTLRGQA
481 QAPPRVICTA RNLYGAKSLE LPFQGAHRLM WAKIGPVGAV VAFAILIAIV CYITQTRRKK
541 NVTESPSFSA GDNPPVLFSS DFRISGAPEK YESERRLGSE RRLLGLRGEP PELDLSYSHS
601 DLGKRPTKDS YTLTEELAEY AEIRVK
```

Database accession

EMBL/GenBank   M29273
SwissProt   P20916

References

[1] Sato, S. et al. (1989) Biochem. Biophys. Res. Commun. 163, 1473–1480.
[2] Kelm, S. et al. (1994) Curr. Biol. 4, 965–972.
[3] Collins, B.E. et al. (1997) J. Biol. Chem. 272, 1248–1255.
[4] Tropak, M.B. and Roder, J.C. (1997) J. Neurochem. 68, 1753–1763.
[5] Fahrig, T. et al. (1987) EMBO J. 6, 2875–2883.
[6] Li, C. et al. (1994) Nature 369, 747–750.
[7] Montag D. et al. (1994) Neuron 13, 229–346.
[8] Fruttiger M. et al. (1995) Eur. J. Neurosci. 7, 511–515.
[9] Li, C. et al. (1998) J. Neurosci. Res. 51, 210–217.

MUC18

Family

Immunoglobulin superfamily

Structure

Molecular weights
Amino acids	646
Polypeptide	71 793

SDS-PAGE reduced   115–130 kDa

Carbohydrates
N-linked sites	8
O-linked sites	

Gene location   11q23.3

Gene structure   14 kb, 16 exons

Alternative forms

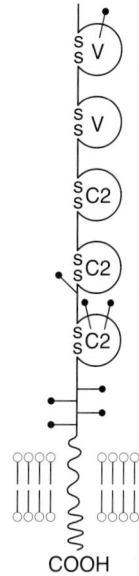

Structure

MUC18 is a highly N-glycosylated type I transmembrane protein with five Ig domains followed by a small membrane proximal stalk region and a 63 amino acid cytoplasmic domain[1,2].

Ligands

None identified.

Function

Owing to its homology to other IgSF members, MUC18 is proposed to act as an adhesion receptor. At least in part, this putative adhesive function may be mediated by a MUC18 signalling function as MUC18 engagement results in tyrosine phosphorylation of focal adhesion kinase, possibly by the MUC18 recruitment of the tyrosine kinase Fyn[3].

Distribution

MUC18 was originally identified as an endothelial antigen where it localizes to cell–cell contact areas. Is is also expressed on smooth muscle, intermediate trophoblast and a subpopulation of activated T cells. Much interest in MUC18 is that it is a marker associated with tumour progression and metastatic development of malignant melanomas[4].

Disease association

OMIM 155735
MUC18 expression is restricted to advanced primary and metastatic melanomas, and has been implicated to play a role in tumour growth and tumour dissemination[5-7].

Knockout

Amino acid sequence of human MUC18

```
  1 MGLPRLVCAF LLAACCCCPR VAGVPGEAEQ PAPELVEVEV GSTALLKCGL SQSQGNLSHV
 61 DWFSVHKEKR TLIFRVRQGQ GQSEPGEYEQ RLSLQDRGAT LALTQVTPQD ERIFLCQGKR
121 PRSQEYRIQL RVYKAPEEPN IQVNPLGIPV NSKEPEEVAT CVGRNGYPIP QVIWYKNGRP
181 LKEEKNRVHI QSSQTVESSG LYTLQSILKA QLVKEDKDAQ FYCELNYRLP SGNHMKESRE
241 VTVPVFYPTE KVWLEVEPVG MLKEGDRVEI RCLADGNPPP HFSISKQNPS TREAEEETTN
301 DNGVLVLEPA RKEHSGRYEC QAWNLDTMIS LLSEPQELLV NYVSDVRVSP AAPERQEGSS
361 LTLTCEAESS QDLEFQWLRE ETDQVLERGP VLQLHDLKRE AGGGYRCVAS VPSIPGLNRT
421 QLVKLAIFGP PWMAFKERKV WVKENMVLNL SCEASGHPRP TISWNVNGTA SEQDQDPQRV
481 LSTLNVLVTP ELLETGVECT ASNDLGKNTS ILFLELVNLT TLTPDSNTTT GLSTSTASPH
541 TRANSTSTER KLPEPESRGV VIVAVIVCIL VLAVLGAVLY FLYKKGKLPC RRSGKQEITL
601 PPSRKTELVV EVKSDKLPEE MGLLQGSSGD KRAPGDQGEK YIDLRH
```

Database accession[1]

EMBL/GenBank   M29277
Swissprot   P43121

References

[1] Lehmann, J.M. et al. (1989) Proc. Natl Acad. Sci. USA 86, 9891–9895.
[2] Sers, C. et al. (1993) Proc. Natl Acad. Sci. USA 90, 8514–8518.
[3] Anfosso, F. et al. (1998) J. Biol. Chem. 273, 26852–26856.
[4] Johnson, J.P et al. (1996) Curr. Top. Microbiol. Immunol. 213, 95–105.
[5] Pickl, W.F. et al. (1997) J. Immunol. 158, 2107–2115.
[6] Jean, D. et al. (1998) J. Biol. Chem. 273, 16501–16508.
[7] Shih, I.M. (1999) J. Pathol. 189, 4–11.

NCAM	Neural cell adhesion molecule, CD56, fascilin II (*Drosophila*), apCAM (*Aplysia*)

Family

Immunoglobulin superfamily

Structure

Molecular weights

Amino acids	NCAM-120	761
	NCAM-140	848
	NCAM-180	1116
Polypeptide	NCAM-120	83 770
	NCAM-140	93 360
	NCAM-180	119 240
SDS-PAGE reduced	NCAM-120	120 kDa
	NCAM-140	140 kDa
	NCAM-180	180 kDa

Carbohydrates

N-linked sites	6
O-linked sites	

Gene location   11q23.1

Gene structure   >100 kb, > 20 exons (mouse)

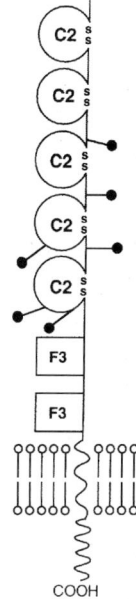

Alternative forms

NCAM is subject to extensive alternative splicing resulting in 20–30 different isoforms. The three major isoforms are NCAM-180, -140 and -120. NCAM-180 differs from NCAM-140 by the inclusion of exon 18, producing a longer cytoplasmic domain. NCAM-120 arises from the inclusion of exon 15, which produces a GPI-anchored isoform. In addition, there is variable insertion of the ten amino acid VASE exon, which converts the fourth Ig domain from a C2 to a V type, and of the MSD1 exons between the two fibronectin type III (F3) repeats[1,2].

Structure

NCAM is a type I transmembrane protein containing five C2 Ig domains followed by two fibronectin type III (F3) domains. In addition to different isoforms generated by alternatively splicing, NCAM is subject to post-translational modification by the addition of varying length of α2–8–linked sialic acid (polysialic acid), which accounts for up to 30% of the molecular weight in the NCAM-180 isoform[2]. The solution structure of the first two Ig domains has been reported[3].

Ligands

NCAM mediates homophilic binding, which involves all five Ig domains pairing in an antiparallel alignment[4]. In neurons, NCAM can bind to

133

chondroitin sulphate proteoglycans in the cells matrix[5]. In addition NCAM can interact in *cis* with other cell adhesion molecules, such as L1, and with signalling receptors, such as the fibroblast growth factor receptor[2,6].

Function

NCAM has been shown to be involved in a diverse range of contact-mediated interactions, but most notably between neurons, neurons and astrocytes, neurons and oligodendrocytes, and neuronal processes and myotubes. These events have implicated a role for NCAM in axonal outgrowth, fasciculation, cell migration, learning and memory[7].

Distribution

NCAM is expressed in adult neural tissue and muscle, on human NK cells, a subpopulation of T cells and in a number of embryonic tissues[8].

Disease association

OMIM 116930
NCAM expression is up-regulated in a number of tumour cell types including neuroblastomas, small cell lung carcinoma, myelomas and Wilm's tumours[8].

Knockout

MGI:97281
The Gene Knockout FactsBook[9], p783
NCAM −/− mice are viable and fertile but with a 10% reduction in overall brain weight, a 36% reduction in olfactory bulb size leading to defects in the migration of olfactory neurons, and deficits in spatial learning[10]. By eliminating only NCAM-180, an equivalent defect is observed[11], indicating that NCAM deficiencies result from loss of the highly polysialylated NCAM form. Interestingly, heterozygous mice with a secreted form of NCAM died *in utero* at E8.5–E9.5[12].

Amino acid sequence of human NCAM-120

```
  1 MLQTKDLIWT LFFLGTAVSL QVDIVPSQGE ISVGESKFFL CQVAGDAKDK DISWFSPNGE
 61 KLTPNQQRIS VVWNDDSSST LTIYNANIDD AGIYKCVVTG EDGSESEATV NVKIFQKLMF
121 KNAPTPQEFR EGEDAVIVCD VVSSLPPTII WKHKGRDVIL KKDVRFIVLS NNYLQIRGIK
181 KTDEGTYRCE GRILARGEIN FKDIQVIVNV PPTIQARQNI VNATANLGQS VTLVCDAEGF
241 PEPTMSWTKD GEQIEQEEDD EKYIFSDDSS QLTIKKVDKN DEAEYICIAE NKAGEQDATI
301 HLKVFAKPKI TYVENQTAME LEEQVTLTCE ASGDPIPSIT WRTSTRNISS EEKTLDGHMV
361 VRSHARVSSL TLKSIQYTDA GEYICTASNT IGQDSQSMYL EVQYAPKLQG PVAVYTWEGN
421 QVNITCEVFA YPSATISWFR DGQLLPSSNY SNIKIYNTPS ASYLEVTPDS ENDFGNYNCT
481 AVNRIGQESL EFILVQADTP SSPSIDQVEP YSSTAQVQFD EPEATGGVPI LKYKAEWRAV
541 GEEVWHSKWY DAKEASMEGI VTIVGLKPET TYAVRLAALN GKGLGEISAA SEFKTQPVHS
601 PPPPASASSS TPVPLSPPDT TWPLPALATT EPAKGEPSAP KLEGQMGEDG NSIKVNLIKQ
661 DDGGSPIRHY LVRYRALSSE WKPEIRLPSG SDHVMLKSLD WNAEYEVYVV AENQQGKSKA
721 AHFVFRTSAQ PTAIPATLGG NSASYTFVSL LFSAVTLLLL C
```

Amino acid sequence of human NCAM-140

```
  1 MLQTKDLIWT LFFLGTAVSL QVDIVPSQGE ISVGESKFFL CQVAGDAKDK DISWFSPNGE
 61 KLTPNQQRIS VVWNDDSSST LTIYNANIDD AGIYKCVVTG EDGSESEATV NVKIFQKLMF
121 KNAPTPQEFR EGEDAVIVCD VVSSLPPTII WKHKGRDVIL KKDVRFIVLS NNYLQIRGIK
181 KTDEGTYRCE GRILARGEIN FKDIQVIVNV PPTIQARQNI VNATANLGQS VTLVCDAEGF
241 PEPTMSWTKD GEQIEQEEDD EKYIFSDDSS QLTIKKVDKN DEAEYICIAE NKAGEQDATI
301 HLKVFAKPKI TYVENQTAME LEEQVTLTCE ASGDPIPSIT WRTSTRNISS EEKTLDGHMV
361 VRSHARVSSL TLKSIQYTDA GEYICTASNT IGQDSQSMYL EVQYAPKLQG PVAVYTWEGN
421 QVNITCEVFA YPSATISWFR DGQLLPSSNY SNIKIYNTPS ASYLEVTPDS ENDFGNYNCT
481 AVNRIGQESL EFILVQADTP SSPSIDQVEP YSSTAQVQFD EPEATGGVPI LKYKAEWRAV
541 GEEVWHSKWY DAKEASMEGI VTIVGLKPET TYAVRLAALN GKGLGEISAA SEFKTQPVQG
601 EPSAPKLEGQ MGEDGNSIKV NLIKQDDGGS PIRHYLVRYR ALSSEWKPEI RLPSGSDHVM
661 LKSLDWNAEY EVYVVAENQQ GKSKAAHFVF RTSAQPTAIP ANGSPTSGLS TGAIVGILIV
721 IFVLLLVVVD ITCYFLNKCG LFMCIAVNLC GKAGPGAKGK DMEEGKAAFS KDESKEPIVE
781 VRTEEERTPN HDGGKHTEPN ETTPLTEPEK GPVEAKPECQ ETETKPAPAE VKTVPNDATQ
841 TKENESKA
```

The NCAM 180 isoform results from insertion of exon 18 at **E**809 (shown in bold)

Amino acid sequence of mouse NCAM-180 (transmembrane and cytoplasmic domain

```
  1 NGSPTAGLST GAIVGILIVI FVLLLVVMDI TCYFLNKCGL LMCIAVNLCG KAGPGAKGKD
 61 MEEGKAAFSK DESKEPIVEV RTEEERTPNH DGGKHTEPNE TTPLTEPELP ADTTATVEDM
121 LPSVTTVTTN SDTITETFAT AQNSPTSETT TLTSSIAPPA TTVPDSNSVP AGQATPSKGV
181 TASSSSPASA PKVAPLVDLS DTPTSAPSAS NLSSTVLANQ GAVLSPSTPA SAGETSKAPP
241 ASKASPAPTP TPAGAASPLA AVAAPATDAP QAKQEAPSTK GPDPEPTQPG TVKNPPEAAT
301 APASPKSKAA TTNPSQGEDL KMDEGNFKTP DIDLAKDVFA ALGSPRPATG ASGQASELAP
361 SPADSAVPPA PAKTEKGPVE TKSEPPESEA KPAPTEVKTV PNDATQTKEN ESKA
```

Database accession

	EMBL/GenBank	SwissProt
NCAM 120	X16841	P13592
NCAM 140	S71824	P13591
NCAM 180	X15052 (mouse)	—

References
1 Walsh, F.S. and Doherty, P. (1991) Semin. Neurosci. 3, 271–284.
2 Walsh, F.S. and Doherty, P. (1997) Annu. Rev. Cell Dev. Biol. 13, 425–456.
3 Jensen, P.H. et al. (1999) Nature Struct. Biol. 6, 486–493.
4 Ranheim, T.S. et al. (1996) Proc. Natl Acad. Sci. USA 93, 4071–4075.
5 Friedlander, D.R. et al. (1994) J. Cell Biol. 125, 669–680.
6 Saffell, J.L. et al. (1997) Neuron 18, 231–242.
7 Goodman, C.S. (1996) Ann. Rev. Neurosci. 19, 341–377.
8 Goridis, C. and Brunet, J.F. (1992) Semin. Cell Biol. 3, 189–197.
9 Mak, T.W. (1998) The Gene Knockout FactsBook. Academic Press, London.
10 Cremer, H. et al. (1994) Nature 367, 455–459.
11 Tomasiewicz, H. et al. (1993) Neuron 11, 1163–1174.
12 Rabinowitz, J.E. et al. (1996) Proc. Natl Acad. Sci. USA 93, 6421–6424.

Family

Immunoglobulin superfamily

Structure

Molecular weights

Amino acids	258
Polypeptide	27 555
SDS-PAGE reduced	28 kDa

Carbohydrate

N-linked sites	
O-linked sites	
Other	Complex, sulphated

Gene location 1q22

Gene structure 7 kb, 6 exons

Alternative forms

Structure

P(0) is a type I integral membrane protein, containing one V-type Ig domain[1,2]. The cytoplasmic tail contains potential phosphorylation sites. The crystal structure of the Ig domain has been determined[3]. One of the two modes of molecular interaction found provides a model for the role of P(0) in homophilic interactions. Individual monomers on the cell membrane are predicted to associate in an antiparallel orientation with monomers on the opposing cell membrane. This would fit a role for P(0) holding adjacent membranes together in the individual wraps of the myelin sheet. Some mutations in Charcot–Marie–Tooth Disease (see below) would disrupt predicted structure.

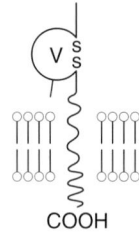

Ligands

P(0) homophilic interaction[4].

Function

Guiding, linking and compaction of adjacent lamellae of myelin sheath.

Distribution

P(0) is the major structural protein of peripheral nerve myelin sheaths, produced by Schwann cells. It is absent from the central nervous system.

Disease association

OMIM 159440

Mutations in the MPZ gene are associated with the autosomal dominant form of Charcot–Marie–Tooth disease (Type 1b), which is due to neuronal demyelination leading to various forms of peripheral neuropathy and distal muscular atrophy, and some other inherited peripheral neurological syndromes[5,6].

Knockout

MGI:103177

A knockout mouse model for P(0) function has been reported to show peripheral nerve myelin degeneration similar to that seen in human mutations[7].

Amino acid sequence of human P(0)

```
  1 MLRAPAPAPA MAPGAPSSSP SPILAVLLFS SLVLSPAQAI VVYTDREVHG AVGSRVTLHC
 61 SFWSSEWVSD DISFTWRYQP EGGRDAISIF HYAKGQPYID EVGTFKERIQ WVGDPRWKDG
121 SIVIHNLDYS DNGTFTCDVK NPPDIVGKTS QVTLYVFEKV PTRYGVVLGA VIGGVLGVVL
181 LLLLLFYVVR YCWLRRQAAL QRRLSAMEKG KLHKPGKDAS KRGRQTPVLY AMLDHSRSTK
241 AVSEKKAKGL GESRKDKK
```

Database accession

EMBL/GenBank   D10537
SwissProt   P25189

References

[1] Hayasaka, K. et al. (1991) Biochem. Biophys. Res. Commun. 180, 515–518.
[2] Hayasaka, K. et al. (1993) Genomics 17, 755–758.
[3] Shapiro, L. et al. (1996) Neuron 17, 435–449.
[4] Filbin, M.T. et al. (1990) Nature 344, 871–872.
[5] Hayasaka, K. et al. (1993) Nature Genet. 5, 266–268.
[6] Hayasaka, K. et al. (1993) Nature Genet. 5, 31–34.
[7] Martini, R. et al. (1995) Nature Genet. 11, 281–286.

Sialoadhesin Siglec-1, CD169

Family

Immunoglobulin superfamily (siglec subfamily)

Structure

Molecular weights

Amino acids	1694
Polypeptide	182 973
SDS-PAGE reduced	185 kDa

Carbohydrate

N-linked sites	15
O-linked sites	

Gene location

20p13[1]

Gene structure

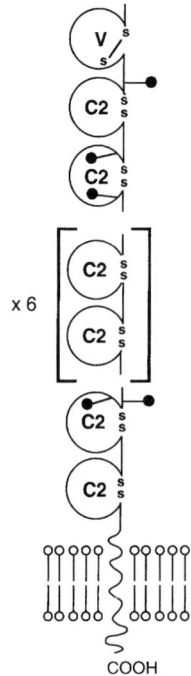

Alternative forms

Two splice variants have been identified, which encode soluble proteins truncated after the third and 16th IgSF domains.

Structure

Sialoadhesin is a member of the siglec family of sialic acid binding lectins (CD33, CD22 and MAG). Structurally sialoadhesin contains 17 Ig domains of which domains 4–17 are made up of seven homologous tandem repeats consisting of a short and long Ig domain[2]. Like other members of the family, sialoadhesin is predicted to have an unusual disulphide bond between β strands B and E in domain 1, and a disulphide bond between domains 1 and 2[3]. The sialic acid binding site has been localized to the N-terminal Ig domain[4] and the crystal structure of this domain in complex with sialyllactose has been obtained[5,6].

Ligands

Sialoadhesin binds to the sialoglycoconjugates NeuAcα2–3Galβ1–3(4)GlcNAc and NeuAcα2–3Galβ1–3GalNAc on glycoproteins and glycolipids[7]. Like other siglecs, sialylation of sialoadhesin negatively regulates ligand binding[8].

Function

Sialoadhesin functions as a macrophage-restricted adhesion receptor mediating preferential recognition of cells of the granulocyte lineage[4,9].

Distribution

Sialoadhesin is expressed on subsets of macrophages in bone marrow and secondary lymphoid organs[2]. Expression in resident peritoneal macrophages is down-regulated by IL-4[10].

Disease association

OMIM 600751

Knockout

MGI:99668

Amino acid sequence of mouse sialoadhesin

```
   1  MCVLFSLLLL ASVFSLGQTT WGVSSPKNVQ GLSGSCLLIP CIFSYPADVP VSNGITAIWY
  61  YDYSGKRQVV IHSGDPKLVD KRFRGRAELM GNMDHKVCNL LLKDLKPEDS GTYNFRFEIS
 121  DSNRWLDVKG TTVTVTTDPS PPTITIPEEL REGMERNFNC STPYLCLQEK QVSLQWRGQD
 181  PTHSVTSSFQ SLEPTGSYHQ TTLHMALSWQ DHGRTLLCQF SLGAHSSRKE VYLQVPHAPK
 241  GVEILLSSSG RNILPGDPVT LTCRVNSSYP AVSAVQWARD GVNLGVTGHV LRLFSAAWND
 301  SGAYTCQATN DMGSLVSSPL SLHVFMAEVK MNPAGPVLEN ETVTLLCSTP KEAPQELRYS
 361  WYKNHILLED AHASTLHLPA VTRADTGFYF CEVQNAQGSE RSSPLSVVVR YPPLTPDLTT
 421  FLETQAGLVG ILHCSVVSEP LATVVLSHGG LTLASNSGEN DFNPRFRISS APNSLRLEIR
 481  DLQPADSGEY TCLAVNSLGN STSSLDFYAN VARLLINPSA EVVEGQAVTL SCRSGLSPAP
 541  DTRFSWYLNG ALLLEGSSSS LLLPAASSTD AGSYYCRTQA GPNTSGPSLP TVLTVFYPPR
 601  KPTFTARLDL DTSGVGDGRR GILLCHVDSD PPAQLRLLHK GHVVATSLPS RCGSCSQRTK
 661  VSRTSNSLHV EIQKPVLEDE GVYLCEASNT LGNSSAAASF NAKATVLVIT PSNTLREGTE
 721  ANLTCNGNQE VAVSPANFSW FRNGVLWTQG SLETVRLQLL ARTDAAVYAC RLLTEDGAQL
 781  SAPVVLSVLY APDPPKLSAL LDVGQGHMAV FICTVDSYPL AHLSLFRGDH LLATNLEPQR
 841  PSHGRIQAKA TANSLQLEVR ELGLVDSGNY HCEATNILGS ANSSLFFQVR GAWVRFTITE
 901  LREGQAVVLS CQVPTGVSEG TSYSWYQDGR PLQESTSSTL RIAAISLRQA GAYHCQAQAP
 961  DTAIASLAAP VSLHVSYTPR HVTLSALLST DPERLGHLVC SVQSDPPAQL QLFHRNRLVA
1021  STLQGADELA GSNPRLHVTV LPNELRLQIH FPELEDDGTY TCEASNTLGQ ASAAADFDAQ
1081  AVRVTVWPNA TVQEGQQVNL TCLVWSTHQD SLSYTWYKGG QQLLGARSIT LPSVKVLDAT
1141  SYRCGVGLPG HAPHLSRPVT LDVLHAPRNL RLTYLLETQG RQLALVLCTV DSRPPAQLTL
1201  SHGDQLVASS TEASVPNTLR LELQDPRPSN EGLYSCSAHS PLGKANTSLE LLLEGVRVKM
1261  NPSGSVPEGE PVTVTCEDPA ALSSALYAWF HNGHWLQEGP ASSLQFLVTT RAHAGAYFCQ
1321  VHDTQGTRSS RPASLQILYA PRDAVLSSFR DSRTRLMVVI QCTVDSEPPA EMVLSHNGKV
1381  LAASHERHSS ASGIGHIQVA RNALRLQVQD VTLGDGNTYV CTAQHTLGSI STTQRLLTET
1441  DIRVTAEPGL DWPEGTALNL SCLLPGGSGP TGNSSFTWFW NRHRLHSAPV PTLSFTPVVR
1501  AQAGLYHCRA DLPTGATTSA PVMLRVLYPP KTPTLIVFVE PQGGHQGILD CRVDSEPLAI
1561  LTLHRGSQLV ASNQLHDAPT KPHIRVTAPP NALRVDIEEL GPSNQGEYVC TASNTLGSAS
1621  ASAYFGTRAL HQLQLFQRLL WVLGFLAGFL CLLLGLVAYH TWRKKSSTKL NEDENSAEMA
1681  TKKNTIQEEV VAAL
```

Database accession

EMBL/GenBank   Z36293 (mouse)
TrEMBL   Q62230

References

1 Mucklow, S. et al. (1995) Genomics 28, 344–346.
2 Crocker, P.R. et al. (1994) EMBO J. 13, 4490–4503.
3 Crocker, P.R. et al. (1996) Biochem. Soc. Trans. 24, 150–156.
4 Vinson, M. et al. (1996) J. Biol. Chem. 271, 9267–9272.
5 May, A.P. et al. (1998) Mol. Cell 1, 719–728.
6 Crocker, P.R. et al. (1999) Biochem. J. 341, 355–361.
7 Kelm, S. et al. (1994) Curr. Biol. 4, 965–972.
8 Barnes, Y.C. et al. (1999) Blood 93, 1245–1252.
9 van den Berg, T.K. et al. (1992) J. Exp. Med. 176, 647–655.
10 McWilliam, A.S. et al. (1992) Proc. Natl Acad. Sci. USA 89, 10522–10526.

TAG-1

Family

Immunoglobulin superfamily

Structure

Molecular weights
Amino acids	1040
Polypeptide	113 393
SDS-PAGE reduced	135 kDa

Carbohydrate
N-linked sites	10
O-linked sites	

Gene location   1q32.1

Gene structure   40 kb, 23 exons

Alternative forms   Soluble TAG-1 is released from cultured neurons.

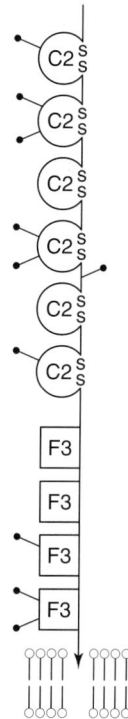

Structure

Type I membrane glyoprotein of the C2 subset of Ig superfamily with six C2 type Ig domains and four membrane-proximal fibronectin type III repeats[1-3]. TAG-1 has no transmembrane sequence and is GPI-anchored.

Ligands

TAG-1 binds to L1 family members. Recent data suggest that this interaction occurs in *cis* in the plane of the membrane rather than representing a heterophilic interaction between neighbouring cells[4,5].

Function

TAG-1 is involved in neurite outgrowth *in vitro* and pathfinder axonal extenstion in embryogenesis[6,7]. Neuronal migration and TAG-1 expression are up-regulated by netrin-1, a long-range diffusible factor[8]. Recent studies indicate that the main function of TAG-1 is to interact in *cis* with L1 family members and thereby regulate L1-mediated adhesive interactions[4,5].

Distribution

Restricted to neural tissue in a subset of central and peripheral neurons in the developing embryo, with retention in adult brain and spinal cord[4,6,7,9–11].

Disease association

OMIM 190197

Knockout

MGI:104518

Amino acid sequence from human TAG-1

```
   1 MGTATRRKPH LLLVAAVALV SSSAWSSALG SQTTFGPVFE DQPLSVLFPE ESTEEQVLLA
  61 CRARASPPAT YRWKMNGTEM KLEPGSRHQL VGGNLVIMNP TKAQDAGVYQ CLASNPVGTV
 121 VSREAILRFG FLQEFSKEER DPVKAHEGWG VMLPCNPPAH YPGLSYRWLL NEFPNFIPTD
 181 GRHFVSQTTG NLYIARTNAS DLGNYSCLAT SHMDFSTKSV FSKFAQLNLA AEDTRLFAPS
 241 IKARFPAETY ALVGQQVTLE CFAFGNPVPR IKWRKVDGSL SPQWTTAEPT LQIPSVSFED
 301 EGTYECEAEN SKGRDTVQGR IIVQAQPEWL KVISDTEADI GSNLRWGCAA AGKPRPTVRW
 361 LRNGEPLASQ NRVEVLAGDL RFSKLSLEDS GMYQCVAENK HGTIYASAEL AVQALAPDFR
 421 LNPVRRLIPA ARGGEILIPC QPRAAPKAVV LWSKGTEILV NSSRVTVTPD GTLIIRNISR
 481 SDEGKYTCFA ENFMGKANST GILSVRDATK ITLAPSSADI NLGDNLTLQC HASHDPTMDL
 541 TFTWTLDDFP IDFDKPGGHY RRTNVKETIG DLTILNAQLR HGGKYTCMAQ TVVDSASKEA
 601 TVLVRGPPGP PGGVVVRDIG DTTIQLSWSR GFDNHSPIAK YTLQARTPPA GKWKQVRTNP
 661 ANIEGNAETA QVLGLTPWMD YEFRVIASNI LGTGEPSGPS SKIRTREAAP SVAPSGLSGG
 721 GGAPGELIVN WTPMSREYQN GDGFGYLLSF RRQGSTHWQT ARVPGADAQY FVYSNESVRP
 781 YTPFEVKIRS YNRRGDGPES LTALVYSAEE EPRVAPTKVW AKGVSSSEMN VTWEPVQQDM
 841 NGILLGYEIR YWKAGDKEAA ADRVRTAGLD TSARVSGLHP NTKYHVTVRA YNRAGTGPAS
 901 PSANATTMKP PPRRPPGNIS WTFSSSSLSI KWDPVVPFRN ESAVTGYKML YQNDLHLTPT
 961 LHLTGKNWIE IPVPEDIGHA LVQIRTTGPG GDGIPAEVHI VRNGGTSMMV EN*MAVRPAPH*
1021 *PGTVISHSVA MLILIGSLEL*
```

The sequences underlined and in italics are cleaved off to form mature TAG-1 and a GPI anchor is added.

Database accession

EMBL/GenBank   X68274
SwissProt   Q02246

References

[1] Hasler, T.H. et al. (1993) Eur. J. Biochem. 211, 329–339.

[2] Tsiotra, C.P. et al. (1993) Genomics 18, 562–567.

[3] Kozlov, S.V. et al. (1995) Genomics 30, 141–148.

[4] Walsh, F.S. and Doherty, P. (1997) Annu. Rev. Cell Dev. Biol. 13, 425–456.

[5] Brummendorf, T. et al. (1998) Curr. Opin. Neurobiol. 8, 87–97.

[6] Furley, A.J. et al. (1990) Cell 61, 157–170.

[7] Kuhn, T.B. et al. (1992) J. Cell Biol. 115, 1113–1126.

[8] Alcantara, S. et al. (2000) Development 127, 1359–1372.

[9] Dodd, J. et al. (1988) Neuron 1, 105–116.

[10] Stoeckli, E.T. et al. (1991) J. Cell Biol. 112, 449–455.

[11] Karagogeos, D. et al. (1991) Development 112, 51–67.

Family

Immunoglobulin superfamily

Structure

Molecular weights

Amino acids	739
Polypeptide	81 276
SDS-PAGE reduced	100–110 kDa

Carbohydrate

N-linked sites	6
O-linked sites	

Gene location	1p32–1p31
Gene structure	~25 kb, 9 exons

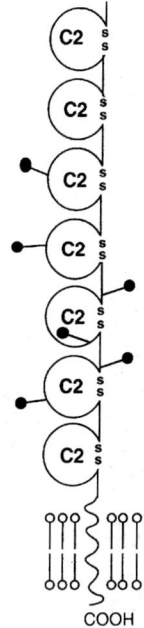

Alternative forms

A variant lacking the fourth Ig domain is produced by alternative splicing[1]. In the mouse, there is a GPI-anchored form of VCAM-1 containing the N terminal three Ig domains[2,3].

Structure

VCAM-1 is a type I transmembrane protein with seven C2-type Ig domains. Ig domains 1 and 2, and domains 4 and 5 show homology to domains 1 and 2 of the other integrin-binding IgSF molecules: MAdCAM, ICAM-1 (CD54), ICAM-2 (CD102) and ICAM-3 (CD50)[4,5]. The integrin binding domain of VCAM-1 has been crystallized[5,6].

Ligands

VCAM-1 binds integrins $\alpha_4\beta_1$ and $\alpha_4\beta_7$[7] with a preference for a $\alpha_4\beta_1$. Binding domains for $\alpha_4\beta_1$ have been mapped to domains 1 and 4[3]. More recently, integrin $\alpha_D\beta_2$ on eosinophils has been shown to act as a VCAM-1 ligand[8].

Function

VCAM-1 mediates the adhesion of lymphocytes, monocytes and eosinophils to inflamed endothelium by promoting both the initial tethering and rolling of lymphocytes, and the subsequent firm adhesion. In non-vascular cells VCAM-1 has been implicated in the interaction of haematopoietic progenitors with bone marrow stromal cells, B cell adhesion to dendritic cells, co-stimulation of T cells, embryonic development and metastatic spread of tumour cells[2,3,9,10].

Distribution

VCAM-1 is present predominantly on activated endothelial cells but is also expressed by tissue macrophages, dendritic cells, bone marrow fibroblasts, myoblasts and myotubes[2,3,9,11].

Disease association

OMIM 192225
There are studies indicating that deficiencies in VCAM-1 can result in epicardial dissolution and subsequent myocardial thinning[12]. The ability of VCAM-1 to mediate adhesion of melanoma cells to the endothelium may play an important role in mestatic dissemination[13].

Knockout

MGI:98926
The Gene Knockout FactsBook[14], p1065
The majority of VCAM-1 $-/-$ mice die at 9.5–11.5 days of gestation due to failure of the allantois to fuse to the chorion. Mice that survive the placental defect subsequently display severe heart abnormalities. However, a few mice develop to become healthy fertile adults, indicating that VCAM-1 is not essential for normal organ development[15,16].

Amino acid sequence of human VCAM-1

```
  1 MPGKMVVILG ASNILWIMFA ASQAFKIETT PESRYLAQIG DSVSLTCSTT GCESPFFSWR
 61 TQIDSPLNGK VTNEGTTSTL TMNPVSFGNE HSYLCTATCE SRKLEKGIQV EIYSFPKDPE
121 IHLSGPLEAG KPITVKCSVA DVYPFDRLEI DLLKGDHLMK SQEFLEDADR KSLETKSLEV
181 TFTPVIEDIG KVLVCRAKLH IDEMDSVPTV RQAVKELQVY ISPKNTVISV NPSTKLQEGG
241 SVTMTCSSEG LPAPEIFWSK KLDNGNLQHL SGNATLTLIA MRMEDSGIYV CEGVNLIGKN
301 RKEVELIVQE KPFTVEISPG PRIAAQIGDS VMLTCSVMGC ESPSFSWRTQ IDSPLSGKVR
361 SEGTNSTLTL SPVSFENEHS YLCTVTCGHK KLEKGIQVEL YSFPRDPEIE MSGGLVNGSS
421 VTVSCKVPSV YPLDRLEIEL LKGETILENI EFLEDTDMKS LENKSLEMTF IPTIEDTGKA
481 LVCQAKLHID DMEFEPKQRQ STQTLYVNVA PRDTTVLVSP SSILEEGSSV NMTCLSQGFP
541 APKILWSRQL PNGELQPLSE NATLTLISTK MEDSGVYLCE GINQAGRSRK EVELIIQVTP
601 KDIKLTAFPS ESVKEGDTVI ISCTCGNVPE TWIILKKKAE TGDTVLKSID GAYTIRKAQL
661 KDAGVYECES KNKVGSQLRS LTLDVQGREN NKDYFSPELL VLYFASSLII PAIGMIIYFA
721 RKANMKGSYS LVEAQKSKV
```

Database accession
EMBL/GenBank	M73255
SwissProt	P19320

References
1 Cybulsky, M.I. et al. (1991) Proc. Natl Acad. Sci. USA 88, 7859–7863.
2 Bevilacqua, M.P. (1993) Ann. Rev. Immunol. 11, 767–804.
3 Vonderheide, R.H. et al. (1994) J. Cell Biol. 125, 215–222.

[4] Jones, E.Y. et al. (1995) Nature 373, 539–544.

[5] Shyjan, A.M. et al. (1996) J. Immunol. 156, 2851–2857.

[6] Wang, J.H. et al. (1995) Proc. Natl Acad. Sci. USA 92, 5714–5718.

[7] Berlin, C. et al. (1993) Cell 74, 185–195.

[8] Grayson, M.H. et al. (1998) J. Exp. Med. 188, 2187–2191.

[9] Carlos, T.M. and Harlan, J.M. (1994) Blood 84, 2068–2101.

[10] Butcher, E.C. and Picker, L.J. (1996) Science 272, 60–66.

[11] Masinovsky, B. et al. (1990) J. Immunol. 145, 2886–2895.

[12] Olson, E. and Srivastava, D. (1996) Science, 272, 671–676.

[13] Taichman, D.B. et al. (1991) Cell Regul. 23, 47–55.

[14] Mak, T.W. (1998) The Gene Knockout FactsBook. Academic Press, London.

[15] Gunter, G.C. et al. (1995) Genes Dev. 9, 1–4.

[16] Kwee, L. et al. (1995) Development 121, 489–503.

Integrins

Integrin α_L CD11a, LFA-1

Family

Integrin (LFA/CD11 subfamily)

Structure

Molecular weights

Amino acids	1170
Polypeptide	128 819

SDS-PAGE reduced	180 kDa (170 kDa unreduced)

Carbohydrate

N-linked sites	12
O-linked sites	

Gene location   16p11.2 (in a cluster with the other CD11 genes)

Gene structure   >32 kb, >6 exons (intron-exon arrangement is similar for all of the CD11a chains).

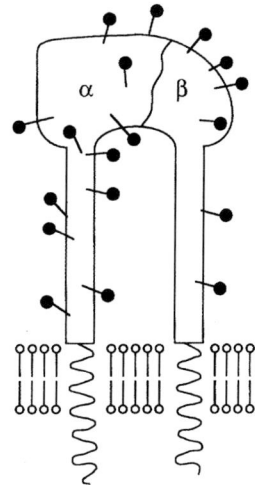

CD11a/CD18

Alternative forms

Structure

Integrin α_L is a type I membrane glycoprotein integrin α chain[1] that dimerizes with β_2 to form $\alpha_L\beta_2$; the β_2 chain is shared with the other integrin α chains of the CD11 family (see entries). α_L contains seven repetitive domains, approximately 50 amino acid long, in its extracellular part; these repeat regions have been modelled to a beta-propeller fold as found in the β subunits of trimeric G proteins[2]. The membrane-proximal three (V–VII) contain calmodulin EF hand-like sequences and are involved in binding divalent cations. The α_L chain is phosphorylated[3]. The α_L chain belongs to the subfamily [with the other leucocyte functional antigen (LFA)/CD11 group of leuco-integrin α chains] of integrin α chains with an inserted or 'I' domain and lacks post-translational proteolytic cleavage[1]. The I domain of α_L (CD11a) shows sequence homology with von Willebrand factor; it has been crystallized and a structural model is available that shows structural homology with an α–β dinucleotide binding fold[4,5]. It also contains an imperfect 'metal ion-dependent adhesion site' (MIDAS) with one uncoordinated site that is thought to be introduced from the integrin ligand, for example, from Glu34 of ICAM-1[6]. The amino acid sequence VGFFKR at the membrane proximal part of the cytoplasmic tail of α_L is involved in receptor dimerization.

Ligands

'Counter-receptors': ICAM-1 (CD54), -2 (CD102), -3 (CD50), which contain immunoglobulin superfamily (IgSF) domains (see individual entries for ICAMs)[7,8]. Interactions occur through the I domain[9]. Bacterial lipopolysaccharides (LPS)[10].

Function

Involved in a variety of immune functions. Cell–cell interactions during lymphocyte transit through endothelium to sites of inflammation and during recirculation through lymph nodes[7,8,11]. Co-stimulatory function in T lymphocyte activation, and effector functions during inflammation and immune responses to T cell antigens. Leucocyte intercelluar adhesion/aggregation. Adhesion of cytotoxic T cells to targets cells. Interaction with ICAM ligand is energy dependent, requires an intact actin cytoskeleton and is divalent cation-dependent (stimulated by magnesium, inhibited by calcium)[12]. Constitutively expressed but 'activation' of the $\alpha_L\beta_2$ receptor is required for it to bind ligand; this occurs via intracellular signalling and under the influence of cations.

Distribution

Lymphocytes, monocyte lineage cells and other leucocytes. Transiently upregulated during T cell activation[11,13]. Absent from non-haemopoeitic tissues and platelets.

Disease association

OMIM 153370
Deficiency of $\alpha_L\beta_2$ leads to lifelong recurrent bacterial infections and is due to mutations in its common β_2 chain (see 'leucocyte adhesion deficiency', LAD-1, under entry for integrin β_2; OMIM 116920). Polymorphic determinants exist on the α_L chain[14]. Antibodies to β_2 family of integrins (to α or β chains and ICAM ligand) block leucocyte immune functions, and are being evaluated clinically for use in a variety of inflammatory diseases and transplant rejection.

Knockout

MGI:96606
The Gene Knockout FactsBook[16], pp. 140–142.
No obvious phenotype at birth. Mice show defective T lymphocyte proliferative responses to T cell mitogens and viruses. Normal thymic development and peripheral blood T cell subsets indicate that α_L is not required for T cell development[15].

Amino acid sequence of human integrin α_L

```
   1 MKDSCITVMA MALLSGFFFF APASSYNLDV RGARSFSPPR AGRHFGYRVL QVGNGVIVGA
  61 PGEGNSTGSL YQCQSGTGHC LPVTLRGSNY TSKYLGMTLA TDPTDGSILA CDPGLSRTCD
 121 QNTYLSGLCY LFRQNLQGPM LQGRPGFQEC IKGNVDLVFL FDGSMSLQPD EFQKILDFMK
 181 DVMKKLSNTS YQFAAVQFST SYKTEFDFSD YVKRKDPDAL LKHVKHMLLL TNTFGAINYV
 241 ATEVFREELG ARPDATKVLI IITDGEATDS GNIDAAKDII RYIIGIGKHF QTKESQETLH
 301 KFASKPASEF VKILDTFEKL KDLFTELQKK IYVIEGTSKQ DLTSFNMELS SSGISADLSR
 361 GHAVVGAVGA KDWAGGFLDL KADLQDDTFI GNEPLTPEVR AGYLGYTVTW LPSRQKTSLL
 421 ASGAPRYQHM GRVLLFQEPQ GGGHWSQVQT IHGTQIGSYF GGELCGVDVD QDGETELLLI
 481 GAPLFYGEQR GGRVFIYQRR QLGFEEVSEL QGDPGYPLGR FGEAITALTD INGDGLVDVA
 541 VGAPLEEQGA VYIFNGRHGG LSPQPSQRIE GTQVLSGIQW FGRSIHGVKD LEGDGLADVA
 601 VGAESQMIVL SSRPVVDMVT LMSFSPAEIP VHEVECSYST SNKMKEGVNI TICFQIKSLY
 661 PQFQGRLVAN LTYTLQLDGH RTRRRGLFPG GRHELRRNIA VTTSMSCTDF SFHFPVCVQD
 721 LISPINVSLN FSLWEEEGTP RDQRAQGKDI PPILRPSLHS ETWEIPFEKN CGEDKKCEAN
 781 LRVSFSPARS RALRLTAFAS LSVELSLSNL EEDAYWVQLD LHFPPGLSFR KVEMLKPHSQ
 841 IPVSCEELPE ESRLLSRALS CNVSSPIFKA GHSVALQMMF NTLVNSSWGD SVELHANVTC
 901 NNEDSDLLED NSATTIIPIL YPINILIQDQ EDSTLYVSFT PKGPKIHQVK HMYQVRIQPS
 961 IHDHNIPTLE AVVGVPQPPS EGPITHQWSV QMEPPVPCHY EDLERLPDAA EPCLPGALFR
1021 CPVVFRQEIL VQVIGTLELV GEIEASSMFS LCSSLSISFN SSKHFHLYGS NASLAQVVMK
1081 VDVVYEKQML YLYVLSGIGG LLLLLLIFIV LYKVGFFKRN LKEKMEAGRG VPNGIPAEDS
1141 EQLASGQEAG DPGCLKPLHE KDSESGGGKD
```

I domain position 170–349.

Database accession

EMBL/GenBank   Y00796
SwissProt   P20701

References

[1] Larson, R.S. et al. (1989) J. Cell Biol. 108, 703–712.

[2] Springer, T.A. (1997) Proc. Natl Acad. Sci. USA 94, 65–72.

[3] Hibbs, M.L. et al. (1991) J. Exp. Med. 174, 1227–1238.

[4] Qu, A. and Leahy, D.J. (1995) Proc. Natl Acad. Sci. USA 92, 10277–10281.

[5] Qu, A. and Leahy, D.J. (1996) Structure 4, 931–942.

[6] Lee, J.O. et al. (1995) Structure 3, 1333–1340.

[7] Springer, T.A. (1990) Nature 346, 425–433.

[8] Springer, T.A. (1994) Cell 76, 301–314.

[9] Landis, R.C. et al. (1994) J. Cell. Biol. 126, 529–537.

[10] Wright, S.D. and Jong, M.T.C. (1986) J. Exp. Med. 164, 1876–1888.

[11] Patarroyo, M. et al. (1990) Immunol. Rev. 114, 67–108.

[12] Dransfield, I. et al. (1992) J. Cell Biol. 116, 219–226.

[13] Larson, R.S. and Springer, T.A. (1990) Immunol. Rev. 114, 181–217.

[14] Pischel, K.D. et al. (1987) J. Clin. Invest. 79, 1607–1687.

[15] Schmits, R. et al. (1996) J. Exp. Med. 183, 1415–1426.

[16] Mak, T.W. (1998) The Gene Knockout FactsBook. Academic Press, London.

Integrin α_M CD11b, Mac-1, CR3

Family

Integrin (LFA/CD11 subfamily)

Structure

Molecular weights

Amino acids	1153
Polypeptide	127 178

SDS-PAGE reduced	170 kDa (165 kDa unreduced)

Carbohydrate

N-linked sites	19
O-linked sites	

Gene location	16p11.2

Gene structure

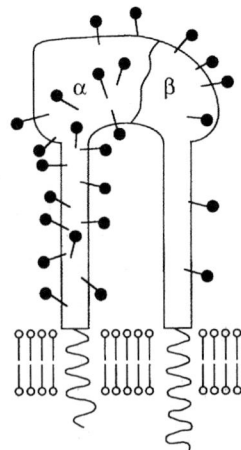

CD11b/CD18

Alternative forms

An extra codon is found at bp 1580 between exons 13 and 14 in some α_M cDNAs (there are no functional effects due to insertion).

Structure

Integrin α_M is a type I membrane glycoprotein[1-3] forming non-covalent dimer with the β_2 integrin chain, forming $\alpha_M\beta_2$. The α_M chain contains an N-terminal inserted 'I' domain and is not proteolytically cleaved. Two crystal structures of the 'I' domain have been obtained, possibly representing different 'activation' states[4-6]; most ligands (for example, ICAMs, iC3b, fibrinogen) bind via the I domain[6a]. α_M contains seven repeat regions, the membrane proximal three being involved in divalent cation binding and receptor activation[7,8].

Ligands

Complement fragment iC3b (possibly via its RGD sequence), ICAM-1 (CD54) via its third Ig domain[9], ICAM-2 (CD102) and ICAM-4, fibrinogen, clotting factor X, CD23, hookworm 'neutrophil inhibitory factor' and heparin. β-Glucan and bacterial LPS and other microbial products, denatured proteins such as albumen[10-20]. The I domain plays a major role in ligand recognition[8,12,21].

Function

Integrin α_M has several roles in inflammation. It mediates adhesion to, and transmigration[22,23] through vascular endothelium of monocytes and granulocytes. It also participates in leucocyte aggregation, and is involved in chemotaxis, apoptosis and other phagocytic activities[10,11,24,25] (uptake of complement-coated particles, hence its original description as 'complement receptor 3';CR3) and subsequent oxidative respiratory burst. It interacts with the FcγR receptor in immune complex-stimulated neutrophil function. Several associated membrane proteins involved with α$_M$β$_2$ in immune functions have been identified: CD14, CD87, Fcγ receptors CD16 and CD32, and uPAR[26,27].

Distribution

Leucocytes, such as neutrophils, monocytes, eosinophils and basophils, NK cells[28,29]. Upregulated in myeloid cells during inflammation (via rapid mobilization from intracellular granules upon exposure to chemoattractants, such as fMLP, complement components and leukotrienes) and differentiation, but reduced in tissue macrophages (see integrin α$_L$ entry).

Disease association

OMIM 120980
See entry for integrin β$_2$ for details of 'leucocyte adhesion deficiency' syndrome (LAD-1, OMIM 116920).

Knockout

MGI:96607
The Gene Knockout FactsBook[30], pp. 143–144.
Outwardly normal at birth, the mutant mice are not severely immunocompromised. The mice have normal leucocyte counts, but these are functionally abnormal and respond poorly to some inflammatory stimuli (response to chemoattractants and reduced phagocytosis but normal peritoneal accumulation with thioglycollate)[31].

Amino acid sequence of human integrin α$_M$

```
   1 MALRVLLLTA LTLCHGFNLD TENAMTFQEN ARGFGQSVVQ LQGSRVVVGA PQEIVAANQR
  61 GSLYQCDYST GSCEPIRLQV PVEAVNMSLG LSLAATTSPP QLLACGPTVH QTCSENTYVK
 121 GLCFLFGSNL RQQPQKFPEA LRGCPQEDSD IAFLIDGSGS IIPHDFRRMK EFVSTVMEQL
 181 KKSKTLFSLM QYSEEFRIHF TFKEFQNNPN PRSLVKPITQ LLGRTHTATG IRKVVRELFN
 241 ITNGARKNAF KILVVITDGE KFGDPLGYED VIPEADREGV IRYVIGVGDA FRSEKSRQEL
 301 NTIASKPPRD HVFQVNNFEA LKTIQNQLRE KIFAIEGTQT GSSSSFEHEM SQEGFSAAIT
 361 SNGPLLSTVG SYDWAGGVFL YTSKEKSTFI NMTRVDSDMN DAYLGYAAAI ILRNRVQSLV
 421 LGAPRYQHIG LVAMFRQNTG MWESNANVKG TQIGAYFGAS LCSVDVDSNG STDLVLIGAP
 481 HYYEQTRGGQ VSVCPLPRGQ RARWQCDAVL YGEQGQPWGR FGAALTVLGD VNGDKLTDVA
 541 IGAPGEEDNR GAVYLFHGTS GSGISPSHSQ RIAGSKLSPR LQYFGQSLSG GQDLTMDGLV
 601 DLTVGAQGHV LLLRSQPVLR VKAIMEFNPR EVARNVFECN DQVVKGKEAG EVRVCLHVQK
 661 STRDRLREGQ IQSVVTYDLA LDSGRPHSRA VFNETKNSTR RQTQVLGLTQ TCETLKLQLP
 721 NCIEDPVSPI VLRLNFSLVG TPLSAFGNLR PVLAEDAQRL FTALFPFEKN CGNDNICQDD
 781 LSITFSFMSL DCLVVGGPRE FNVTVTVRND GEDSYRTQVT FFFPLDLSYR KVSTLQNQRS
 841 QRSWRLACES ASSTEVSGAL KSTSCSINHP IFPENSEVTF NITFDVDSKA SLGNKLLLKA
 901 NVTSENNMPR TNKTEFQLEL PVKYAVYMVV TSHGVSTKYL NFTASENTSR VMQHQYQVSN
 961 LGQRSLPISL VFLVPVRLNQ TVIWDRPQVT FSENLSSTCH TKERLPSHSD FLAELRKAPV
1021 VNCSIAVCQR IQCDIPFFGI QEEFNATLKG NLSFDWYIKT SHNHLLIVST AEILFNDSVF
1081 TLLPGQGAFV RSQTETKVEP FEVPNPLPLI VGSSVGGLLL LALITAALYK LGFFKRQYKD
1141 MMSEGGPPGA EPQ
```

I domain position 164–349.

Database accession

EMBL/GenBank   J03925
SwissProt   P1121

References

1 Corbi, A.L. et al. (1988) J. Biol. Chem. 263, 12403–12411.

2 Arnaout, M.A. et al. (1988) J. Cell Biol. 106, 2153–2158.

3 Hickstein, D.D. et al. (1989) Proc. Natl Acad. Sci. USA 86, 257–261.

4 Lee, J.O. et al. (1995a) Cell 80, 631–638.

5 Lee, J.O. et al. (1995b) Structure 3, 1333–1340.

6 Baldwin, E.T. et al. (1998) Structure 6, 923–935.

6a Zhang, L. and Plow, E.F. (1999) Biochemistry 38, 8064–8071.

7 Oxvig, C. et al. (1999) Proc. Natl Acad. Sci. USA 96, 2215–2220.

8 Michishita, M. et al. (1993) Cell 72, 857–867.

9 Sanchez-Madrid, F. and Corbi, A.L. (1993) Semin. Cell Biol. 3, 199–210.

10 Beller, D.I. et al. (1982) Exp. Med. 156, 1000–1009.

11 Wright, S.D. et al. (1983) Proc. Natl Acad. Sci. USA 80, 5699–5703.

12 Zhang, L. and Plow, E.F. (1996) J. Biol. Chem. 271, 18211–18216.

13 Lecoanet-Henchoz, S. et al. (1995) Immunity 3, 119–125.

14 Diamond, M.S. et al. (1995) J. Cell Biol. 130, 1473–1482.

15 Thornton, B.P. et al. (1996) Immunol. 156, 1235–1246.

16 Wright, S.D. and Jong, M.T.C. (1986) J. Exp. Med. 164, 1876–1888.

17 Davis, G.E. (1992) Exp. Cell Res. 200, 242–252.

18 Diamond, M.S. et al. (1991) Cell 65, 961–971.

19 Alteri, D.C. et al. (1988) Proc. Natl Acad. Sci. USA 85, 7462–7466.

20 Wright, S.D. et al. (1988) Proc. Natl Acad. Sci. USA 85, 7734–7738.

21 Diamond, M.S. et al. (1993) J. Cell Biol. 120, 1031–1043.

22 Springer, T.A. (1994) Cell 76, 301–314.

23 Arnaout, M.A. et al. (1988) J. Cell Physiol. 137, 305–309.

24 Gresham, H.D. et al. (1991) J. Clin. Invest. 88, 588–596.

25 Carlos, T.M. and Harlan, J.M. (1990) Immunol. Rev. 114, 1–24.

26 Petty, H.R. and Todd, R.F. III. (1996) Immunol. Today 17, 209–212.

27 Annendov, A. et al. (1996) Eur. J. Immunol. 26, 207–212.

28 Larson, R.S. and Springer, T.A. (1990) Immunol. Rev. 114, 181–217.

29 Patarroyo, M. et al. (1990) Immunol. Rev. 114, 67–108.

30 Mak, T.W. (1998) The Gene Knockout FactsBook. Academic Press, London.

31 Coxon, A. et al. (1996) Immunity 5, 653–666.

Integrin α_X

Family

Integrin (LFA/CD11 subfamily)

Structure

Molecular weights

Amino acids	1163
Polypeptide	127 885

SDS-PAGE reduced	150 kDa (145 kDa unreduced)

Carbohydrate

N-linked sites	8
O-linked sites	

Gene location	16p11.2

Gene structure	25 kb. 31 exons

Alternative forms

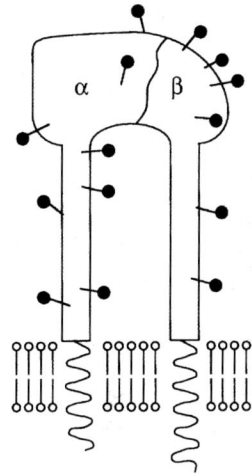

CD11c/CD18

Structure

Integrin α_X is a type I membrane glycoprotein[1] forming a non-covalent dimer with the β_2 integrin chain, $\alpha_X\beta_2$. The α_X chain contains an N-terminal inserted I domain and is not proteolytically cleaved. It contains seven N-terminal repeat regions, the membrane proximal three being involved in divalent cation binding.

Ligands

α_X has similar ligands to α_M but they are less well characterized: complement fragment iC3b, fibrinogen, ICAM-1 (CD54), CD23 and bacterial LPS[2-7].

Function

It has a similar function to $\alpha_M\beta_2$ (CD11b), including cytotoxic T cell killing, and monocyte and granulocyte adhesion to endothelium[8,9]. Major β_2 receptor of tissue macrophages.

Distribution

Tissue macrophages and monocytes (high levels, inverse distribution to $\alpha_M\beta_2$ during monocyte–macrophage maturation)[10] and less on other leucocytes[10,11].

Disease association

OMIM 151510
See entry for integrin β$_2$ for details of 'leucocyte adhesion deficiency' syndrome (LAD-1, OMIM 116920).

Knockout

Amino acid sequence of human integrin α$_X$

```
   1  MTRTRAALLL FTALATSLGF NLDTEELTAF RVDSAGFGDS VVQYANSWVV VGAPQKITAA
  61  NQTGGLYQCG YSTGACEPIG LQVPPEAVNM SLGLSLASTT SPSQLLACGP TVHHECGRNM
 121  YLTGLCFLLG PTQLTQRLPV SRQECPRQEQ DIVFLIDSGS SISSRNFATM MNFVRAVISQ
 181  FQRPSTQFSL MQFSNKFQTH FTFEEFRRTS NPLSLLASVH QLQGFTYTAT AIQNVVHRLF
 241  HASYGARRDA TKILIVITDG KKEGDSLDYK DVIPMADAAG IIRYAIGVGL AFQNRNSWKE
 301  LNDIASKPSQ EHIFKVEDFD ALKDIQNQLK EKIFAIEGTE TTSSSSFELE MAQEGFSAVF
 361  TPDGPVLGAV GSFTWSGGAF LYPPNMSPTF INMSQENVDM RDSYLGYSTE LALWKGVQSL
 421  VLGAPRYQHT GKAVIFTQVS RQWRMKAEVT GTQIGSYFGA SLCSVDVDTD GSTDLVLIGA
 481  PHYYEQTRGG QVSVCPLPRG WRRWWCDAVL YGEQGHPWGR FGAALTVLGD VNGDKLTDVV
 541  IGAPGEEENR GAVYLFHGVL GPSISPSHSQ RIAGSQLSSR LQYFGQALSG GQDLTQDGLV
 601  DLAVGARGQV LLLRTRPVLW VGVSMQFIPA EIPRSAFECR EQVVSEQTLV QSNICLYIDK
 661  RSKNLLGSRD LQSSVTLDLA LDPGRLSPRA TFQETKNRSL SRVRVLGLKA HCENFNLLLP
 721  SCVEDSVTPI TLRLNFTLVG KPLLAFRNLR PMLAALAQRY FTASLPFEKN CGADHICQDN
 781  LGISFSFPGL KSLLVGSNLE LNAEVMVWND GEDSYGTTIT FSHPAGLSYR YVAEGQKQGQ
 841  LRSLHLTCDS APVGSQGTWS TSCRINHLIF RGGAQITFLA TFDVSPKAVL GDRLLLTANV
 901  SSENNTPRTS KTTFQLELPV KYAVYTVVSS HEQFTKYLNF SESEEKESHV AMHRYQVNNL
 961  GQRDLPVSIN FWVPVELNQE AVWMDVEVSH PQNPSLRCSS EKIAPPASDF LAHIQKNPVL
1021  DCSIAGCLRF RCDVPSFSVQ EELDFTLKGN LSFGWVRQIL QKKVSVVSVA EITFDTSVYS
1081  QLPGQEAFMR AQTTTVLEKY KVHNPTPLIV GSSIGGLLLL ALITAVLYKV GFFKRQYKEM
1141  MEEANGQIAP ENGTQTPSPP SEK
```

I domain position 165–351.

Database accession

EMBL/GenBank   M811695
SwissProt   P20702

References

[1] Corbi, A.L. et al. (1987) EMBO J. 6, 4023–4028.

[2] Myones, B.L. et al. (1988) J. Clin. Invest. 82, 640–651.

[3] de Fougerolles, A.R. et al. (1995) Eur. J. Immunol. 24, 1008–1012.

[4] Loike, J.D. et al. (1991) Proc. Natl Acad. Sci. USA 88, 1044–1048.

[5] Postigo, A.A. et al. (1991) J. Exp. Med. 174, 1313–1322.

[6] Wright, S.D. and Jong, M.T.C. (1986) J. Exp. Med. 164, 1876–1888.

[7] Ingalls, R.R. and Golenbock, D.T. (1995) J. Exp. Med. 181, 1473–1479.

[8] Keizer, G.D. et al. (1987) J. Immunol. 138, 3130–3136.

[9] Stacker, S.A. and Springer, T.A. (1991) J. Immunol. 146, 648–655.

[10] Larson, R.S. and Springer, T.A. (1990) Immunol. Rev. 114, 181–217.

[11] Patarroyo, M. et al. (1990) Immunol. Rev. 114, 67–108.

Integrin α_D

Family

Integrin (LFA/CD11 subfamily)

Structure

Molecular weights

Amino acids	1162
Polypeptide	126 886

SDS-PAGE reduced   150 kDa

Carbohydrate

N-linked sites	10
O-linked sites	
Other	

Gene location   16p11.2 (CD11 family gene cluster on chromosome 16)[1]

Gene structure

Alternative forms

Structure

Integrin α_D forms a typical integrin heterodimer with integrin β$_2$ (CD18) to form α_Dβ$_2$. It contains an N-terminal I domain as with other β$_2$ (CD11) subfamily members[1,2].

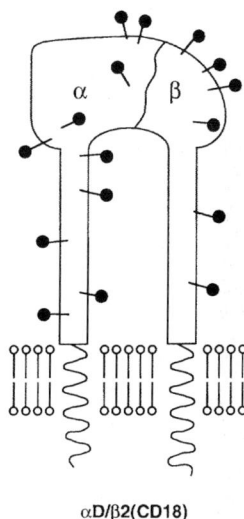

αD/β2(CD18)

Ligands

It binds ICAM-3[3] (CD50), but not ICAM-1 (CD54). α_D interacts with VCAM-1 (CD106)[3] on eosinophils.

Function

Currently undefined though likely to be similar to other β$_2$ integrins, including a role in phagocytosis by macrophages.

Distribution

Moderate levels on peripheral leucocytes[4], and higher expression on tissue macrophages such as in spleen[2].

Disease association

OMIM 602453

Knockout

Amino acid sequence of human α_D

```
   1 MTFGTVLLLS VLASYHGFNL DVEEPTIFQE DAGGFGQSVV QFGGSRLVVG APLEVVAANQ
  61 TGRLYDCAAA TGMCQPIPLH IRPEAVNMSL GLTLAASTNG SRLLACGPTL HRVCGENSYS
 121 KGSCLLLGSR WEIIQTVPDA TPECPHQEMD IVFLIDGSGS IDQNDFNQMK GFVQAVMGQF
 181 EGTDTLFALM QYSNLLKIHF TFTQFRTSPS QQSLVDPIVQ LKGLTFTATG ILTVVTQLFH
 241 HKNGARKSAK KILIVITDGQ KYKDPLEYSD VIPQAEKAGI IRYAIGVGHA FQGPTARQEL
 301 NTISSAPPQD HVFKVDNFAA LGSIQKQLQE KIYAVEGTQS RASSSFQHEM SQEGFSTALT
 361 MDGLFLGAVG SFSWSGGAFL YPPNMSPTFI NMSQENVDMR DSYLGYSTEL ALWKGVQNLV
 421 LGAPRYQHTG KAVIFTQVSR QWRKKAEVTG TQIGSYFGAS LCSVDVDSDG STDLILIGAP
 481 HYYEQTRGGQ VSVCPLPRGQ RVQWQCDAVL RGEQGHPWGR FGAALTVLGD VNEDKLIDVA
 541 IGAPGEQENR GAVYLFHGAS ESGISPSHSQ RIASSQLSPR LQYFGQALSG GQDLTQDGLM
 601 DLAVGARGQV LLLRSLPVLK VGVAMRFSPV EVAKAVYRCW EEKPSALEAG DATVCLTIQK
 661 SSLDQLGDIQ SSVRFDLALD PGRLTSRAIF NETKNPTLTR RKTLGLGIHC ETLKLLLPDC
 721 VEDVVSPIIL HLNFSLVREP IPSPQNLRPV LAVGSQDLFT ASLPFEKNCG QDGLCEGDLG
 781 VTLSFSGLQT LTVGSSLELN VIVTVWNAGE DSYGTVVSLY YPAGLSHRRV SGAQKQPHQS
 841 ALRLACETVP TEDEGLRSSR CSVNHPIFHE GSNGTFIVTF DVSYKATLGD RMLMRASASS
 901 ENNKASSSKA TFQLELPVKY AVYTMISRQE ESTKYFNFAT SDEKKMKEAE HRYRVNNLSQ
 961 RDLAISINFW VPVLLNGVAV WDVVMEAPSQ SLPCVSERKP PQHSDFLTQI SRSPMLDCSI
1021 ADCLQFRCDV PSFSVQEELD FTLKGNLSFG WVRETLQKKV LVVSVAEITF DTSVYSQLPG
1081 QEAFMRAQME MVLEEDEVYN AIPIIMGSSV GALLLLALIT ATLYKLGFFK RHYKEMLEDK
1141 PEDTATFSGD DFSCVAPNVP LS
```

I domain position 164–349.

Database accession
EMBL/GenBank   U37028
SwissProt   Q13349

References
[1] Wong, D.A. et al. (1996) Gene 171, 291–294.
[2] Van Der Vieren, M. et al. (1995) Immunity 3, 683–690.
[3] Grayson, M.H. et al. (1998) J. Exp. Med. 188, 2187–2191.
[4] Danilenko, D.M. et al. (1995) J. Immunol. 155, 35–44.

Integrin α_1 CD49a, VLA1

Family

Integrin

Structure

Molecular weights
Amino acids	1151
Polypeptide	127 837

SDS-PAGE reduced   210 kDa (200 kDa unreduced)

Carbohydrate
N-linked sites	26
O-linked sites	

Gene location   Chromosome 5

Gene structure

Alternative forms

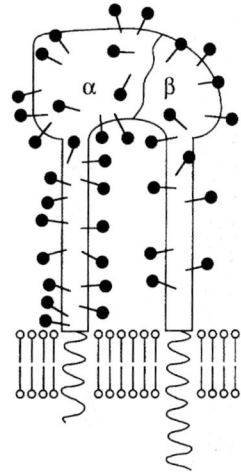

CD49a/CD29

Structure
The type I membrane glycoprotein integrin α_1 chain associates with β_1 to form the non-covalently linked $\alpha_1\beta_1$ heterodimer[1]. The α_1 chain falls into the subclass of integrin α chains with an N-terminal inserted 206 amino acid sequence, the I domain, and is not post-translationally proteolytically cleaved. It is heavily N-glycosylated compared with other integrin α chains.

Ligands

Laminin (E1 fragment) and collagen[2,3].

Function

Cell lines (for example, melanoma) have been shown to bind laminin and collagen via $\alpha_1\beta_1$[2,3], but adhesion of stimulated T cells to collagen does not utilize $\alpha_1\beta_1$[4]. A role in embryonic nervous system development and placentation is suggested by its early expression, but this is not supported by mouse knockout data – see below. The finding of $\alpha_1\beta_1$ on T cells in certain immune diseases indicates a functional role in inflammation[5].

Distribution

Widespread distribution, including in embryogenesis (placenta, nervous system). Key collagen receptor in smooth muscle and liver[6,7]. Expressed by some haemopoietic cells including monocytes and lymphocytes[8]. Resting T lymphocytes express low levels of $\alpha_1\beta_1$, and this is increased substantially upon prolonged stimulation *in vitro*[8,9] and in some inflammatory conditions, such as in rheumatoid arthritis and inflammatory bowel disease. This feature led to its initial name, 'very late T cell antigen' (or VLA).

Disease association

OMIM 192968

Knockout

MGI:96599
The Gene Knockout FactsBook[11], p. 619.
Live, fertile offspring have no obvious phenotype. Fibroblasts from α_1 $-/-$ mice show defective adhesion and migration on collagen type IV and laminin, though adhesion to collagen type I is normal[10].

Amino acid sequence of human integrin α_1

```
   1 FNVDVKNSMT FSGPVEDMFG YTVQQYENEE GKWVLIGSPL VGQPKNRTGD VYKCPVGRGE
  61 SLPCVKLDLP VNTSIPNVTE VKENMTFGST LVTNPNGGFL ACGPLYAYRC GHLHYTTGIC
 121 SDVSPTFQVV NSIAPVQECS TQLDIVIVLD GSNSIYPWDS VTAFLNDLLK RMDIGPKQTQ
 181 VGIVQYGENV THEFNLNKYS STEEVLVAAK KIVQRGGRQT MTALGTDTAR KEAFTEARGA
 241 RRGVKKVMVI VTDGESHDNH RLKKVIQDCE DENIQRFSIA ILGSYNRGNL STEKFVEEIK
 301 SIASEPTEKH FFNVSDELAL VTIVKTLGER IFALEATADQ SAASFEMEMS QTGFSAHYSQ
 361 DWVMLGAVGA YDWNGTVVMQ KASQIIIPRN TTFNVESTKK NEPLASYLGY TVNSATASSG
 421 DVLYIAGQPR YNHTGQVIIY RMEDGNIKIL QTLSGEQIGS YFGSILTTTD IDKDSNTDIL
 481 LVGAPMYMGT EKEEQGKVYV YALNQTRFEY QMSLEPIKQT CCSSRQHNSC TTENKNEPCG
 541 ARFGTAIAAV KDLNLDGFND IVIGAPLEDD HGGAVYIYHG SGKTIRKEYA QRIPSGGDGK
 601 TLKFFGQSIH GEMDLNGDGL TDVTIGGLGG AALFWSRDVA VVKVTMNFEP NKVNIQKKNC
 661 HMEGKETVCI NATVCFEVKL KSKEDTIYEA DLQYRVTLDS LRQISRSFFS GTQERKVQRN
 721 ITVRKSECTK HSFYMLDKHD FQDSVRITLD FNLTDPENGP VLDDSLPNSV HEYIPFAKDC
 781 GNKEKCISDL SLHVATTEKD LLIVRSQNDK FNVSLTVKNT KDSAYNTRTI VHYSPNLVFS
 841 GIEAIQKDSC ESNHNITCKV GYPFLRRGEM VTFKILFQFN TSYLMENVTI YLSATSDSEE
 901 PPETLSDNVV NISIPVKYEV GLQFYSSASE YHISIAANET VPEVINSTED IGNEINIFYL
 961 IRKSGSFPMP ELKLSISFPN MTSNGYPVLY PTGLSSSENA NCRPHIFEDP FSINSGKKMT
1021 TSTDHLKRGT ILDCNTCKFA TITCNLTSSD ISQVNVSLIL WKPTFIKSYF SSLNLTIRGE
1081 LRSENASLVL SSSNQKRELA IQISKDGLPG RVPLWVILLS AFAGLLLLML LILALWKIGF
1141 FKRPLKKKME K
```

I domain position 147–360.

Database accession

EMBL/GenBank   X68742
SwissProt   P56199

References

[1] Briesewitz, R. et al. (1993) J. Biol. Chem. 268, 2989–2996.
[2] Hall, D.E. et al. (1990) J. Cell Biol. 110, 2175–2184.
[3] Kramer, R.H. and Marks, N. (1989) J. Biol. Chem. 264, 4684–4688.
[4] Goldman, R. et al. (1992) Eur. J. Immunol. 22, 1109–1114.
[5] MacDonald, T.T. et al. (1990) J. Clin. Pathol. 43, 313–315.
[6] Ignatius, M. and Reichardt, L.F. (1988) Neuron 1, 713–725.
[7] Belkin, V.M. et al. (1990) J. Cell Biol. 111, 2159–2170.
[8] Hemler, M.E. (1990) Annu. Rev. Immunol. 114, 365–400.
[9] Hemler, M.E. et al. (1984) J. Immunol. 132, 3011–3018.
[10] Gardner, H. et al. (1996) Dev. Biol. 175, 301–313.
[11] Mak, T.W. (1998) The Gene Knockout FactsBook. Academic Press, London.

Integrin α₂

Family

Integrin

Structure

Molecular weights

Amino acids	1181
Polypeptide	126 378

SDS-PAGE reduced	165 kDa (160 kDa unreduced)

Carbohydrate

N-linked sites	10
O-linked sites	

Gene location

5q23–31

Gene structure

Alternative forms

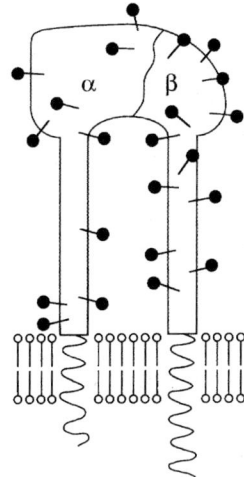

CD49b/CD29

Structure

Integrin α_2 is a type I membrane glyoprotein integrin α chain forming a heterodimer with β_1 to form $\alpha_2\beta_1$[1]. The α_2 chain falls into the subclass of integrin α chains with an N-terminal inserted or I domain (between the second and third of the seven α chain repeat domains), and is not post-translationally proteolytically cleaved. There are three cation-binding domains in the α2 chain. The I domain from α2 has been crystallized and its structure solved[2].

Ligands

Collagen types I–IV and laminin[3]. Echovirus-1 receptor[4].

Function

Magnesium-dependent platelet adhesion to collagen[5] is mediated by $\alpha_2\beta_1$; this has been suggested to occur via the amino acid sequence DGEA in collagen[6] and involve the integrin I domain[7]. Other cells also use $\alpha_2\beta_1$ to interact with collagen[8–10]. The interaction between $\alpha_2\beta_1$ and laminin is only found in certain tissues; thus, cell lines, but not platelets utilize α_2 to bind laminin. Fibroblasts may utilize $\alpha_2\beta_1$ in tissue collagen remodelling during wound repair[11,12]. The level of $\alpha_2\beta_1$ is modified in neoplastic tissues and may be involved in tumour progression[12–14].

Distribution

Originally identified as one of the 'very late antigens' (VLA) of stimulated T cells[3], α₂β₁ was subsequently identified in several other tissues–fibroblasts, endothelium, peripheral nerves–and several blood cell types including platelets[3,15,16].

Disease association

OMIM 19274

Inherited deficiency of platelet α₂β₁ has been described in Japanese populations and leads to a mild bleeding diathesis[17]. A point mutation at amino acid position 534 (Lys to Glu transformation) leads to the platelet antigen Zav/Hc/Br(a), and is involved in neonatal and transfusion alloimmune thrombocytopenia[18].

Knockout

MGI 96600

Amino acid sequence of human integrin α₂

```
    1 MGPERTGAAP LPLLLVLALS QGILNCCLAY NVGLPEAKIF SGPSSEQFGY AVQQFINPKG
   61 NWLLVGSPWS GFPENRMGDV YKCPVDLSTA TCEKLNLQTS TSIPNVTEMK TNMSLGLILT
  121 RNMGTGGFLT CGPLWAQQCG NQYYTTGVCS DISPDFQLSA SFSPATQPCP SLIDVVVVCD
  181 ESNSIYPWDA VKNFLEKFVQ GLDIGPTKTQ VGLIQYANNP RVVFNLNTYK TKEEMIVATS
  241 QTSQYGGDLT NTFGAIQYAR KYAYSAASGG RRSATKVMVV VTDGESHDGS MLKAVIDQCN
  301 HDNILRFGIA VLGYLNRNAL DTKNLIKEIK AIASIPTERY FFNVSDEAAL LEKAGTLGEQ
  361 IFSIEGTVQG GDNFQMEMSQ VGFSADYSSQ NDILMLGAVG AFGWSGTIVQ KTSHGHLIFP
  421 KQAFDQILQD RNHSSYLGYS VAAISTGEST HFVAGAPRAN YTGQIVLYSV NENGNITVIQ
  481 AHRGDQIGSY FGSVLCSVDV DKDTITDVLL VGAPMYMSDL KKEEGRVYLF TIKKGILGQH
  541 QFLEGPEGIE NTRFGSAIAA LSDINMDGFN DVIVGSPLEN QNSGAVYIYN GHQGTIRTKY
  601 SQKILGSDGA FRSHLQYFGR SLDGYGDLNG DSITDVSIGA FGQVVQLWSQ SIADVAIEAS
  661 FTPEKITLVN KNAQIILKLC FSAKFRPTKQ NNQVAIVYNI TLDADGFSSR VTSRGLFKEN
  721 NERCLQKNMV VNQAQSCPEH IIYIQEPSDV VNSLDLRVDI SLENPGTSPA LEAYSETAKV
  781 FSIPFHKDCG EDGLCISDLV LDVRQIPAAQ EQPFIVSNQN KRLTFSVTLK NKRESAYNTG
  841 IVVDFSENLF FASFSLPVDG TEVTCQVAAS QKSVACDVGY PALKREQQVT FTINFDFNLQ
  901 NLQNQASLSF QALSESQEEN KADNLVNLKI PLLYDAEIHL TRSTNINFYE ISSDGNVPSI
  961 VHSFEDVGPK FIFSLKVTTG SVPVSMATVI IHIPQYTKEK NPLMYLTGVQ TDKAGDISCN
 1021 ADINPLKIGQ TSSSVSFKSE NFRHTKELNC RTASCSNVTC WLKDVHMKGE YFVNVTTRIW
 1081 NGTFASSTFQ TVQLTAAAEI NTYNPEIYVI EDNTVTIPLM IMKPDEKAEV PTGVIIGSII
 1141 AGILLLLALV AILWKLGFFK RKYEKMTKNP DEIDETTELS S
```

I domain position 188–378.

Database accession

EMBL/GenBank   X17033
SwissProt   P17301

References

[1] Takada, Y. and Hemler, M.E. (1989) J. Cell Biol. 109, 397–407.

[2] Emsley, J. et al. (1997) J. Biol. Chem. 272, 28512–29517.

[3] Hemler, M.E. (1990) Annu. Rev. Immunol. 114, 365–400.

[4] Bergelson, J.M. et al. (1992) Science 255, 1718–1720.

[5] Staatz, W.D. et al. (1989) J. Cell Biol. 108, 1917–1924.

⁶ Staatz, W.D. et al. (1991) J. Biol. Chem. 266, 7363–7367.

⁷ Kamata, T. et al. (1994) J. Biol. Chem. 269, 9659–9663.

⁸ Goldman, R. et al. (1992) Eur. J. Immunol. 22, 1109–1114.

⁹ Elices, M.J. and Hemler, M.E. (1989) Proc. Natl Acad. Sci. USA 86, 9906–9910.

¹⁰ Wayner, E.A. and Carter, W.G. (1987) J. Cell Biol. 105, 1873–1884.

¹¹ Schiro, J.A. et al. (1991) Cell 67, 403–410.

¹² Klein, C.E. et al. (1991) J. Cell Biol. 115 1427–1436.

¹³ Chen, F.A. et al. (1991) J. Exp. Med. 173, 1111–1119.

¹⁴ Chan, B.M.C. et al. (1991) Science 251, 1600–1602.

¹⁵ Hemler, M.E. et al. (1984) J. Immunol. 132, 3011–3018.

¹⁶ Perez-Villar, J.J. et al. (1996) Eur. J. Immunol. 26, 2023–2029.

¹⁷ Nieuwenhuis, H.K. et al. (1985) Nature 318, 470–472.

¹⁸ Santoso, S. et al. (1993) J. Clin. Invest. 92, 2427–2432.

Integrin α_3

Family

Integrin

Structure

Molecular weights

Amino acids	1051
Polypeptide	116 612

SDS-PAGE reduced	130, 25 kDa (150 kDa unreduced)

Carbohydrate

N-linked sites	14
O-linked sites	

Gene location	17q

Gene structure	26 exons

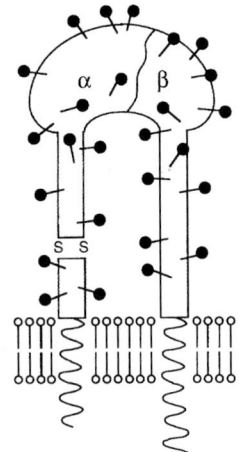

CD49c/CD29

Alternative forms

An α_3 form with a variant cytoplasmic tail has been detected in brain and heart, and arises by alternate splicing.

Structure

The type I membrane glycoprotein integrin α_3 chain associates with β_1 to form the non-covalently linked $\alpha_3\beta_1$ heterodimer[1-3]. The α_3 chain falls into the subclass of integrin α chains without an inserted I domain and is post-translationally proteolytically cleaved into a heavy and a light chain, which are disulphide bonded. The α_3 chain contains seven homologous repeat domains, the three most membrane proximal having divalent cation binding sequences.

Ligands

$\alpha_3\beta_1$ has complex ligand binding specificity[4,5]. Thus, $\alpha_3\beta_1$ will bind fibronectin in an RGD-dependent manner but only if the $\alpha_5\beta_1$ fibronectin receptor is not expressed[4]. A second site is involved in interaction with collagen and laminin 5 (E3 fragment)[4,5] and binding is not sensitive to inhibition by RGD peptides. $\alpha_3\beta_1$ in epithelium also binds the epithelial basement membrane protein, epiligrin[6,7]. The bacterial coat protein, invasin, is an additional ligand[8]. Fibronectin binding is inhibited by calcium, but not collagen and laminin adhesion.

Function

Epithelial–basement membrane interaction and structural integrity in, for example, the epidermal–dermal junction. Expression in transformed cells (as 'galactoprotein β3') suggests a possible role in oncogenesis. It is one of the integrin dimers expressed on T lymphocytes after prolonged stimulation ('very late antigen', VLA). $\alpha_3\beta_1$ is also found at sites of cell–cell contact and thus may also be involved in intercellular as well as cell–matrix adhesion[9].

Distribution

Low levels on monocytes, B and T cells, kidney, thyroid[10] and most non-lymphoid adherent cell lines[11]. Expression levels in lymphocytes are increased upon culture *in vitro*.

Disease association

Knockout

MGI:96602
The Gene Knockout FactsBook[14], p. 620.
Mutant mice die at birth with abnormalities in kidney and lung, and skin blistering due to defective cell–basement interaction[12,13].

Amino acid sequence of human integrin α₃

```
   1 MGPGPSRAPR APRLMLCALA LMVAAGGCVV SAFNLDTRFL VVKEAGNPGS LFGYSVALHR
  61 QTERQQRYLL LAGAPRELAV PDGYTNRTGA VYLCPLTAHK DDCERMNITV KNDPGHHIIE
 121 DMWLGVTVAS QGPAGRVLVC AHRYTQVLWS GSEDQRRMVG KCYVRGNDLE LDSSDDWQTY
 181 HNEMCNSNTD YLETGMCQLG TSGGFTQNTV YFGAPGAYNW KGNSYMIQRK EWDLSEYSYK
 241 DPEDQGNLYI GYTMQVGSFI LHPKNITIVT GAPRHRHMGA VFLLSQEAGG DLRRRQVLEG
 301 SQVGAYFGSA IALADLNNDG WQDLLVGAPY YFERKEEVGG AIYVFMNQAG TSFPAHPSLL
 361 LHGPSGSAFG LSVASIGDIN QDGFQDIAVG APFEGLGKVY IYHSSSKGLL RQPQQVIHGE
 421 KLGLPGLATF GYSLSGQMDV DENFYPDLLV GSLSDHIVLL RARPVINIVH KTLVPRPAVL
 481 DPALCTATSC VQVELCFAYN QSAGNPNYRR NITLAYTLEA DRDRRPPRLR FAGSESAVFH
 541 GFFSMPEMRC QKLELLLMDN LRDKLRPIII SMNYSLPLRM PDRPRLGLRS LDAYPILNQA
 601 QALENHTEVQ FQKECGPDNK CESNLQMRAA FVSEQQQKLS RLQYSRDVRK LLLSINVTNT
 661 RTSERSGEDA HEALLTLVVP PALLLSSVRP PGACQANETI FCELGNPFKR NQRMELLIAF
 721 EVIGVTLHTR DLQVQLQLST SSHQDNLWPM ILTLLVDYTL QTSLSMVNHR LQSFFGGTVM
 781 GESGMKTVED VGSPLKYEFQ VGPMGEGLVG LGTLVLGLEW PYEVSNGKWL LYPTEITVHG
 841 NGSWPCRPPG DLINPLNLTL SDPGDRPSSP QRRRRQLDPG GGQGPPPVTL AAAKKAKSET
 901 VLTCATGRAH CVWLECPIPD APVVTNVTVK ARVWNSTFIE DYRDFDRVRV NGWATLFLRT
 961 SIPTINMENK TTWFSVDIDS ELVEELPAEI ELWLVLVAVG AGLLLLGLII LLLWKCGFFK
1021 RARTRALYEA KRQKAEMKSQ PSETERLTDD Y
```

Proteolytic cleavage site at position 872.

Database accession

EMBL/GenBank   M59911
SwissProt   P26006

References

[1] Takada, Y. et al. (1991) J. Cell Biol. 115, 257–266.

[2] Tsuji, T. et al. (1991) J. Biochem. 109, 659–665.

[3] Takada, Y. et al. (1987) Proc. Natl Acad. Sci. USA 84, 3239–3243.

[4] Elices, M.J. et al. (1991) J. Cell Biol. 112, 169–181.

[5] Gehlsen, K.R. et al. (1989) J. Biol. Chem. 264, 19034–19038.

[6] Carter, W.G. et al. (1991) Cell 65, 599–610.

[7] Weitzman, J.B. et al. (1993) J. Biol. Chem. 268, 8651–8657.

[8] Isberg, R.R. and Leong, J.M. (1990) Cell 60, 861–871.

[9] Kaufmann, R. et al. (1989) J. Cell Biol. 109, 1807–1815.

[10] Hemler, M.E. (1990) Annu. Rev. Immunol. 114, 365–400.

[11] Rettig, W.J. and Old, L.J. (1992) Annu. Rev. Immunol. 7, 481–511.

[12] Kreidberg, J.A. et al. (1996) Development 122, 3537–3547.

[13] DiPersio, C.M. et al. (1997) J. Cell Biol. 137, 729–742.

[14] Mak, T.W. (1998) The Gene Knockout FactsBook. Academic Press, London.

Integrin α₄   CD49d, VLA4

Family

Integrin

Structure

Molecular weights
Amino acids	1038
Polypeptide	111 228

SDS-PAGE reduced: 150 kDa (180, 150 kDa unreduced)

Carbohydrate
N-linked sites	11
O-linked sites	

Gene location   2q31–q32

Gene structure

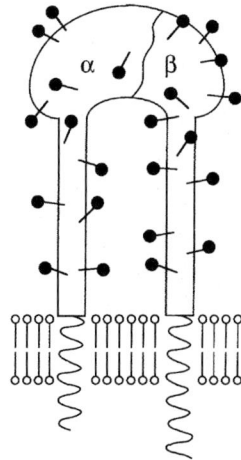

CD49d/CD29

Alternative forms
An additional 180 kDa form is also found (α_4 /180) in non-reducing SDS-PAGE and may be the main functional form of α_4[1,2].

Structure
Integrin α_4 is a type I membrane glycoprotein[3,4] that forms non-covalently linked dimers with two integrin β chains, β_1 and β_7, to form $\alpha_4\beta_1$ and $\alpha_4\beta_7$. The α_4 chain does not belong to the two main groups of integrin α chains[3]. It neither contains an I domain nor is it proteolytically cleaved in a standard α chain manner; instead there is variable cleavage in some cell types of the 145 kDa protein into two, similarly sized, non-disulphide linked fragments of 80 and 70 kDa (at Arg597)[1,5,6]. The larger 180 kDa form is functionally active[7,8]. The extracellular portion of α_4 contains three EF-hand loop-like domains that bind divalent cations and are involved in modifying adhesion.

Ligands

VCAM-1 (CD106) in activated endothelium[9] and alternately spliced forms of fibronectin in matrix (via EILDV sequence in the CS-1 region of fibronectin)[10-13] bind to both integrin α_4 dimers, $\alpha_4\beta_1$ and $\alpha_4\beta_7$. MAdCAM-1 is an additional counter-receptor for $\alpha_4\beta_7$[12,14,15] and thrombospondin for $\alpha_4\beta_1$. α_4 is a further integrin cellular receptor for the bacterial coat protein, invasin.

Function

α_4 integrins are key molecules maintaining the structural integrity of the placenta and heart during embryogenesis (see below). In the adult α_4 integrins are involved in lymphocyte homing to Peyer's patch high venular

167

endothelium via interaction between $\alpha_4\beta_7$ and MAdCAM-1 (see entry for β_7[14,16]). Initial adhesion and rolling, and subsequent tight adhesion of T cells on vascular endothelium after endothelial activation is mediated by α_4 integrins[17], and is independent of selectins (CD62) (see entry). Mediates homotypic cell–cell aggregation of α_4-positive lymphocytes. Fibronectin binding by $\alpha_4\beta_1$ induces T cell activation and proliferation through the T cell receptor-CD3 complex via focal adhesion kinase and the MAP kinase signalling cascade. Mediates adhesion of haemopoeitic stem cells to marrow stroma and hence blood cell differentiation.

Distribution

Most leucoytes including T lymphocytes (but not platelets and neutrophils), haemopoeitic precursors[12,14,15], muscle and some other non-haemopoeitic tissues[18].

Disease association

OMIM 192975

Knockout

MGI:96603
The Gene Knockout FactsBook[19], p. 621.
Embryonic lethal mutation with failure of placentation in half of the embryos due to failure of fusion of the VCAM-1+ve allantois with the integrin α_4+ve chorion. Surviving embryos die due to disruption of the integrin α_4+ve epicardium and underlying coronary vessels[20,22]. Experiments with α_4 $-/-$ cells have confirmed the requirement for this integrin in lymphoid homing to Peyer's patches[21].

Amino acid sequence of human integrin α₄

```
   1 MFPTESAWLG KRGANPGPEA AVRETVMLLL CLGVPTGRPY NVDTESALLY QGPHNTLFGY
  61 SVVLHSHGAN RWLLVGAPTA NWLANASVIN PGAIYRCRIG KNPGQTCEQL QLGSPNGEPC
 121 GKTCLEERDN QWLGVTLSRQ PGENGSIVTC GHRWKNIFYI KNENKLPTGG CYGVPPDLRT
 181 ELSKRIAPCY QDYVKKFGEN FASCQAGISS FYTKDLIVMG APGSSYWTGS LFVYNITTNK
 241 YKAFLDKQNQ VKFGSYLGYS VGAGHFRSQH TTEVVGGAPQ HEQIGKAYIF SIDEKELNIL
 301 HEMKGKKLGS YFGASVCAVD LNADGFSDLL VGAPMQSTIR EEGRVFVYIN SGSGAVMNAM
 361 ETNLVGSDKY AARFGESIVN LGDIDNDGFE DVAIGAPQED DLQGAIYIYN GRADGISSTF
 421 SQRIEGLQIS KSLSMFGQSI SGQIDADNNG YVDVAVGAFR SDSAVLLRTR PVVIVDASLS
 481 HPESVNRTKF DCVENGWPSV CIDLTLCFSY KGKEVPGYIV LFYNMSLDVN RKAESPPRFY
 541 FSSNGTSDVI TGSIQVSSRE ANCRTHQAFM RKDVRDILTP IQIEAAYHLG PHVISKRSTE
 601 EFPPLQPILQ QKKEKDIMKK TINFARFCAH ENCSADLQVS AKIGFLKPHE NKTYLAVGSM
 661 KTLMLNVSLF NAGDDAYETT LHVKLPVGLY FIKILELEEK QINCEVTDNS GVVQLDCSIG
 721 YIYVDHLSRI DISFLLDVSS LSRAEEDLSI TVHATCENEE EMDNLKHSRV TVAIPLKYEV
 781 KLTVHGFVNP TSFVYGSNDE NEPETCMVEK MNLTFHVINT GNSMAPNVSV EIMVPNSFSP
 841 QTDKLFNILD VQTTTGECHF ENYQRVCALE QQKSAMQTLK GIVRFLSKTD KRLLYCIKAD
 901 PHCLNFLCNF GKMESGKEAS VHIQLEGRPS ILEMDETSAL KFEIRATGFP EPNPRVIELN
 961 KDENVAHVLL EGLHHQRPKR YFTIVIISSS LLLGLIVLLL ISYVMWKAGF FKRQYKSILQ
1021 EENRRDSWSY INSKSNDD
```

Proteolytic cleavage site at position 597.

Database accession

EMBL/GenBank   X16983
SwissProt   P13612

References

[1] Szabo, M. and McIntyre, B.W. (1995) Mol. Immunol. 32, 1453–1454.

[2] Rubio, M. et al. (1992) Eur. J. Immunol. 22, 1099–1102.

[3] Takada, Y. et al. (1989) EMBO J. 8, 1361–1368.

[4] Takada, Y. et al. (1987) Proc. Natl Acad. Sci. USA 84, 3239–3243.

[5] Hemler, M.E. et al. (1987) J. Biol. Chem. 262, 11478–11485.

[6] Teixido, J. et al. (1992) J. Biol. Chem. 267, 1786–1791.

[7] Parker, C.M. et al. (1990) J. Biol. Chem. 268, 7028–7035.

[8] Pujades, C. et al. (1996) Biochem. J. 313, 899–908.

[9] Elices, M.J. et al. (1990) Cell 60, 577–584.

[10] Wayner, E.A. et al. (1989) J. Cell Biol. 109, 1321–1326.

[11] Mould, A.P. and Humphries, M.J. (1991) EMBO J. 10, 4089–4095.

[12] Lobb, R.R. and Hemler, M.E. (1994) J. Clin. Invest. 94, 1722–1728.

[13] Kilger, G. et al. (1995) J. Biol. Chem. 270, 5979–5984.

[14] Rott, L.S. et al. (1996) J. Immunol. 156, 3727–3736.

[15] Erle, D.J. et al. (1994) J. Immunol. 153, 517–528.

[16] Berlin, C. et al. (1993) Cell 74, 185–195.

[17] Springer, T.A. (1994) Cell 76, 301–314.

[18] Rosen, G.D. et al. (1992) Cell 69, 1107–1119.

[19] Mak, T.W. (1998) The Gene Knockout FactsBook. Academic Press, London.

[20] Yang, J.T. et al. (1995) Development 121, 549–560.

[21] Arroyo, A.G. et al. (1996) Cell 85, 997–835.

[22] Yang, J.T. et al. (1996) J. Cell Biol. 135, 829–835.

Integrin α_5

Family

Integrin

Structure

Molecular weights

Amino acids	1049
Polypeptide	114 508

SDS-PAGE reduced	135, 25 kDa (155 kDa unreduced)

Carbohydrate

N-linked sites	14
O-linked sites	

Gene location   12q11–q13

Gene structure

Alternative forms

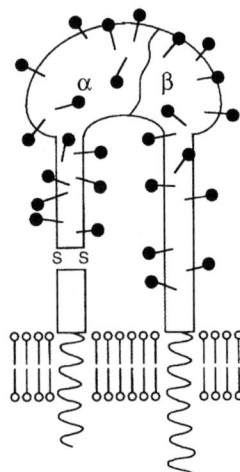

CD49e/CD29

Structure

The type I membrane glycoprotein integrin α_5 chain[1-4] associates with β_1 to form the non-covalently linked $\alpha_5\beta_1$ heterodimer. The α_5 chain falls into the subclass of integrin α chains without an inserted I domain, and is post-translationally proteolytically cleaved into heavy and light chains that are disulphide bonded. Rotary shadowing electron microscopy of purified $\alpha_5\beta_1$ has provided a structural model for all integrin dimers in the absence of crystallography or NMR data[5]. The α_5 chain contains five potential divalent cation binding sites.

Ligands

Major receptor for fibronectin, via RGD sequence and alternative ancillary domains[6,7]. $\alpha_5\beta_1$ also interacts with the neural adhesion molecule, L1[8] and the bacterial coat protein, invasin[9].

Function

Mediates fibronectin assembly into the extracellular matrix, and fibronectin-dependent cell adhesion, spreading, migration and signal transduction[7]. It acts as a co-stimulatory molecule to the T cell receptor–CD3 complex in T cell activation and enhances phagocytosis of opsonized particles by monocytes[10]. $\alpha_5\beta_1$ is also involved in monocyte migration[11] and $\alpha_2\beta_1$-mediated adhesion to collagen is enhanced by $\alpha_5\beta_1$[12]. There is extensive evidence for $\beta1$ integrin dimers being involved in signal transduction and several general reviews are provided in the 'Integrins' section in Chapter 2 (p. 16).

Distribution

Widespread in embryonic and adult tissues[13]. Expressed by blood cells, including monocytes, platelets and lymphocytes, especially in B and T cells after activation[14-16].

Disease association

OMIM 135620

Knockout

MGI:96604
The Gene Knockout FactsBook[17], p. 622.
Embryos die at day 10–11 of gestation with vascular defects in the embryonic and extra-embryonic circulation; the posterior end of the embryo fails to develop properly. Integrin α₅ null fibroblasts function remarkably normally *in vitro*, adhering and spreading on fibronectin, and assembling fibronectin-rich extracellular matrix, etc. Thus, the phenotype is milder than seen with fibronectin null mice, implying integrin receptor redundancy for interaction with fibronectin[18,19].

Amino acid sequence of human integrin α₅

```
   1 MGSRTPESPL HAVQLRWGPR RRPPLVPLLL LLVPPPPRVG GFNLDAEAPA VLSGPPGSFF
  61 GFSVEFYRPG TDGVSVLVGA PKANTSQPGV LQGGAVYLCP WGASPTQCTP IEFDSKGSRL
 121 LESSLSSSEG EEPVEYKSLQ WFGATVRAHG SSILACAPLY SWRTEKEPLS DPVGTCYLST
 181 DNFTRILEYA PCRSDFSWAA GQGYCQGGFS AEFTKTGRVV LGGPGSYFWQ GQILSATQEQ
 241 IAESYYPEYL INLVQGQLQT RQASSIYDDS YLGYSVAVGE FSGDDTEDFV AGVPKGNLTY
 301 GYVTILNGSD IRSLYNFSGE QMASYFGYAV AATDVNGDGL DDLLVGAPLL MDRTPDGRPQ
 361 EVGRVYVYLQ HPAGIEPTPT LTLTGHDEFG RFGSSLTPLG DLDQDGYNDV AIGAPFGGET
 421 QQGVVFVFPG GPGGLGSKPS QVLQPLWAAS HTPDFFGSAL RGGRDLDGNG YPDLIVGSFG
 481 VDKAVVYRGR PIVSASASLT IFPPAMFNPEE RSCSLEGNPV ACINLSFCLN ASGKHVADSI
 541 GFTVELQLDW QKQKGGVRRA LFLASRQATL TQTLLIQNGA REDCREMKIY LRNESEFRDK
 601 LSPIHIALNF SLDPQAPVDS HGLRPALHYQ SKSRIEDKAQ ILLDCGEDNI CVPDLQLEVF
 661 GEQNHVYLGD KNALNLTFHA QNVGEGGAYE AELRVTAPPE AEYSGLVRHP GNFSSLSCDY
 721 FAVNQSRLLV CDLGNPMKAG ASLWGGLRFT VPHLRDTKKT IQFDFQILSK NLNNSQSDVV
 781 SFRLSVEAQA QVTLNGVSKP EAVLFPVSDW HPRDQPQKEE DLGPAVHHVY ELINQGPSSI
 841 SQGVLELSCP QALEGQQLLY VTRVTGLNCT TNHPINPKGL ELDPEGSLHH QQKREAPSRS
 901 SASSGPQILK CPEAECFRLR CELGPLHQQE SQSLQLHFRV WAKTFLQREH QPFSLQCEAV
 961 YKALKMPYRI LPRQLPQKER QVATAVQWTK AEGSYG͟V͟P͟L͟W͟ ͟I͟I͟I͟L͟A͟I͟L͟F͟G͟L͟ ͟L͟L͟L͟G͟L͟L͟I͟Y͟I͟L͟
1021 ͟Y͟K͟L͟G͟F͟F͟K͟R͟S͟L͟ PYGTAMEKAQ LKPPATSDA
```

Proteolytic cleavage site at position 894.

Database accession

EMBL/GenBank   X06256
SwissProt   P08648

References

[1] Argraves, W.S. et al. (1987) J. Cell Biol. 105, 1183–1190.

[2] Fitzgerald, L.A. et al. (1987) Biochem. 26, 8158–8165.

[3] Argarves, W.S. et al. (1986) J. Biol. Chem. 261, 12922–12924.

4 Takada, Y. et al. (1987) Proc. Natl Acad. Sci. USA 84, 3239–3243.

5 Nermut, M.N. et al. (1988) EMBO J. 7, 4093–4099.

6 Hemler, M.E. (1990) Annu. Rev. Immunol. 114, 365–400.

7 Hynes, R.O. (1992) Cell 69, 11–25.

8 Ruppert, M. et al. (1995) J. Cell Biol. 131, 1881–1891.

9 Isberg, R.R. and Leong, J.M. (1990) Cell 60, 861–871.

10 Wright, S.D. et al. (1984) J. Cell Biol. 99, 336–339.

11 Weber, C. et al. (1996) J. Cell Biol. 134, 1063–1073.

12 Pacifici, R. et al. (1994) J. Immunol. 153, 2222–2233.

13 Sanchez-Madrid, F. and Corbi, A.L. (1993) Semin. Cell Biol. 3, 199–210.

14 Hemler, M.E. (1990) Annu. Rev. Immunol. 8, 365–400.

15 Shimizu, Y. et al. (1990) Nature 345, 250–253.

16 Ballard, L.L. et al. (1991) Clin. Exp. Immunol. 84, 336–346.

17 Mak, T.W. (1998) The Gene Knockout FactsBook. Academic Press, London.

18 Yang, J.T. et al. (1993) Development 119, 1093–1105.

19 Yang, J.T. et al. (1996) Mol. Biol. Cell 7, 1737–1748.

Integrin α_6 CD49f, VLA6, platelet gpIc

Family

Integrin

Structure

Molecular weights
Amino acids	1073
Polypeptide	117 263 (A form), 119 869 (B form)
SDS-PAGE reduced	120, 30 kDa (140 kDa unreduced)

Carbohydrate
N-linked sites	8
O-linked sites	

Gene location
Chromosome 2

Gene structure

CD49f/CD29

Alternative forms
Two alternatively spliced RNA forms with differing cytploplasmic tails $(\alpha_6 A$ and B)[1,2].

Structure
Integrin α_6[2–4] exists as alternate dimers, $\alpha_6\beta_1$ (VLA6) and $\alpha_6\beta_4$ (see entry for β_4/CD104). The C-terminus is phosphorylated. α_6 is most homologogous to the integrin α_3 chain. It is cleaved post-translationally to produce a disulphide-bonded α chain, and does not contain an inserted I domain. The α_6 chain contains seven homologous repeat domains, the three most membrane proximal having divalent cation binding sequences; a fourth site is located between repeat domains III and IV.

Ligands

α_6 integrin dimers, $\alpha_6\beta_1$ and $\alpha_6\beta_4$, are receptors for laminin[5–9].

Function

$\alpha_6\beta_1$ mediates platelet interaction with subendothelial basement membrane laminin exposed on blood vessel damage. Platelet binding of laminin via $\alpha_6\beta_1$ requires magnesium but not calcium cations. On T lymphocytes $\alpha_6\beta_1$ is a co-stimulatory molecule for T cell activation. $\alpha_6\beta_4$ forms specialized adhesive interactions in the hemidesmosomes of stratified epithelia, Schwann cells, etc., where it is essential for the structural integrity of cell–matrix interactions (see below).

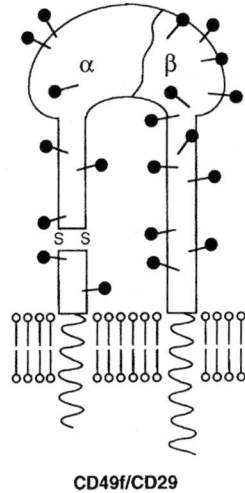

Distribution

$α_6β_1$ is expressed widely in epithelia[10], and in haemopoeitic tissues is found on monocytes, platelets and lymphocytes[5,6,11]. Activation increases expression levels in T cells[2,5,6]. The A form is found in lung, liver, spleen and cervix, whereas the B form is expressed in brain, kidney and ovary. In other tissues, both forms are co-expressed. $α_6β_4$ is found in stratified epithelia, Schwann cells and some endothelia.

Disease association

OMIM 147556

Mutations in integrin $α_6$ give rise to junctional epidermolysis bullosa with pyloric atresia[12].

Knockout

MGI:96605

The Gene Knockout FactsBook[13], p. 623.

Embryos die neonatally due to detachment of epithelia (epidermis and simple epithelium of mouth and oesophagus) from the basement membrane. Exhibits a similar phenotype to integrin $β_4$ knockout mice and to the human blistering skin disease epidermiolysis bullosa with pyloric atresia (as with mutations in integrin $β_4$)[14-16].

Amino acid sequence of human integrin α₆

```
   1 MAAAGQLCLL YLSAGLLSRL GAAFNLDTRE DNVIRKYGDP GSLFGFSLAM HWQLQPEDKR
  61 LLLVGAPRGE ALPLQRANRT GGLYSCDITA RGPCTRIEFD NDADPTSESK EDQWMGVTVQ
 121 SQGPGGKVVT CAHRYEKRQH VNTKQESRDI FGRCYVLSQN LRIEDDMDGG DWSFCDGRLR
 181 GHEKFGSCQQ GVAATFTKDF HYIVFGAPGT YNWKGIVRVE QKNNTFFDMN IFEDGPYEVG
 241 GETEHDESLV PVPANSYLGF SLDSGKGIVS KDEITFVSGA PRANHSGAVV LLKRDMKSAH
 301 LLPEHIFDGE GLASSFGYDV AVVDLNKDGW QDIVIGAPQY FDRDGEVGGA VYVYMNQQGR
 361 WNNVKPIRLN GTKDSMFGIA VKNIGIDINQD GYPDIAVGAP YDDLGKVFIY HGSANGINTK
 421 PTQVLKGISP YFGYSIAGNM DLDRNSYPDV AVGSLSDSVT IFRSRPVINI QKTITVTPNR
 481 IDLRQKTACG APSGICLQVK SCFEYTANPA GYNPSISIVG TLEAEKERRK SGLSSRVQFR
 541 NQGSEPKYTQ ELTLKRQKQK VCMEETLWLQ DNIRDKLRPI PITASVEIQE PSSRRRVNSL
 601 PEVLPILNSD EPKTAHIDVH FLKEGCGDDN VCNSNLKLEY KFCTREGNQD KFSYLPIQKG
 661 VPELVLKDQK DIALEITVTN SPSNPRNPTK DGDDAHEAKL IATFPDTLTY SAYRELRAFP
 721 EKQLSCVANQ NGSQADCELG NPFKRNSNVT FYLVLSTTEV TFDTPYLDIN LKLETTSNQD
 781 NLAPITAKAK VVIELLLSVS GVAKPSQVYF GGTVVGEQAM KSEDEVGSLI EYEFRVINLG
 841 KPLTNLGTAT LNIQWPKEIS NGKWLLYLVK VESKGLEKVT CEPQKEINSL NLTESHNSRK
 901 KREITEKQID DNRKFSLFAE RKYQTLNCSV NVNCVNIRCP LRGLDSKASL ILRSRLWNST
 961 FLEEYSKLNY LDILMRAFID VTAAAENIRL PNAGTQVRVT VFPSKTVAQY SGVPWWIILV
1021 AILAGILMLA LLVFILWKCG FFKRNKKDHY DATYHKAEIH AQPSDKERLT SDA
```

Proteolytic cleavage site at position 899.

The above sequence is for integrin $α_6$A, and the sequence in italics and underlined is substituted as below to make the B form:

```
B; CGFFKRSRYD DSVPRYHAVR IRKEEREIKD EKYIDNLEKK QWITKWNRNE SYS
```

Database accession

EMBL/GenBank   X53586
SwissProt   P23229

References

[1] Tamura, R.N. et al. (1991) Proc. Natl Acad. Sci. USA 88, 10183–10187.

[2] Hogervorst, F. et al. (1991) Eur. J. Biochem. 199, 425–433.

[3] Hemler, M.E. et al. (1989) J. Biol. Chem. 264, 6529–6535.

[4] Tamura, R.N. et al. (1990) J. Cell Biol. 111, 1593–1604.

[5] Shimizu, Y. et al. (1990) J. Immunol. 145, 59–67.

[6] Shimizu, Y. et al. (1990) Nature 345, 250–253.

[7] Tobias, J.W. et al. (1987) Blood, 69, 1265–1268.

[8] Sonnenberg, A. et al. (1988) Nature 336, 487–489.

[9] Sonnenberg, A. et al. (1990) J. Cell Biol. 110, 2145–2155.

[10] Natali, P.G. et al. (1992) J. Cell Sci. 103, 1243–1247.

[11] Hemler, M.E. (1990) Annu. Rev. Immunol. 114, 365–400.

[12] Ruzzi, L. et al. (1997) J. Clin. Invest. 99, 2826–2831.

[13] Mak, T.W. (1998) The Gene Knockout FactsBook. Academic Press, London.

[14] Georges-Labouesse, E.N. et al. (1996) Nature Genet. 13, 370–373.

[15] Van der Neut, R. et al. (1996) Nature Genet. 13, 367–369.

[16] Dowling, J. et al. (1996) J. Cell Biol. 134, 559–572.

Integrin α_7

Family

Integrin

Structure

Molecular weights
Amino acids	1137
Polypeptide	120957

SDS-PAGE reduced   100, 30 kDa (120 kDa)

Carbohydrate
N-linked sites	5
O-linked sites	

Gene location   12q13

Gene structure

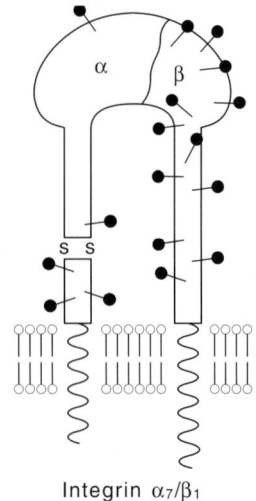

Integrin α_7/β_1

Alternative forms
Two extracellular and three cytoplasmic variants have been described[1].

Structure
Typical proteolytically cleaved, non-I domain containing integrin α chain that forms dimer with β_1 to form $\alpha_7\beta_1$.

Ligands

Receptor for basement membrane laminin types 1, 2 and 4 (E8 region)[2-4].

Function

Integrin α_7 is involved in muscle development[5]. α_7 is also involved in the formation and function of muscle fibre to extracellular matrix linkage, distinct from that mediated by the dystrophin–dystroglycan complex (see entry on dystroglycan), at the muscle basement membrane, as exemplified by the phenotype seen in human and mouse mutations (see below).

Distribution

Developmentally regulated expression and splice variant usage during myogenesis[5-8], with integrin α_7 expressed in skeletal, smooth and cardiac muscle[2,4,9,10]. Additionally, α_7 is expressed in developing nervous system[1]. The α_7A and B spliced forms of α_7 are concentrated in myotendinous and neuromuscular junctions[1,11].

Disease association

OMIM 600536
Human mutations lead to congenital myopathy and variable mental retardation[12].

Knockout

MGI:102700
Viable mice at birth despite its apparent role in myogenesis, with later development of pathological and clinical features of muscular dystrophy[13].

Amino acid sequence of human integrin α₇

```
   1 MAGARSRDPW GASGICYLFG SLLVELLFSR AVAFNLDVMG ALRKEGEPGS LFGFSVALHR
  61 QLQPRPQSWL LVGAPQALAL PGQQANRTGG LFACPLSLEE TDCYRVDIDQ GADMQKESKE
 121 NQWLGVSVRS QGPGGKIVTC AHRYEARQRV DQILETRDMI GRCFVLSQDL AIRDELDGGE
 181 WKFCEGRPQG HEQFGFCQQG TAAAFSPDSH YLLFGAPGTY NWKGLLFVTN IDSSDPDQLV
 241 YKTLDPADRL PGPAGDLALN SYLGFSIDSG KGLVRAEELS FVAGAPRANH KGAVVILRKD
 301 SASRLVPEVM LSGERLTSGF GYSLAVADLN SDGWPDLIVG APYFFERQEE LGGAVYVYLN
 361 QGGHWAGISP LRLCGSPDSM FGISLAVLGD LNQDGFPDIA VGAPFDGDGK VFIYHGSSLG
 421 VVAKPSQVLE GEAVGIKSFG YSLSGSLDMD GNQYPDLLVG SLADTAVLFR ARPILHVSHE
 481 VSIAPRSIDL EQPNCAGGHS VCVDLRVCFS YIAVPSSYSP TVALDYVLDA DTDRRLRGQV
 541 PRVTFLSRNL EEPKHQASGT VWLKHQHDRV CGDAMFQLQE NVKDKLRAIV VTLSYSLQTP
 601 RLRRQAPGQG LPPVAPILNA HQPSTQRAEI HFLKQGCGED KICQSNLQLV HARFCTRVSD
 661 TEFQPLPMDV DGTTALFALS GQPVIGLELM VTNLPSDPAQ PQADGDDAHE AQLLVMLPDS
 721 LHYSGVRALD PAEKPLCLSN ENASHVECEL GNPMKRGAQV TFYLILSTSG ISIETTELEV
 781 ELLLATISEQ ELHPVSARAR VFIELPLSIA GMAIPQQLFF SGVVRGERTM QSERDVGSKV
 841 KYEVTVSNQG QSLRTLGSAF LNIMWPHEIA NGKWLLYPMQ VELEGGQGPG QKGLCSPRPN
 901 ILHLDVDSRD RRRRELEPPE QQEPGERQEP SMSWWPVSSA EKKKNITLDC ARGTANCVVF
 961 ICPLYSFDRA AVLHVWGRLW NSTFLEEYSA VKSLEVIVRA NITVKSSIKN LMLRDASTVI
1021 PVMVYLDPMA VVAEGVPWWV ILLAVLAGLL VLALLVLLLW KMGFFKRAKH PEATVPQYHA
1081 VKIPREDRQQ FKEEKTGTIL RNNWGSPRRE GPDAHPILAA DCHPELGPDG HPGPGTA
```

Proteolytic cleavage site at position 882.

Database accession

EMBL/GenBank	L23423 (mouse)
TrEMBL	Q61738

References

1 Velling, T. et al (1996) Dev. Dyn. 207, 355–371.
2 Kramer, R.H. et al. (1991) Cell Regul. 2, 805–817.
3 Yao, C.C. et al. (1996) J. Biol. Chem. 271, 25598–25603.
4 Yao, C.C. et al. (1997) J. Cell Sci. 110, 1477–1487.
5 Song, W.K. et al. (1992) J. Cell Biol. 117, 643–657.
6 Song, W.K. et al. (1993) J. Cell Sci. 106, 1139–1152.
7 Zieber, B.L. et al. (1993) J. Biol. Chem. 268, 26773–26783.
8 Leung, E. et al. (1998) Biochem. Biophys. Res. Commun. 243, 317–325.
9 Kramer, R.H. et al. (1989) J. Biol. Chem. 264, 15642–15649.
10 von der Mark, H. et al. (1991) J. Biol. Chem. 266, 23593–23601.
11 Martin, P.T. et al. Dev. Biol. (1996) 171, 125–139.
12 Hayashi, Y.K. et al. (1998) Nature Genet. 19, 94–97.
13 Mayer, U. et al. (1997) Nature Genet. 17, 318–323.

Integrin α_8

Family
Integrin

Structure

Molecular weights
Amino acids	1025
Polypeptide	113612

SDS-PAGE reduced

Carbohydrate
N-linked sites	15
O-linked sites	

Gene location

Gene structure

Alternative forms

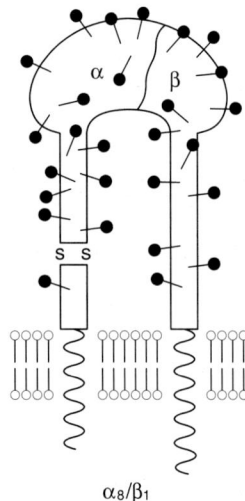

α_8/β_1

Structure
Integrin α_8 is expressed as a heterodimer with integrin β_1 chain to form $\alpha_8\beta_1$[1]. The α_8 chain is disulphide bonded at post-translational proteolytic cleavage site and has no I domain.

Ligands
Osteopontin[2], and fibronectin, vetronectin and tenascin[3].

Function
Location of integin α_8 in smooth muscle and kidney is suggestive of a functional role. Osteopontin interacts with $\alpha_8\beta_1$ during kidney morphogenesis[2], where $\alpha_8\beta_1$ appears to be involved in epithelial–mesenchymal interactions[4]. In chick embryo it is predominantly expressed neural tissue[5].

Distribution
Predominantly expressed in smooth muscle and kidney.

Disease association
OMIM 604063

Knockout
MGI:109442

Amino acid sequence of human integrin α_8

```
   1 FNLDVEKLAV YSGPKGSYFG YAVDFHIPDA RTASVLVGAP KANTSQPDIV EGGAVYYCPW
  61 PAEGSAQCRQ IPFDTTNNRK IRVNGTKEPI EFKSNQWFGA TVKAHKGKVV ACAPLYHWRT
 121 LKPTPEKGPV GTCYVAIQNF SAYAEFSPCG NSNADPEGQG YCQAGFSLDF YKNGDLIVGG
 181 PGSFYWQGQV ITASVADIIA NYSFKDILRK LAGEKQTEVA PASYDDSYLG YSVAAGEFTG
 241 DSQQELVAGI PRGAQNFGYV SIINSYDMTF IQNFTGEQMA SYFGYTVVVS DVNSDGLDDV
 301 LVGAPLFMER EFESNPREVG QIYLYLQVSS LLFRDPQILT GTETFGRFGS AMAHLGDLNQ
 361 DGYNDIAIGV PFAGKDQRGK VLIYNGNKDG LNTKPSQVLQ GVWASHAVPS GFGFTLRGDS
 421 DIDKNDYPDL IVGAFGTGKV AVYRARPVVT VDAQLLLHPM IINLENKTCQ VPDSMTSAAC
 481 FSLRVCASVT GQSIANTIVL MAEVQLDSLK QKGAIKRTLF LDNHQAHRVF PLVIKRQKSH
 541 QCQDFIVYLR DETEFRDKLS PINISLNYSL DESTFKEGLE VKPILNYYRE NIVSEQAHIL
 601 VDCGEDNLCV PDLKLSARPD KHQVIIGDEN HLMLIINARN EGEGAYEAEL FVMIPEEADY
 661 VGIERNNKGF RPLSCEYKME NVTRMVVCDL GNPMVSGTNY SLGLRFAVPR LEKTNMSINF
 721 DLQIRSSNKD NPDSNFVSLQ INITAVAQVE IRGVSHPPQI VLPIHNWEPE EEPHKEEEVG
 781 PLVEHIYELH NIGPSTISDT ILEVGWPFSA RDEFLLYIFH IQTLGPLQCQ PNPNINPQDI
 841 KPAASPEDTP ELSAFLRNST IPHLVRKRDV HVVEFHRQSP AKILNCTNIE CLQISCAVGR
 901 LEGGESAVLK VRSRLWAHTF LQRKNDPYAL ASLVSFEVKK MPYTDQPAKL PEGSIAIKTS
 961 VIWATPNVSF SIPLWVIILA ILLGLLVLAI LTLALWKCGF FDRARPPQED MTDREQLTND
1021 KTPEA
```

Proteolytic cleavage site at position 868.

Database accession

EMBL/GenBank   L36531
SwissProt   P53708

References

[1] Schnapp, L.M. et al. (1995) J. Cell Sci. 108, 537–544.
[2] Denda, S. et al. (1998) Mol. Biol. Cell 9, 1425–1535.
[3] Schnapp, L.M. et al. (1995) J. Biol. Chem. 270, 23196–23202.
[4] Muller, U. et al. (1997) Cell 88, 603–613.
[5] Bossy, B. et al. (1991) EMBO J. 10, 2375–2385.

179

Integrin α_9

Family

Integrin

Structure

Molecular weights
Amino acids	1035
Polypeptide	114 560

SDS-PAGE reduced

Carbohydrate
N-linked sites	11
O-linked sites	

Gene location   3p21.3

Gene structure

Alternative forms

α_9/β_1

Structure

Integrin α_9 forms heterodimer with integrin β_1 to form $\alpha_9\beta_1$[1]. It is post-translationally cleaved to form disulphide-bonded heavy and light α chains; the cleavage site is non-standard as in the α_4 chain[1].

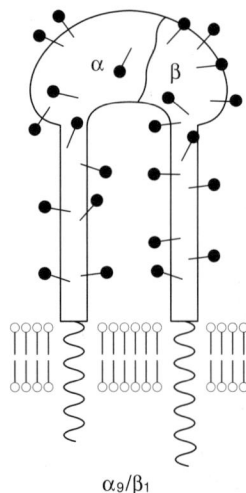

Ligands

Type III fibronectin repeat of tenascin-C[2], the N-terminal fragment of osteopontin through a cryptic site[3] and VCAM-1[4]. Together with the related integrin, $\alpha_4\beta_1$, $\alpha_9\beta_1$, recognises tissue transglutaminase, coagulation factor XIII and von Willebrand factor pro-peptide[5].

Function

The function of α_9 remains to be fully clarified, but it appears to be involved in a number of events in the epithelium of gut[6] and respiratory tract.

Distribution

Widely distributed in epithelia (skin, airway), muscle (smooth, skeletal) and liver[1]. Upregulated in lung and colon carcinoma[7]. Its expression is regulated in embryonic airway and intestinal development[6,8].

Disease association

OMIM 603963

Knockout

MGI:104756

The Gene Knockout FactsBook[9], p. 624.

Live offspring are produced at birth but these die in the first postnatal week with respiratory failure due to chylothorax as integrin α9 expression is critical for thoracic duct development (D. Sheppard, unpublished).

Amino acid sequence of human integrin α9

```
   1 MGGPAAPRGA GRLRALLLAL VVAGIPAGAY NLDPQRPVHF QGPADSFFGY AVLEHFHDNT
  61 RWVLVGAPKA DSKYSPSVKS PGAVFKCRVH TNPDRRCTEL DMARGKNRGT SCGKTCREDR
 121 DDEWMGVSLA RQPKADGRVL ACAHRWKNIY YEADHILPHG FCYIIPSNLQ AKGRTLIPCY
 181 EEYKKKYGEE HGSCQAGIAG FFTEELVVMG APGSFYWAGT IKVLNLTDNT YLKLNDEVIM
 241 NRRYTYLGYA VTAGHFSHPS TIDVVGGAPQ DKGIGKVYIF RADRRSGTLI KIFQASGKKM
 301 GSYFGSSLCA VDLNGDGLSD LLVGAPMFSE IRDEGQVTVY INRGNGALEE QLALTGDGAY
 361 NAHFGESIAS LDDLDNDGFP DVAIGAPKED DFAGAVYIYH GDAGGIVPQY SMKLSGQKIN
 421 PVLRMFGQSI SGGIDMDGNG YPDVTVGAFM SDSVVLLRAR PVITVDVSIF LPGSINITAP
 481 QCHDGQQPVN CLNVTTCFSF HGKHVPEEIG LNYVLMADVA KKEKGQMPRV YFVLLGETMG
 541 QVTEKLQLTY MEETCRHYVA HVKRRVQDVI SPIVFEAAYS LSEHVTGEEE RELPPLTPVL
 601 RWKKGQKIAQ KNQTVFERNC RSEDCAADLQ LQGKLLLSSM DEKTLYLALG AVKNISLNIS
 661 ISNLGDDAYD ANVSFNVSRE LFFINMWQKE EMGISCELLE SDFLKCSVGF PFMRSKSKYE
 721 FSVIFDTSHL SGEEEVLSFI VTAQSGNTER SESLHDNTLV LMVPLMHEVD TSITGIMSPT
 781 SFVYGESVDA ANFIQLDDLE CHFQPINITL QVYNTGPSTL PGSSVSISFP NRLSSGGAEM
 841 FHVQEMVVGQ EKGNCSFQKN PTPCIIPQEQ ENIFHTIFAF FTKSGRKVLD CEKPGISCLT
 901 AHCNFSALAK EESRTIDIYM LLNTEILKKD SSSVIQFMSR AKVKVDPALR VVEIAHGNPE
 961 EVTVVFEALH NLEPRGYVVG WIIAISLLVG ILIFLLLAVL LWKMGFFRRR YKEIIEAEKN
1021 RKENEDSWDW VQKNQ
```

Proteolytic cleavage site at position 565.

Database accession

EMBL/GenBank	L24158
SwissProt	Q13797

References

[1] Palmer, E.L. et al. (1993) J. Cell Biol. 123, 1289–1297.

[2] Yokosaki,Y. et al. (1998) J. Biol. Chem. 273, 11423–11428.

[3] Smith, L.L. et al. (1996) J. Biol. Chem. 271, 28485–28491.

[4] Taooka, Y. et al. (1999) J. Cell Biol. 145, 413–420.

[5] Takahashi, N. et al. (2000) J. Biol. Chem., in press.

[6] Desloges, N. et al. (1998) J. Cell Biochem. 71, 536–545.

[7] Basora, N. et al. (1998) Int. J. Cancer 75, 738–743.

[8] Wang, A. et al. (1995) Dev. Dyn. 204, 421–431.

[9] Mak, T.W. (1998) The Gene Knockout FactsBook. Academic Press, London.

Integrin α_{10}

■ **Family**

Integrin

■ **Structure**

Molecular weights

Amino acids	1167
Polypeptide	127 574

SDS-PAGE reduced	160 kDa

Carbohydrate

N-linked sites	10
O-linked sites	

Gene location	1q21

Gene structure	17 exons

Alternative forms

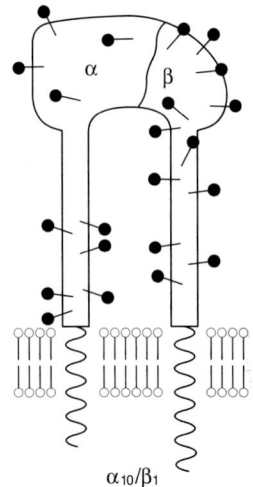

α_{10}/β_1

Structure

The integrin α_{10} chain forms a dimer with β_1 to form $\alpha_{10}\beta_1$. Integrin α_{10} is most homologous to integrin chains $\alpha 1$ and $\alpha 2$, and to the uncleaved, I domain-containing subgroup of integrin α chains[1].

■ **Ligands**

Collagen type II building integrin in chondrocytes[1].

■ **Function**

Unknown.

■ **Distribution**

Chondrocytes. Widely expressed with high levels in skeletal muscle and heart[2].

■ **Disease association**

OMIM 604042

■ **Knockout**

Amino acid sequence of human integrin α_{10}

```
   1 MELPFVTHLF LPLVFLTGLC SPFNLDEHHP RLFPGPPEAE FGYSVLQHVG GGQRWMLVGA
  61 PWDGPSGDRR GDVYRCPVGG AHNAPCAKGH LGDYQLGNSS HPAVNMHLGM SLLETDGDGG
 121 FMACAPLWSR ACGSSVFSSG ICARVDASFQ PQGSLAPTAQ RCPTYMDVVI VLDGSNSIYP
 181 WSEVQTFLRR LVGKLFIDPE QIQVGLVQYG ESPVHEWSLG DFRTKEEVVR AAKNLSRREG
 241 RETKTAQAIM VACTEGFSQS HGGRPEAARL LVVVTDGESH DGEELPAALK ACEAGRVTRY
 301 GIAVLGHYLR RQRDPSSFLR EIRTIASDPD ERFFFNVTDE AALTDIVDAL GDRIFGLEGS
 361 HAENESSFGL EMSQIGFSTH RLKDGILFGM VGAYDWGGSV LWLEGGHRLF PPRMALEDEF
 421 PPALQNHAAY LGYSVSSMLL RGGRRLFLSG APRFRHRGKV IAFQLKKDGA VRVAQSLQGE
 481 QIGSYFGSEL CPLDTDRDGT TDVLLVAAPM FLGPQNKETG RVYVYLVGQQ SLLTLQGTLQ
 541 PEPPQDARFG FAMGALPDLN QDGFADVAVG APLEDGHQGA LYLYHGTQSG VRPHPAQRIA
 601 AASMPHALSY FGRSVDGRLD LDGDDLVDVA VGAQGAAILL SSRPIVHLTP SLEVTPQAIS
 661 VVQRDCRRRG QEAVCLTAAL CFQVTSRTPG RWDHQFYMRF TASLDEWTAG ARAAFDGSGQ
 721 RLSPRRLRLS VGNVTCEQLH FHVLDTSDYL RPVALTVTFA LDNTTKPGPV LNEGSPTSIQ
 781 KLVPFSKDCG PDNECVTDLV LQVNMDIRGS RKAPFVVRGG RRKVLVSTTL ENRKENAYNT
 841 SLSIIFSRNL HLASLTPQRE SPIKVECAAP SAHARLCSVG HPVFQTGAKV TFLLEFEFSC
 901 SSLLSQVFGK LTASSDSLER NGTLQENTAQ TSAYIQYEPH LLFSSESTLH RYEVHPYGTL
 961 PVGPGPEFKT TLRVQNLGCY VVSGLIISAL LPAVAHGGNY FLSLSQVITN NASCIVQNLT
1021 EPPGPPVHPE ELQHTNRLNG SNTQCQVVRC HLGQLAKGTE VSVGLLRLVH NEFFRRAKFK
1081 SLTVVSTFEL GTEEGSVLQL TEASRWSESL LEVVQTRPIL ISLWILIGSV LGGLLLLALL
1141 VFCLWKLGFF AHKKIPEEEK REEKLEQ
```

I domain position 140–337.

Database accession

EMBL/GenBank   AF074015
SwissProt   O75578

Reference

[1] Camper, L. et al. (1998) J. Biol. Chem. 273, 20383–20389.
[2] Lehnert, K. et al. (1999) Cytogenet. Cell Genet. 87, 238–244.

Integrin α$_{IIb}$ CD41, platelet glycoprotein gpIIb

Family

Integrin

Structure

Molecular weights

Amino acids	1039
Polypeptide	110 019

SDS-PAGE reduced	125, 22 kDa (140 kDa unreduced)

Carbohydrate

N-linked sites	7
O-linked sites	

Gene location	17q21.32, in the same region as β3

Gene structure	17 kb, 30 exons

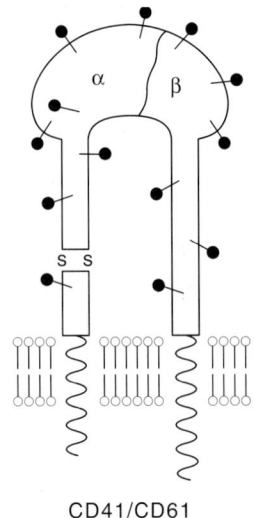

CD41/CD61

Alternative forms

Alternatively spliced variant with extracellular domain 34 amino acid deletion, which stops surface expression[1].

Structure

Integrin α$_{IIb}$ is a type I membrane glycoprotein integrin α chain[2-4] forming a non-covalently linked dimer with β$_3$ (CD61), α$_{IIb}$β$_3$. α$_{IIb}$ is proteolytically cleaved and disulphide bonded. Conformational changes in α$_{IIb}$ occur on co-activation through signals mediated via the cytoplasmic tail leading to high-affinity ligand binding and exposure of antibody-defined neo-epitopes (see entry for integrin β$_3$)[5-10]. Rotary shadowing electron microscopy has led to the development of a consensus structural model for integrins[11]. A model for the distribution of intramolecular disulphide bonding has been proposed[12].

Ligands

The α$_{IIb}$β$_3$ dimer (CD41/61, platelet gpIIbIIIa) is the platelet receptor for the adhesive proteins fibrinogen, fibronectin, vitronectin, thrombospondin and von Willebrand factor[6,13]. Binding occurs through an Arg-Gly-Asp (RGD) sequence in matrix/serum proteins (see also integrin α$_v$)[6,13,14]; interaction with fibrinogen is via an alternate sequence, HHLGGAKQAGDV (the H12 sequence) in the fibrinogen γ chain[15,16]. Adhesion is enhanced by co-stimulation, triggered by the agonists thrombin, ADP or collagen, leading to integrin receptor activation, high-affinity adhesion to soluble ligand and platelet aggregation[6,13,14].

Function

$\alpha_{IIb}\beta_3$ is the major integrin in platelets. Essential for normal platelet adhesion to damaged vascular endothelium and platelet aggregation. RGD peptide-based inhibitory drugs have been developed for clinical use in a variety of cardiovascular diseases where there is excessive platelet aggregation and thrombosis[17,18].

Distribution

Expression is limited to platelets and megakaryocytes[19].

Disease association

OMIM 273800

Mutations in α_{IIb}[20–24] cause one form of Glanzmann's thrombasthenia (type A) (see also OMIM 173470, entry for integrin β_3/CD61), which causes an inherited disorder of mucocutaneous bleeding. Polymorphism at the position 874 (ser874 to ile) forms the HPA-3/Bak/Lek platelet alloantigen system and is involved in neonatal thrombocytopenia[25,26].

Knockout

MGI:96601

Amino acid sequence of human integrin α_{IIb}

```
   1 MARALCPLQA LWLLEWVLLL LGPCAAPPAW ALNLDPVQLT FYAGPNGSQF GFSLDFHKDS
  61 HGRVAIVVGA PRTLGPSQEE TGGVFLCPWR AEGGQCPSLL FDLRDETRNV GSQTLQTFKA
 121 RQGLGASVVS WSDVIVACAP WQHWNVLEKT EEAEKTPVGS CFLAQPESGR RAEYSPCRGN
 181 TLSRIYVEND FSWDKRYCEA GFSSVVTQAG ELVLGAPGGY YFLGLLAQAP VADIFSSYRP
 241 GILLWHVSSQ SLSFDSSNPE YFDGYWGYSV AVGEFDGDLN TTEYVVGAPT WSWTLGAVEI
 301 LDSYYQRLHR LRAEQMASYF GHSVAVTDVN GDGRHDLLVG APLYMESRAD RKLAEVGRVY
 361 LFLQPRGPHA LGAPSLLLTG TQLYGRFGSA IAPLGDLDRD GYNDIAVAAP YGGPSGRGQV
 421 LVFLGQSEGL RSRPSQVLDS PFPTGSAFGF SLRGAVDIDD NGYPDLIVGA YGANQVAVYR
 481 AQPVVKASVQ LLVQDSLNPA VKSCVLPQTK TPVSCFNIQM CVGATGHNIP QKLSLNAELQ
 541 LDRQKPRQGR RVLLLGSQQA GTTLNLDLGG KHSPICHTTM AFLRDEADFR DKLSPIVLSL
 601 NVSLPPTEAG MAPAVVLHGD THVQEQTRIV LDCGEDDVCV PQLQLTASVT GSPLLVGADN
 661 VLELQMDAAN EGEGAYEAEL AVHLPQGAHY MRALSNVEGF ERLICNQKKE NETRVVLCEL
 721 GNPMKKNAQI GIAMLVSVGN LEEAGESVSF QLQIRSKNSQ NPNSKIVLLD VPVRAEAQVE
 781 LRGNSFPASL VVAAEEGERE QNSLDSWGPK VEHTYELHNN GPGTVNGLHL SIHLPGQSQP
 841 SDLLYILDIQ PQGGLQCFPQ PPVNPLKVDW GLPIPSPSPI HPAHHKRDRR QIFLPEPEQP
 901 SRLQDPVLVS CDSAPCTVVQ CDLQEMARGQ RAMVTVLAFL WLPSLYQRPL DQFVLQSHAW
 961 FNVSSLPYAV PPLSLPRGEA QVWTQLLRAL EERAIPIWWV LVGVLGGLLL LTILVLAMWK
1021 VGFFKRNRPP LEEDDEEGE
```

Proteolytic cleavage site at position 902.

Database accession

EMBL/GenBank   J02764
SwissProt   P08514

References

[1] Bray, P.F. et al. (1990) J. Biol. Chem. 265, 9587–9590.

[2] Poncz, M. et al. (1987) J. Biol. Chem. 262, 8476–8482.

[3] Frachet, P. et al. (1990) Mol. Biol. Rep. 14, 27–33.

[4] Charo, I.F. et al. (1986) Proc. Natl Acad. Sci. USA 83, 8351–8355.

[5] Frelinger, A.L. et al. (1991) J. Biol. Chem. 266, 17106–17111.

[6] Du, X. et al. (1991) Cell 65, 409–416.

[7] Loftus, J.C. et al. (1990) Science 249, 915–918.

[8] Frelinger, A.L. et al. (1990) J. Biol. Chem. 265, 6346–6352.

[9] Frelinger, A.L. et al. (1988) J. Biol. Chem. 263, 12397–12402.

[10] Mondoro, T.H. et al. (1996) Blood 88, 3824–3830.

[11] Carrell, N.A. et al. (1985) J. Biol. Chem. 260, 1743–1749.

[12] Calvete, J.J. et al. (1989) Biochem. J. 261, 561–568.

[13] Phillips, D.R. et al. (1991) Cell 65, 359–362.

[14] Ginsberg, M.H. et al. (1993) Thromb. Haemost. 70, 87–93.

[15] D'Souza, S.E. et al. (1991) Nature 350, 66–68.

[16] Savage, B. et al. (1995) J. Biol. Chem. 270, 28812–28817.

[17] Ferguson, J.J. and Zaqqa, M. (1999) Drugs 58, 965–982.

[18] Wang, W. et al. (2000) Curr. Med. Chem. 7, 437–453.

[19] Kieffer, N. and Phillips, D.R. (1990) Annu. Rev. Cell Biol. 6, 329–357.

[20] Bray, P.F. (1994) Thromb. Haemost. 72, 492–502.

[21] Poncz, M. et al. (1994) J. Clin. Invest. 93, 172–179.

[22] Wilcox, D.A. et al. (1994) J. Biol. Chem. 269, 4450–4457.

[23] Wilcox, D.A. et al. (1995) J. Clin. Invest. 95, 1553–1560.

[24] Tadokoro, S. et al. (1998) Blood 92, 2750–2758.

[25] Calvete, J.J. and Muniz-Diaz, E. (1993) FEBS Lett. 328, 30–34.

[26] Lyman, S. et al. (1990) Blood 75, 2343–2348.

Integrin α_V

Family

Integrin

Structure

Molecular weights

Amino acids	1048
Polypeptide	116 051

SDS-PAGE reduced	125, 25 kDa (165 kDa unreduced)

Carbohydrate

N-linked sites	13
O-linked sites	

Gene location

2q31–q32

Gene structure

Alternative forms

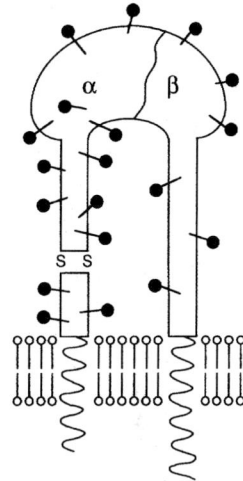

CD51/CD61

Structure

Integrin α_V is a type I membrane glycoprotein of the integrin subclass with no I (inserted) domain[1,2]. The α_V chain is post-translationally cleaved into a heavy and light chain, which is expressed on the cell surface as a disulphide-bonded polypeptide. α_V is a promiscuous integrin α chain forming dimers with many integrin β chains: β_1 (CD29), β_3 (CD61), β_5, β_6 and β_8.

Ligands

Most α_V integrins recognize the Arg-Gly-Asp (RGD) amino acid sequence. RGD-dependent interactions have been shown with many extracellular matrix and serum proteins[3-5], the specificity depending upon which α_V integrin is utilized. These include, for example, for $\alpha_V\beta_3$ (see also entries for integrins β_5, β_6, β_8), vitronectin[6,7], fibronectin, fibrinogen, thrombospondin, osteopontin, bone sialoprotein and von Willebrand factor. Unlike the other β_3 dimer, platelet $\alpha_{IIb}\beta_3$, ligand binding is constitutively activated. In contrast, $\alpha_V\beta_1$ is a fibronectin/vitronectin receptor[8]. Several proteins when in their native conformation fail to interact, for example laminin[9] and collagen type I, but do so if 'denatured'. Other matrix interactions with $\alpha_V\beta_3$ are not dependent on RGD, such as with the neural cell adhesion molecule L1 (see entry)[10]; $\alpha_V\beta_3$ also mediates intercellular (leucocyte–endothelium) adhesion via interaction with CD31 (see entry)[11,12].

Function

A role in normal bone resorption and hence bone turnover has been extensively demonstrated for $\alpha_V\beta_3$[13], a finding that is being exploited pharmaceutically. Likewise $\alpha_V\beta_3$ integrins play an important regulatory role in angiogenesis in various pathological conditions such as in tumours[14,15]. Recent evidence shows reciprocal involvement of $\alpha_V\beta_5$ and $\alpha_V\beta_3$ dimers in angiogenesis in experimental models, with $\alpha_V\beta_5$ being regulated by vascular factors such as VEGF, whereas $\alpha_V\beta_3$ is modulated by bFGF[16]. Experiments in mice have shown that $\alpha_V\beta_3$ can act as a co-stimulatory molecule for T cell activation. It is upregulated on inflammatory monocytes, where it is involved, with CD36, in the recognition of apoptotic neutrophils (see CD36 entry). There are a number of membrane molecules associated with $\alpha_V\beta_3$ that are involved in signal transduction and regulating $\alpha_V\beta_3$ activation, for example, CD47[17,18]. Melanoma cells express high levels of $\alpha_V\beta_3$ *in vivo*, and this integrin dimer is probably involved in tumour progression and invasion[19] (for example, via cell–endothelial adhesion via CD31). It thus has a role in melanoma invasion in the clinical situation due to upregulation of $\alpha_V\beta_3$ in neoplastic versus normal melanocytes[20].

Distribution

Low levels of $\alpha_V\beta_3$ integrin are found in a number of tissues, including endothelium, smooth muscle, some activated macrophages and T cells, and platelets[3,5]. Very high amounts are expressed constitutively by osteoclasts in bone[13]. Most cultured, adherent cell lines are α_V positive.

Disease association

OMIM:193210

Knockout

MGI:96608
The Gene Knockout FactsBook[21], p. 625
Eighty per cent of embryos die at E10.5–12.5. The remainder survive to birth but rapidly die with brain haemorrhage due to malformation of the cerebral vasculature. The unexpectedly 'mild' phenotype points to a complex role of α_V integrins in development and angiogenesis (leading to fatal intracerebral and intestinal haemorrhage)[22].

Amino acid sequence of human integrin α_V

```
  1 MAFPPRRRLR LGPRGLPLLL SGLLLPLCRA FNLDVDSPAE YSGPEGSYFG FAVDFFVPSA
 61 SSRMFLLVGA PKANTTQPGI VEGGQVLKCD WSSTRRCQPI EFDATGNRDY AKDDPLEFKS
121 HQWFGASVRS KQDKILACAP LYHWRTEMKQ EREPVGTCFL QDGTKTVEYA PCRSQDIDAD
181 GQGFCQGGFS IDFTKADRVL LGGPGSFYWQ GQLISDQVAE IVSKYDPNVY SIKYNNQLAT
241 RTAQAIFDDS YLGYSVAVGD FNGDGIDDFV SGVPRAARTL GMVYIYDGKN MSSLYNFTGE
301 QMAAYFGFSV AATDINGDDY ADVFIGAPLF MDRGSDGKLQ EVGQVSVSLQ RASGDFQTTK
361 LNGFEVFARF GSAIAPLGDL DQDGFNDIAI AAPYGGEDKK GIVYIFNGRS TGLNAVPSQI
421 LEGQWAARSM PPSFGYSMKG ATDIDKNGYP DLIVGAFGVD RAILYRARPV ITVNAGLEVY
481 PSILNQDNKT CSLPGTALKV SCFNVRFCLK ADGKGVLPRK LNFQVELLLD KLKQKGAIRR
541 ALFLYSRSPS HSKNMTISRG GLMQCEELIA YLRDESEFRD KLTPITIFME YRLDYRTAAD
601 TTGLQPILNQ FTPANISRQA HILLDCGEDN VCKPKLEVSV DSDQKKIYIG DDNPLTLIVK
661 AQNQGEGAYE AELIVSIPLQ ADFIGVVRNN EALARLSCAF KTENQTRQVV CDLGNPMKAG
```

```
 721 TQLLAGLRFS VHQQSEMDTS VKFDLQIQSS NLFDKVSPVV SHKVDLAVLA AVEIRGVSSP
 781 DHIFLPIPNW EHKENPETEE DVGPVVQHIY ELRNNGPSSF SKAMLHLQWP YKYNNNTLLY
 841 ILHYDIDGPM NCTSDMEINP LRIKISSLQT TEKNDTVAGQ GERDHLITKR DLALSEGDIH
 901 TLGCGVAQCL KIVCQVGRLD RGKSAILYVK SLLWTETFMN KENQNHSYSL KSSASFNVIE
 961 FPYKNLPIED ITNSTLVTTN VTWGIQPAPM PVPVWVIILA VLAGLLLLAV LVFVMYRMGF
1021 FKRVRPPQEE QEREQLQPHE NGEGNSET
```

Proteolytic cleavage site at position 889.

Database accession

EMBL/GenBank   M14648
SwissProt   P06756

References

[1] Suzuki, S. et al. (1986) Proc. Natl Acad. Sci. USA 83, 8614–8618.
[2] Suzuki, S. et al. (1987) J. Biol. Chem. 262, 14080–14085.
[3] Keiffer, N. and Phillips, D.R. (1990) Annu. Rev. Cell Biol. 6, 329–357.
[4] Ginsberg, M.H. et al. (1993) Thromb. Haemost. 70, 87–93.
[5] Horton, M.A. (1997) Int. J. Biochem. Cell Biol. 29, 721–725.
[6] Pytela, R. et al. (1985) Proc. Natl Acad. Sci. USA 82, 5766–5770.
[7] Cheresh, D.A. (1987) Proc. Natl Acad. Sci. USA 84, 6471–6475.
[8] Vogel,B.E. et al. (1990) J. Biol. Chem. 265, 5934–5937.
[9] Kramer, R.H. et al. (1990) J. Cell Biol. 111, 1233–1243.
[10] Ebeling, O. et al. (1996) Eur. J. Immunol. 26, 2508–2516.
[11] Piali, L. et al. (1995) J. Cell Biol. 130, 451–460.
[12] Buckley, C.D. et al. (1996) J. Cell Sci. 109, 437–455.
[13] Helfrich, M.H. and Horton, M.A. (1999) Dynamics of Bone and Cartilage Metabolism, Seibel, M.J., Robins, S.P. and Bilezikian, J.P., eds. Academic Press, London, pp.111–125.
[14] Varner, J.A. and Cheresh, D.A. (1996) Curr. Opin. Cell Biol. 8, 724–730.
[15] Varner, J.A. et al. (1995) Cell Adhes. Comm. 3, 367–374.
[16] Friedlander, M. et al. (1995) Science 270, 1500–1502.
[17] Lindberg, F.P. et al. (1996) J. Cell Biol. 134, 1313–1322.
[18] Gao, A.G. et al. (1996) J. Cell Biol. 135, 533–544.
[19] Johnson, J.P. (1999) Cancer Metastasis Rev. 18, 345–357.
[19] Seftor, R.E. et al. (1992) Proc. Natl Acad. Sci. USA 89, 1557–1561.
[20] Mak, T.W. (1998) The Gene Knockout FactsBook. Academic Press, London.
[21] Bader, B.L. et al. (1998) Cell 95, 507–519.

Integrin α$_E$ CD103; HML-1

Family

Integrin

Structure

Molecular weights
Amino acids	1178
Polypeptide	129 714

SDS-PAGE reduced	150, 25 kDa (175 kDa unreduced)

Carbohydrate
N-linked sites	11
O-linked sites	

Gene location   17q13

Gene structure

Alternative forms

CD103(αE)/β7

Structure
Integrin α$_E$ (CD103) is a type I membrane glycoprotein[1-4] which exists as a dimer with β$_7$ to form α$_E$β$_7$. It has an N-terminal I or inserted domain characteristic of a subclass of integrin α chains, but, unusually, it is also post-translationally cleaved into two polypeptides (an N-terminal 25 kDa and larger C-terminal 150 kDa fragment), which are disulphide bonded. An extra sequence (the 'X' domain) of 55 amino acids (16 are acidic) is found immediately N-terminal to the I domain and is the site of proteolytic cleavage[1].

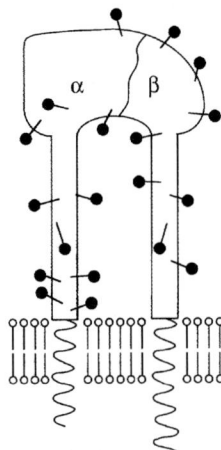

Ligands

E-Cadherin (cadherin-1)[5-8] on epithelium.

Function

Adhesion of intraepithelial lymphocytes, a subset of T cells, to the basolateral surface of intestinal epithelial cells[9]. Presumed role in lymphocyte homing to epithelium[5,9,10].

Distribution

Integrin α$_E$ was defined as the HML-1 antigen by the M290 antibody in the mouse in the early literature. It is expressed on the majority of intraepithelial lymphocytes in the intestine and other epithelia (bronchi, skin, breast, tumours), and a subpopulation of T cells in the lamina propria. Less than 5% of circulating lymphocytes express α$_E$, but this is increased by TGFβ1[3,11].

Disease association

OMIM 604682

Knockout

MGI:1298377

The function of α$_E$ has been defined in β$_7$-deficient mice (see entry) and in α$_E$ knockout mice[12], where intraepithelial lymphocyte numbers are reduced, sparing T cell levels in tissues such as spleen and Peyer's patches.

Amino acid sequence of human integrin α$_E$

```
   1 MWLFHTLLCI ASLALLAAFN VDVARPWLTP KGGAPFVLSS LLHQDPSTNQ TWLLVTSPRT
  61 KRTPGPLHRC SLVQDEILCH PVEHVPIQGE APGSDRCPEP PRCFDMHSSA GPAPHSLSSE
 121 LTGTCSLLGP DLRPQAQANF FDLENLLDPD ARVDTGDCYS NKEGGGEDDV NTARQRRALE
 181 KEEEEDKEEE EDEEEEEAGT EIAIILDGSG SIDPPDFQRA KDFISNMMRN FYEKCFECNF
 241 ALVQYGGVIQ TEFDLRDSQD VMASLARVQN ITQVGSVTKT ASAMQHVLDS IFTSSHGSRR
 301 KASKVMVVLT DGGIFEDPLN LTTVINSPKM QGVERFAIGV GEEFKSARTA RELNLIASDP
 361 DETHAFKVTN YMALDGLLSK LRYNIISMEG TVGDALHYQL AQIGFSAQIL DERQVLLGAV
 421 GAFDWSGGAL LYDTRSRRGR FLNQTAAAAA DAEAAQYSYL GYAVAVLHKT CSLSYVAGAP
 481 QYKHHGAVFE LQKEGREASF LPVLEGEQMG SYFGSELCPV DIDMDGSTDF LLVAAPFYHV
 541 HGEEGRVYVY RLSEQDGSFS LARILSGHPG FTNARFGFAM AAMGDLSQDK LTDVAIGAPL
 601 EGFGADDGAS FGSVYIYNGH WDGLSASPSQ RIRASTVAPG LQYFGMSMAG GFDISGDGLA
 661 DITVGTLGQA VVFRSRPVVR LKVSMAFTPS ALPIGFNGVV NVRLCFEISS VTTASESGLR
 721 EALLNFTLDV DVGKQRRRLQ CSDVRSCLGC LREWSSGSQL CEDLLLMPTE GELCEEDCFS
 781 NASVKVSYQL QTPEGQTDHP QPILDRYTEP FAIFQLPYEK ACKNKLFCVA ELQLATTVSQ
 841 QELVVGLTKE LTLNINLTNS GEDSYMTSMA LNYPRNLQLK RMQKPPSPNI QCDDPQPVAS
 901 VLIMNCRIGH PVLKRSSAHV SVVWQLEENA FPNRTADITV TVTNSNERRS LANETHTLQF
 961 RHGFVAVLSK PSIMYVNTGQ GLSHHKEFLF HVHGENLFGA EYQLQICVPT KLRGLQVAAV
1021 KKLTRTQAST VCTWSQERAC AYSSVQHVEE WHSVSCVIAS DKENVTAAEE ISWDHSEELL
1081 KDVTELQILG EISFNKSLYE GLNAENHRTK ITVVFLKDEK YHSLPTIIKG SVGGLLVLIV
1141 ILVILFKCGF FKRKYQQLNL ESIRKAQLKS ENLLEEEN
```

Proteolytic cleavage site at position 176.
I domain position 199–390 and an 'X' domain position 144–198.

Database accession

EMBL/GenBank   L25851
SwissProt   P38570

References

[1] Shaw, S.K. et al. (1994) J. Biol. Chem. 269, 6016–6025.
[2] Micklem, K.J. et al. (1991) Am. J. Pathol. 139, 1297–1301.
[3] Parker, C.M. et al. (1992) Proc. Natl Acad. Sci. USA 89, 1924–1928.
[4] Kilshaw, P.J. and Murant S.J. (1990) Eur. J. Immunol. 20, 2201.
[5] Cepek, K.L. et al. (1994) Nature 372, 190–193.
[6] Higgins, J.M. et al. (1998) J. Cell Biol. 140, 197–210.
[7] Karecla, P.I. et al. (1995) Eur. J. Immunol. 25, 852–856.
[8] Karecla, P. et al. (1996) J. Biol. Chem. 271, 30909–30915.
[9] Cepek, K.L. et al. (1993) J. Immunol. 150, 3459–3470.
[10] Shaw, S.K. and Brenner, M.B. (1995) Semin. Immunol. 7, 335–342.
[11] Schieferdecker, H.L. et al. (1990) J. Immunol. 144, 2541–2549.
[12] Scon, M.P. et al. (1999) J. Immunol. 162, 6641–6649.

Integrin β_1

Family

Integrin β chain

Structure

Molecular weights

Amino acids	798
Polypeptide	88 465 (4 isoforms with polypeptides varying by approx. 4 kDa)
SDS-PAGE reduced	115 kDa (130 kDa unreduced)

Carbohydrate

N-linked sites	12
O-linked sites	

Gene location   10p11.2

Gene structure   30 kb, 7 exons

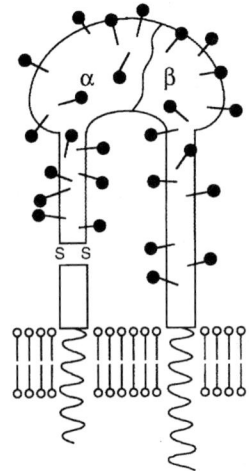

CD49e/CD29

Alternative forms

Four alternate RNA splice variants producing different C-termini with differential tissue expression[1-4].

Structure

β_1 is the prototype integrin β chain (fibronectin receptor, $\alpha_5\beta_1$)[2]. It is a type I membrane glycoprotein that associates with a large number of integrin α chains forming non-covalently linked dimers with α_1, α_2, α_3, α_4, α_5, α_6, α_7, α_8, α_9, α_{10} and α_V. β_1, like all integrin β chains, is cyteine-rich (56 residues in four repeat regions) and internally disulphide bonded. The N-terminal part of the β chain is involved in ligand interaction and shows homology with the von Willebrand factor A-domain[5].

Ligands

Ligand specificity depends upon the exact integrin dimer formed (see individual α-chain entries for details). For example, α_1 and α_6 form dimers with β_1 that bind laminin, $\alpha_2\beta_1$ is a collagen receptor and $\alpha_5\beta_1$ is the major fibronectin receptor.

Function

See individual α-chain entries for details. β_1 integrins mediate a broad range of cell–matrix and intercellular interactions[6]. The β_1 integrin is activated for ligand binding by divalent cation exposure and 'inside-out' cell signalling[7,8].

Distribution

Widespread in most tissues during embryogensis and adult life as the common β chain of the large β_1 subclass of integrins[6] (see individual α-chain entries for details of individual dimer tissue distribution). The β-chain splice forms show different tissue expression: the B variant[1] is found in placenta, lymphoid tissues, liver and endothelium; the C isoform[3] is expressed by some subpopulations of haemopoietic cells; the D variant[4] is found only in skeletal muscle and heart.

Disease association

OMIM 135630

Knockout

MGI:96610
The Gene Knockout FactsBook[9], p. 626.
Embryos implant but die at around E5.5 due to defective inner cell mass endoderm morphogenesis following normal blastocyst development and trophoblast function[10-12].

Amino acid sequence of human integrin β₁

```
  1 MNLQPIFWIG LISSVCCVFA QTDENRCLKA NAKSCGECIQ AGPNCGWCTN STFLQEGMPT
 61 SARCDDLEAL KKKGCPPDDI ENPRGSKDIK KNKNVTNRSK GTAEKLKPED IHQIQPQQLV
121 LRLRSGEPQT FTLKFKRAED YPIDLYYLMD LSYSMKDDLE NVKSLGTDLM NEMRRITSDF
181 RIGFGSFVEK TVMPYISTTP AKLRNPCTSE QNCTTPFSYK NVLSLTNKGE VFNELVGKQR
241 ISGNLDSPEG GFDAIMQVAV CGSLIGWRNV TRLLVFSTDA GFHFAGDGKL GGIVLPNDGQ
301 CHLENNMYTM SHYYDYPSIA HLVQKLSENN IQTIFAVTEE FQPVYKELKN LIPKSAVGTL
361 SANSSNVIQL IIDAYNSLSS EVILENGKLS EGVTISYKSY CKNGVNGTGE NGRKCSNISI
421 GDEVQFEISI TSNKCPKKDS DSFKIRPLGF TEEVEVILQY ICECECQSEG IPESPKCHEG
481 NGTFECGACR CNEGRVGRHC ECSTDEVNSE DMDAYCRKEN SSEICSNNGE CVCGQCVCRK
541 RDNTNEIYSG KFCECDNFNC DRSNGLICGG NGVCKCRVCE CNPNYTGSAC DCSLDTSTCE
601 ASNGQICNGR GICECGVCKC TDPKFQGQTC EMCQTCLGVC AEHKECVQCR AFNKGEKKDT
661 CTQECSYFNI TKVESRDKLP QPVQPDPVSH CKEKDVDDCW FYFTYSVNGN NEVMVHVVEN
721 PECPTGPDII PIVAGVVAGI VLIGLALLLI WKLLMIIHDR REFAKFEKEK MNAKWDTGEN
781 PIYKSAVTTV VNPKYEGK
```

Integrin β1 exists in four cytoplasmic tail splice variants. The A isoform is illustrated in italics and underlined and is replaced by the following after T777:

```
B: VSYKTSKKQS GL
C: SLSVAQPGVQ WCDISSLQPL TSRQQFSCLS LPSTWDYRVK ILFIRVP
D: QENPIYKSPI NNFKNPNYGR KAGL
```

Database accession

EMBL/GenBank	A form: X07979; B form: U33879; C form: U33882; D form: U33880
SwissProt	P05556

References

1 Balzac, F. et al. (1993) J. Cell Biol. 121, 171–178.
2 Argraves, W.S. et al. (1987) J. Cell Biol. 105, 1183–1190.
3 Languino, R.L. and Ruoslahti, E. (1992) J. Biol. Chem. 267, 7116–7120.
4 Belkin, A.M. et al. (1996) J. Cell Biol. 132, 211–216.
5 Tuckwell, D.S. and Humphries, M.J. (1997) FEBS Lett. 400, 297–303.
6 Hemler, M.E. (1990) Annu. Rev. Immunol. 114, 365–400.
7 Shimizu, Y. et al. (1990) Nature 345, 250–253.
8 Schwartz, M.A. et al. (1995) Annu. Rev. Cell Dev. Biol. 11, 549–599.
9 Mak, T.W. (1998) The Gene Knockout FactsBook. Academic Press, London.
10 Stephens, L.E. et al. (1995) Genes Dev. 9, 1883–1895.
11 Fassler, R. and Meyer, M. (1995) Genes Dev. 9, 1896–1908.
12 Brakebusch, C. et al. (1997) J. Cell Sci. 110, 2895–2904.

Integrin β₂

Family

Integrin

Structure

Molecular weights

Amino acids	769
Polypeptide	84 790

SDS-PAGE reduced	95 kDa (90 kDa unreduced)

Carbohydrate

N-linked sites	6
O-linked sites	

Gene location	21q22.3

Gene structure	40 kb

Alternative forms

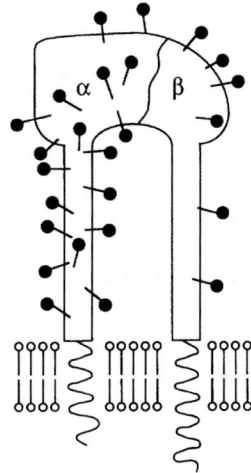

CD11b/CD18

Structure

Integrin β₂ is a type I membrane glycoprotein[1-3] of the general integrin β chain structure (see β₁) that forms dimers with four α chains (CD11a–d). The cytoplasmic tail of β₂ contains eight potential posphorylation sites.

Ligands

See individual entries for α_L, α_M, α_X and α_D (CD11a–d) molecules for details of β₂ ligands. β₂ integrins also recognize several microbial ligands including those from *Histoplasma* sp., *Legionella* sp. and *Leishmania* sp[4-7].

Function

Adhesion and signalling in haemopoietic cells. For details of function, see entries for individual α chains, α_L, α_M, α_X and α_D (CD11a–d) integrins. The cytoplasmic tail of β₂ is important in regulating signal transduction from the individual dimer via cytoskeletal interactions and cytoplasmic tail phosphorylation[8,9]. Ligand binding is regulated by receptor activation as with several other integrins[10,11].

Distribution

Integrin β₂ (CD18) is expressed by all leucocytes[12,13] with the four leuco-integrin dimers being differentially expressed – see individual entries for α_L, α_M, α_X and α_D (CD11a–d) integrins for details.

Disease association

OMIM 116920

Leucocyte adhesion deficiency syndrome (LAD-1) is caused by mutations in the β_2 gene (see also LAD-2, in entry for selectins) in which patients suffer from varying severity of disease owing to recurrent bacterial infections due to defective leucocyte (granulocytes, monocytes, lymphocytes) adhesion[14-17]; in part this relates to the molecular defect and its impact upon the level of β_2 integrin expression in the cell membrane

Knockout

MGI:96611

β_2 (CD18) knockout mice[18] show similar deficits in inflammation as in the human disease, underscoring the importance of β_2 integrins in normal immune function. An inherited disease with similar features was observed in Holstein cattle due to point mutations in the bovine β_2 gene[19,20].

Amino acid sequence of human integrin β_2

```
  1 MLGLRPPLLA LVGLLSLGCV LSQECTKFKV SSCRECIESG PGCTWCQKLN FTGPGDPDSI
 61 RCDTRPQLLM RGCAADDIMD PTSLAETQED HNGGQKQLSP QKVTLYLRPG QAAAFNVTFR
121 RAKGYPIDLY YLMDLSYSML DDLRNVKKLG GDLLRALNEI TESGRIGFGS FVDKTVLPFV
181 NTHPDKLRNP CPNKEKECQP PFAFRHVLKL TNNSNQFQTE VGKQLISGNL DAPEGGLDAM
241 MQVAACPEEI GWRNVTRLLV FATDDGFHFA GDGKLGAILT PNDGRCHLED NLYKRSNEFD
301 YPSVGQLAHK LAENNIQPIF AVTSRMVKTY EKLTEIIPKS AVGELSEDSS NVVHLIKNAY
361 NKLSSRVFLD HNALPDTLKV TYDSFCSNGV THRNQPRGDC DGVQINVPIT FQVKVTATEC
421 IQEQSFVIRA LGFTDIVTVQ VLPQCECRCR DQSRDRSLCH GKGFLECGIC RCDTGYIGKN
481 CECQTQGRSS QELEGSCRKD NNSIICSGLG DCVCGQCLCH TSDVPGKLIY GQYCECDTIN
541 CERYNGQVCG GPGRGLCFCG KCRCHPGFEG SACQCERTTE GCLNPRRVEC SGRGRCRCNV
601 CECHSGYQLP LCQECPGCPS PCGKYISCAE CLKFEKGPFG KNCSAACPGL QLSNNPVKGR
661 TCKERDSEGC WVAYTLEQQD GMDRYLIYVD ESRECVAGPN IAAIVGGTVA GIVLIGILLL
721 VIWKALIHLS DLREYRRFEK EKLKSQWNND NPLFKSATTT VMNPKFAES
```

Database accession

EMBL/GenBank   Y00057
SwissProt   P05107

References

1 Law, S.K.A. et al. (1987) EMBO J. 6, 915–919.

2 Weitzman, J.B. et al. (1991) FEBS Lett. 294, 97–103.

3 Kishimoto, T.K. et al. (1987) Cell 48, 681–690.

4 Aderem, A. and Underhill, D.M. (1999) Annu. Rev. Immunol. 17, 593–623.

5 Mosser, D.M. et al. (1992) J. Cell Biol. 116, 511–520.

6 Rosenthal, L.A. et al. (1996) Infect. Immun. 64, 2206–2215.

7 Talamas-Rohana, P. et al. (1990) J. Immunol. 144, 4817–4824.

8 Pavalko, F.M. and LaRoche, S.M. (1993) J. Immunol. 151, 3795–3807.

9 Sharma, C.P. et al. (1995) J. Immunol. 154, 3461–3470.

10 Ortlepp, S. et al. (1995) Eur. J. Immunol. 25, 637–643.

11 Petruzzelli, L. et al. (1995) J. Immunol. 155, 854–866.

12 Larson, R.S. and Springer, T.A. (1990) Immunol. Rev. 114, 181–217.

[13] Patarroyo, M. et al. (1990) Immunol. Rev. 114, 67–108.

[14] Nelson, C. et al. (1992) J. Biol. Chem. 267, 3351–3357.

[15] Wardlaw, A.J. et al. (1990) J. Exp. Med. 172, 335–345.

[16] Arnaout, M.A. et al. (1990) J. Clin. Invest. 85, 977–981.

[17] Kishimoto, T.K. et al. (1987) Cell 50, 193–202.

[18] Wilson, R.W. (1993) J. Immunol. 151, 1571–1578.

[19] Shuster, D.E. et al. (1992) Proc. Natl Acad. Sci. USA 89, 9225–9229.

[20] Shier, P. et al. (1996) J. Immunol. 157, 5375–5386.

Integrin β₃

CD61, platelet glycoprotein gpIIIa

Family

Integrin

Structure

Molecular weights

Amino acids	788
Polypeptide	87213

SDS-PAGE reduced	105 kDa (90 kDa unreduced)

Carbohydrate

N-linked sites	6
O-linked sites	

Gene location 17q21–23 near α_{IIb}

Gene structure 60 kb, 14 exons

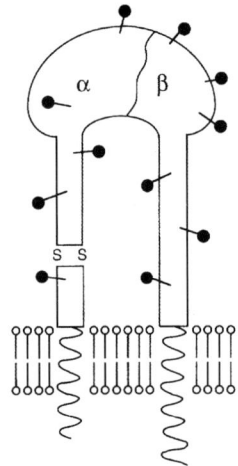

CD41/CD61

Alternative forms

There are at least three cytoplasmic tail alternate splice forms[1], with one form having reduced ligand binding[2].

Structure

Integrin β₃ is a typical integrin β chain type I glyoprotein[3-5] that dimerizes with two alternate α chains, α_{IIb} and α_V, to form $\alpha_{IIb}\beta_3$ and $\alpha_V\beta_3$, respectively. Point mutations lead to polymorphisms, and deletions or mutations to an inherited bleeding disorder (see below). Tyr 773 is potentially phophorylated and is deleted by RNA splicing. A model for intramolecular disulphide bond usage has been proposed[6].

Ligands

The two β₃ integrin dimers, $\alpha_V\beta_3$ and $\alpha_{IIb}\beta_3$, bind different ligands (see individual α_V and α_{IIb} entries for details).

Function

See individual α_V and α_{IIb} entries for details of β₃ function.

Distribution

β₃ expression in tissues follows that of the exact integrin dimer formed. $\alpha_V\beta_3$ (CD51/CD61) is found in several tissues (see entry for integrin α_V). In contrast, $\alpha_{IIb}\beta_3$(CD41/CD61, gpIIbIIIa) expression is restricted to the platelet/megakaryocyte lineage[7], where it plays a key role in platelet aggregation and hence haemostasis (see entry for α_{IIb}).

98

Disease association

OMIM 173470

Glanzmann's thrombasthenia (type B) (see also OMIM 273800, and entry for integrin α$_{IIb}$) which causes an inherited haemorrhagic diathesis, is due to mutation (cytoplasmic tail truncation due to stop codon leading to altered signalling to, and interaction with, platelet actin cytoskeleton) or major deletions (Iraqi-Jewish form) in the β₃ gene[8-11]. The significance of functional regions within the β₃ chain important in ligand binding have been exemplified by certain mutations in Glanzmann's thrombasthenia[12,13], which lead to loss of binding activity.

Alloantigens on the β₃ chain include Zw/Pl (a), Pen and Mo lead to neonatal and post-transfusion purpura[10,14,15].

Knockout

MGI:96612

The Gene Knockout FactsBook[16], p. 627.

Mice are viable at birth with bleeding diathesisis due to a platelet defect, the murine homologue of Glanzmann's thrombasthenia. As predicted from studies of the α$_V$β₃ integrin (vitronectin receptor), there is some evidence of an osteoclast functional defect in the bones of mutant mice[17].

Amino acid sequence of human integrin β₃

```
  1 MRARPRPRPL WVTVLALGAL AGVGVGGPNI CTTRGVSSCQ QCLAVSPMCA WCSDEALPLG
 61 SPRCDLKENL LKDNCAPESI EFPVSEARVL EDRPLSDKGS GDSSQVTQVS PQRIALRLRP
121 DDSKNFSIQV RQVEDYPVDI YYLMDLSYSM KDDLWSIQNL GTKLATQMRK LTSNLRIGFG
181 AFVDKPVSPY MYISPPEALE NPCYDMKTTC LPMFGYKHVL TLTDQVTRFN EEVKKQSVSR
241 NRDAPEGGFD AIMQATVCDE KIGWRNDASH LLVFTTDAKT HIALDGRLAG IVQPNDGQCH
301 VGSDNHYSAS TTMDYPSLGL MTEKLSQKNI NLIFAVTENV VNLYQNYSEL IPGTTVGVLS
361 MDSSNVLQLI VDAYGKIRSK VELEVRDLPE ELSLSFNATC LNNEVIPGLK SCMGLKIGDT
421 VSFSIEAKVR GCPQEKEKSF TIKPVGFKDS LIVQVTFDCD CACQAQAEPN SHRCNNGNGT
481 FECGVCRCGP GWLGSQCECS EEDYRPSQQD ECSPREGQPV CSQRGECLCG QCVCHSSDFG
541 KITGKYCECD DFSCVRYKGE MCSGHGQCSC GDCLCDSDWT GYYCNCTTRT DTCMSSNGLL
601 CSGRGKCECG SCVCIQPGSY GDTCEKCPTC PDACTFKKEC VECKFDREP YMTENTCNRY
661 CRDEIESVKE LKDTGKDAVN CTYKNEDDCV VRFQYYEDSS GKSILYVVEE PECPKGPDIL
721 VVLLSVMGAI LLIGLAALLI WKLLITIHDR KEFAKFEEER ARAKWDTANN PLYKEATSTF
781 TNITYRGT
```

Database accession

EMBL/GenBank   U95204
SwissProt   P05106

References

[1] van Kuppevelt, T.H. et al. (1989) Proc. Natl Acad. Sci. USA 86, 5414–5418.

[2] Kumar, C.S. et al. (1997) J. Biol. Chem. 272, 16390–16397.

[3] Fitzgerald, L.A. et al. (1987) J. Biol. Chem. 262, 3936–3939.

[4] Frachet, P. et al. (1990) Mol. Biol. Rep. 14, 27–33.

[5] Zimrin, A.B. et al. (1988) J. Clin. Invest. 81, 1470–1475.

[6] Calvete, J.J. et al. (1991) Biochem. J. 274, 63–71.

7 Kieffer, N. and Phillips, D.R. (1990) Annu. Rev. Cell Biol. 6, 334–357.

8 Wang, R. et al. (1992) J. Clin. Invest. 89, 1995–2004.

9 Simsek, S. et al. (1993) Blood 81, 2044–2049.

10 Kuijpers, R.W.A.M. et al. (1993) Blood 81, 70–76.

11 Lanza, F. et al. (1990) J. Biol. Chem. 265, 18098–18103.

12 Loftus, J.C. et al. (1990) Science 249, 915–918.

13 Bajt, M.L. et al. (1992) J. Biol. Chem. 267, 3789–3794.

14 Newman, P.J. et al. (1989) J. Clin. Invest. 83, 1778–1781.

15 Wang, R. et al. (1992) J. Clin. Invest. 90, 2038–2043.

16 Mak, T.W. (1998) The Gene Knockout FactsBook. Academic Press, London.

17 Hodivala-Dilke, K.M. et al. (1999) J. Clin. Invest. 103, 229–238.

Integrin β_4 CD104, TSP-180

Family

Integrin

Structure

Molecular weights
Amino acids	1875
Polypeptide	208 024

SDS-PAGE reduced   220 kDa

Carbohydrate
N-linked sites	5
O-linked sites	

Gene location   17q11–qter

Gene structure

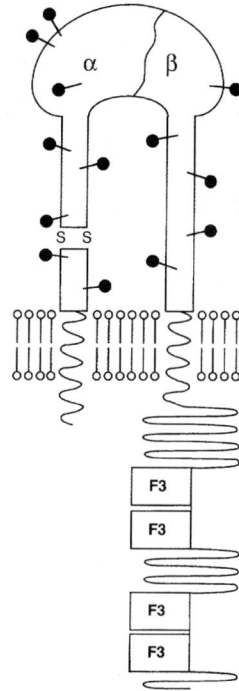

CD49f/CD104

Alternative forms

Three RNA splice variants reported (with two alternate cytoplasmic insertions at positions 1370 or 1520 between the second and third type III fibronectin repeats) or neither[1-3].

Structure

Integrin β_4 forms a heterodimer with integrin α_6 chain[1-5] to form $\alpha_6\beta_4$. Unique amongst integrin β chains, β_4 contains a large 118 kDa (approximately 1000 amino acid) cytoplasmic domain with four fibronectin type III repeats. It has only 48 of the usual 56 cysteine residues conserved across most integrin β chains. Interacts via unique cytoplasmic tail with keratin cytoskeletal filaments, rather than actin as found with other integrins[6-8].

Ligands

Laminin 5[9,10].

Function

Its distribution on the basal surface of keratinocytes and association with hemidesmosomal cytoskeleton in stratified and transitional epithelia indicates that the receptor is involved in adhesion to basement membrane matrix proteins[11,12] (see below for effect of mutation in β₄ upon epithelial integrity and adhesion). *In vitro* experiments and evidence from β₄ knockout mice also suggest a role for β₄ integrin in epithelial proliferation and signal transduction, the latter mediated specifically via its unique cytoplasmic tail. A cytoplasmic insert at position 1370 (insert 1) is found in carcinoma cell lines suggesting a variant-related function[3,13].

Distribution

β₄ (CD104) is expressed by basal keratinocytes in association with hemidesmosomes[4-6] and in simple epithelia, Schwann cells[11,14] and some endothelium, which do not have hemidesmosomes[15].

Disease association

OMIM 147557
Mutations in β₄ lead to a variant of epidermolysis bullosa with pyloric atresia[16-18].

Knockout

MGI:96613
Extensive epithelial detachment at birth and consequent death, features similar to the human disease[6,19].

Amino acid sequence of human integrin β₄

```
   1 MAGPRPSPWA RLLLAALISV SLSGTLANRC KKAPVKSCTE CVRVDKDCAY CTDEMFRDRR
  61 CNTQAELLAA GCQRESIVVM ESSFQITEET QIDTTLRRSQ MSPQGLRVRL RPGEERHFEL
 121 EVFEPLESPV DLYILMDFSN SMSDDLDNLK KMGQNLARVL SQLTSDYTIG FGKFVDKVSV
 181 PQTDMRPEKL KEPWPNSDPP FSFKNVISLT EDVDEFRNKL QGERISGNLD APEGGFDAIL
 241 QTAVCTRDIG WRPDSTHLLV FSTESAFHYE ADGANVLAGI MSRNDERCHL DTTGTYTQYR
 301 TQDYPSVPTL VRLLAKHNII PIFAVTNYSY SYYEKLHTYF PVSSLGVLQE DSSNIVELLE
 361 EAFNRIRSNL DIRALDSPRG LRTEVTSKMF QKTRTGSFHI RRGEVGIYQV QLRALEHVDG
 421 THVCQLPEDQ KGNIHLKPSF SDGLKMDAGI ICDVCTCELQ KEVRSARCSF NGDFVCGQCV
 481 CSEGWSGQTC NCSTGSLSDI QPCLREGEDK PCSGRGECQC GHCVCYGEGR YEGQFCEYDN
 541 FQCPRTSGFL CNDRGRCSMG QCVCEPGWTG PSCDCPLSNA TCIDSNGGIC NGRGHCECGR
 601 CHCHQQSLYT DTICEINYSA IHPGLCEDLR SCVQCQAWGT GEKKGRTCEE CNFKVKMVDE
 661 LKRAEEVVVR CSFRDEDDDC TYSYTMEGDG APGPNSTVLV HKKKDCPPGS FWWLIPLLLL
 721 LLPLLALLLL LCWKYCACCK ACLALLPCCN RGHMVGFKED HYMLRENLMA SDHLDTPMLR
 781 SGNLKGRDVV RWKVTNNMQR PGFATHAASI NPTELVPYGL SLRLARLCTE NLLKPDTREC
 841 AQLRQEVEEN LNEVYRQISG VHKLQQTKFR QQPNAGKKQD HTIVDTVLMA PRSAKPALLK
 901 LTEKQVEQRA FHDLKVAPGY YTLTADQDAR GMVEFQEGVE LVDVRVPLFI RPEDDDEKQL
 961 LVEAIDVPAG TATLGRRLVN ITIIKEQARD VVSFEQPEFS VSRGDQVARI PVIRRVLDGG
1021 KSQVSYRTQD GTAQGNRDYI PVEGELLFQP GEAWKELQVK LLELQEVDSL LRGRQVRRFH
1081 VQLSNPKFGA HLGQPHSTTI IIRDPDELDR SFTSQMLSSQ PPPHGDLGAP QNPNAKAAGS
1141 RKIHFNWLPP SGKPMGYRVK YWIQGDSESE AHLLDSKVPS VELTNLYPYC DYEMKVCAYG
1201 AQGEGPYSSL VSCRTHQEVP SEPGRLAFNV VSSTVTQLSW AEPAETNGEI TAYEVCYGLV
1261 NDDNRPIGPM KKVLVDNPKN RMLLIENLRE SQPYRYTVKA RNGAGWGPER EAIINLATQP
1321 KRPMSIPIIP DIPIVDAQSG EDYDSFLMYS DDVLRSPSGS QRPSVSDDTG CGWKFEPLLG
1381 EELDLRRVTW RLPPELIPRL SASSGRSSDA EAPTAPRTTA ARAGRAAAVP RSATPGPPGE
1441 HLVNGRMDFA FPGSTNSLHR MTTTSAAAYG THLSPHVPHR VLSTSSTLTR DYNSLTRSEH
```

```
1501  SHSTTLPRDY  STLTSVSSHG  LPPIWEHGRS  RLPLSWALGS  RSRAQMKGFP  PSRGPRDSII
1561  LAGRPAAPSW  GPDSRLTAGV  PDTPTRLVFS  ALGPTSLRVS  WQEPRCERPL  QGYSVEYQLL
1621  NGGELHRLNI  PNPAQTSVVV  EDLLPNHSYV  FRVRAQSQEG  WGREREGVIT  IESQVHPQSP
1681  LCPLPGSAFT  LSTPSAPGPL  VFTALSPDSL  QLSWERPRRP  NGDIVGYLVT  CEMAQGGGPA
1741  TAFRVDGDSP  ESRLTVPGLS  ENVPYKFKVQ  ARTTEGFGPE  REGIITIESQ  DGGPFPQLGS
1801  RAGLFQHPLQ  SEYSSITTTH  TSATEPFLVD  GPTLGAQHLE  AGGSLTRHVT  QEFVSRTLTT
1861  SGTLSTHMDQ  QFFQT
```

Two splice variants are marked (underlined and in italics).

Database accession

EMBL/GenBank   X51841
SwissProt   P16144

References

[1] Hogervorst, F. et al. (1990) EMBO J. 9, 765–770.
[2] Suzuki, S. and Naitoh, Y. (1990) EMBO J. 9, 757–763.
[3] Tamura, R.N. et al. (1990) J. Cell Biol. 111, 1593–1604.
[4] Kajiji, S. et al. (1989) EMBO J. 8, 673–680.
[5] Hemler, M.E. et al. (1989) J. Biol. Chem. 264, 6529–6535.
[6] Garrod, D.R. (1993) Curr. Opin. Cell Biol. 5, 30–40.
[7] Dowling, J. et al. (1996) J. Cell Biol. 134, 559–572.
[8] Niessen, C.M. et al. (1994) Exp. Cell Res. 211, 360–367.
[9] Lotz, M.M. et al. (1990) Cell Regul. 1, 249–257.
[10] Lee, E.C. et al. (1992) J. Cell Biol. 117, 671–678.
[11] Sonnenberg, A. et al. (1990) J. Cell Sci. 96, 207–217.
[12] Sonnenberg, A. et al. (1991) J. Cell Biol. 113, 907–917.
[13] Shaw, L.M. et al. (1997) Cell 91, 949–960.
[14] Natali, P.G. et al. (1992) J. Cell Sci. 103, 1243–1247.
[15] Kennel, S.J. et al. (1992) J. Cell Sci. 101, 145–150.
[16] Vidal, F. et al. (1995) Nature Genet. 10, 220–234.
[17] Pulkkinen, L. et al. (1998a) Am. J. Hum. Genet. 63, 1376–1387.
[18] Pulkkinen, L. et al. (1998b) Am. J. Pathol. 152, 157–166.
[19] Murgia, C. et al. (1998) EMBO J. 17, 3940–3951.

Integrin β₅

Family

Integrin

Structure

Molecular weights
Amino acids	799
Polypeptide	88 053

SDS-PAGE reduced	100 kDa

Carbohydrate
N-linked sites	8
O-linked sites	

Gene location

Gene structure

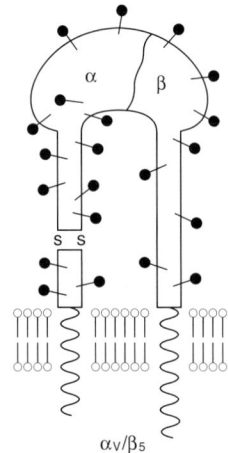

αv/β₅

Alternative forms
Two alternately spliced β₅ transcripts[1] have been described (β₅A and β₅B).

Structure
Integrin β₅ is highly homologous (56%) to the first integrin αᵥ partner discovered, β₃, and forms the αᵥβ₅ dimer[2-5].

Ligands

Vitronectin[2] and penton base of adenovirus[6].

Function

Cell adhesion to, and migration on, vitronectin and other extracellular matrix substrates[4,5]. There is data for a role for αᵥβ₅ in vitronectin endocytosis[7,8]. Recent evidence shows reciprocal involvement of αᵥβ₅ and αᵥβ₃ in angiogenesis in experimental models, with αᵥβ₅ being regulated by vascular factors such as VEGF, whereas αᵥβ₃ is modulated by bFGF[9]

Distribution

αᵥβ₅ is more widely spread in tissues than the related αᵥ integrin, αᵥβ₃. Expression of αᵥβ₅ is regulated in angiogenesis in tumours and at other sites of pathological new vessel formation.

Disease association

OMIM 147561

Knockout

MGI:96614

Amino acid sequence of human integrin β5

```
  1  MPRAPAPLYA  CLLGLCALLP  RLAGLNICTS  GSATSCEECL  LIHPKCAWCS  KEDFGSPRSI
 61  TSRCDLRANL  VKNGCGGEIE  SPASSFHVLR  SLPLSSKGSG  SAGWDVIQMT  PQEIAVNLRP
121  GDKTTFQLQV  RQVEDYPVDL  YYLMDLSLSM  KDDLDNIRSL  GTKLAEEMRK  LTSNFRLGFG
181  SFVDKDISPF  SYTAPRYQTN  PCIGYKLFPN  CVPSFGFRHL  LPLTDRVDSF  NEEVRKQRVS
241  RNRDAPEGGF  DAVLQAAVCK  EKIGWRKDAL  HLLVFTTDDV  PHIALDGKLG  GLVQPHDGQC
301  HLNEANEYTA  SNQMDYPSLA  LLGEKLAENN  INLIFAVTKN  HYMLYKNFTA  LIPGTTVEIL
361  DGDSKNIIQL  IINAYNSIRS  KVELSVWDQP  EDLNLFFTAT  CQDGVSYPGQ  RKCEGLKIGD
421  TASFEVSLEA  RSCPSRHTEH  VFALRPVGFR  DSLEVGVTYN  CTCGCSVGLE  PNSARCNGSG
481  TYVCGLCECS  PGYLGTRCEC  QDGENQSVYQ  NLCREAEGKP  LCSGRGDCSC  NQCSCFESEF
541  GKIYGPFCEC  DNFSCARNKG  VLCSGHGECH  CGECKCHAGY  IGDNCNCSTD  ISTCRGRDGQ
601  ICSERGHCLC  GQCQCTEPGA  FGEMCEKCPT  CPDACSTKRD  CVECLLLHSG  KPDNQTCHSL
661  CRDEVITWVD  TIVKDDQEAV  LCFYKTAKDC  VMMFTYVELP  SGKSNLTVLR  EPECGNTPNA
721  MTILLAVVGS  ILLVGLALLA  IWKLLVTIHD  RREFAKFQSE  RSRARYEMAS  NPLYRKPIST
781  HTVDFTFNKF  NKSYNGTVD
```

Database accession

EMBL/GenBank   X53002
SwissProt   P18084

References

[1] Zhang, J. et al. (1998) Biochem. J. 331, 631–637.

[2] McLean, J.W. et al. (1990) J. Biol. Chem. 265, 17126–17131.

[3] Suzuki, S. et al. (1990) Proc. Natl Acad. Sci. USA 87, 5354–5358.

[4] Ramaswamy, H. and Hemler, M.E. (1990) EMBO J. 9, 1561–1568.

[5] Cheresh, D.A. et al. (1989) Cell 57, 59–69.

[6] Mathias, P. et al. (1998) J. Virol. 72, 8669–8675.

[7] Memmo, L.M. et al. (1998) J. Cell Sci. 111, 425–433.

[8] De Deyne, P.G. et al. (1998) J. Cell Sci. 111, 2729–2740.

[9] Friedlander, M. et al. (1995) Science 270, 1500–1502.

Integrin β₆

Family

Integrin

Structure

Molecular weights

Amino acids	788
Polypeptide	85 975

SDS-PAGE reduced	105 kDa

Carbohydrate

N-linked sites	9
O-linked sites	

Gene location   2q24–q31

Gene structure

Alternative forms

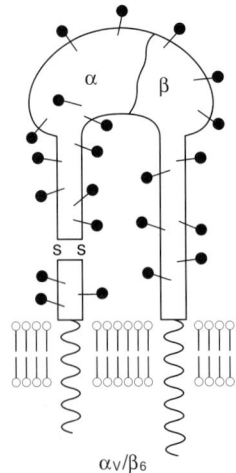

αv/β₆

Structure

Integrin β₆ dimerizes with the integrin αᵥ chain to form αᵥβ₆[1]. It has high degree of homology with other integrin β chains but has a unique 11 amino acid cytoplasmic extension, which is conserved across species[1].

Ligands

αᵥβ₆ binds fibronectin[2–4], tenascin[5] and vitronectin[6]. Recent findings have indicated that TGFβ-associated latency-associated peptide (LAP) is a ligand for αᵥβ₆[7].

Function

αᵥβ₆ is involved in the response of epithelium to inflammatory stimuli and injury, such as in asthma. Regulation of epithelial proliferation *in vitro*[8] and wound healing[9].

Distribution

Epithelial cells, including carcinomas. In view of the phenotype of the knockout mouse (see below) the distribution of αᵥβ₆ in airway epithelium has been extensively studied[5].

Disease association

OMIM 147558

Knockout

MGI:96615

The Gene Knockout FactsBook[10], pp. 628–629.

Mutant mice are normal at birth, but develop alopecia after minor trauma and show airway hyper-responsiveness and 'asthma'. Both are associated with an abnormal inflammatory cell infiltration and subsequent fibrosis[11].

Amino acid sequence of human integrin β₆

```
  1 MGIELLCLFF LFLGRNDSRT RWLCLGGAET CEDCLLIGPQ CAWCAQENFT HPSGVGERCD
 61 TPANLLAKGC QLNFIENPVS QVEILKNKPL SVGRQKNSSD IVQIAPQSLI LKLRPGGAQT
121 LQVHVRQTED YPVDLYYLMD LSASMDDDLN TIKELGSGLS KEMSKLTSNF RLGFGSFVEK
181 PVSPFVKTTP EEIANPCSSI PYFCLPTFGF KHILPLTNDA ERFNEIVKNQ KISANIDTPE
241 GGFDAIMQAA VCKEKIGWRN DSLHLLVFVS DADSHFGMDS KLAGIVIPND GLCHLDSKNE
301 YSMSTVLEYP TIGQLIDKLV QNNVLLIFAV TQEQVHLYEN YAKLIPGATV GLLQKDSGNI
361 LQLIISAYEE LRSEVELEVL GDTEGLNLSF TAICNNGTLF QHQKKCSHMK VGDTASFSVT
421 VNIPHCERRS RHIIIKPVGL GDALELLVSP ECNCDCQKEV EVNSSKCHHG NGSFQCGVCA
481 CHPGHMGPRC ECGEDMLSTD SCKEAPDHPS CSGRGDCYCG QCICHLSPYG NIYGPYCQCD
541 NFSCVRHKGL LCGGNGDCDC GECVCRSGWT GEYCNCTTST DSCVSEDGVL CSGRGDCVCG
601 KCVCTNPGAS GPTCERCPTC GDPCNSKRSC IECHLSAAGQ AGEECVDKCK LAGATISEEE
661 DFSKDGSVSC SLQGENECLI TFLITTDNEG KTIIHSINEK DCPKPPNIPM IMLGVSLATL
721 LIGVVLLCIW KLLVSFHDRK EVAKFEAERS KAKWQTGTNP LYRGSTSTFK NVTYKHREKQ
781 KVDLSTDC
```

Database accession
EMBL/GenBank   M35198
SwissProt   P18564

References
[1] Sheppard, D. et al. (1990) J. Biol. Chem. 265, 11502–11507.
[2] Busk, M. et al. (1992) J. Biol. Chem. 267, 5790–5796.
[3] Chen, J. et al. (1996) Cell Adhes. Commun. 4, 237–250.
[4] Weinacker, A. et al. (1994) J. Biol. Chem. 269, 6940–6948.
[5] Weinacker, A. et al. (1995) Am. J. Respir. Cell Mol. Biol. 12, 547–556.
[6] Huang, X. et al. (1998) J. Cell Sci. 111, 2189–2195.
[7] Munger, J.S. et al. (1999) Cell 96, 319–328.
[8] Agrez, M. et al. (1994) J. Cell Biol. 127, 547–556.
[9] Breuss, J.M. et al. (1995) J. Cell Sci. 108, 241–251.
[10] Mak, T.W. (1998) The Gene Knockout FactsBook. Academic Press, London.
[11] Huang, X. et al. (1996) J. Cell Biol. 133, 921–928.

Integrin β₇ — LPAM-1, β_p

Family

Integrin

Structure

Molecular weights

Amino acids	798
Polypeptide	86 903

SDS-PAGE reduced   110 kDa

Carbohydrate

N-linked sites	8
O-linked sites	

Gene location   12q13.13

Gene structure   10 kb, 14 exons

Alternative forms

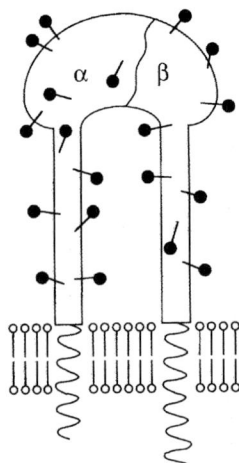

$\alpha 4$(CD49d)/$\beta 7$

Structure

A β_2-integrin-related β chain[1,2] that combines with the integrin α_4 chain (CD49d) to form $\alpha_4\beta_7$ or with α_E (CD103) to form $\alpha_E\beta_7$ (HML-1[3]) (see entries for α_4 and α_E). The cytoplasmic tail contains two potential tyrosine phosphorylation sites. β_7 contains 54 instead of the usual 56 cysteine residues found in most β subunits[1,2].

Ligands

$\alpha_4\beta_7$ binds VCAM-1 (CD106) and the CS1 domain of fibronectin, as does the $\alpha_4\beta_1$ integrin[4,5]; additionally, $\alpha_4\beta_7$ binds MAdCAM-1[6,7]. $\alpha_E\beta_7$ recognizes epithelial E-cadherin (cadherin-1)[8-10].

Function

$\alpha_4\beta_7$ mediates lymphocyte adhesion to MAdCAM-1 on high venular endothelial cells and is involved in lymphocyte homing to intestinal Peyer's patches[6,7]. It is predicted that the interaction of $\alpha_E\beta_7$ with E-cadherin between lymphocytes and epithelium is likewise involved in intestinal targeting of lymphocyte subpopulations[8,11].

Distribution

β_7 was isolated from T lymphocytes. In its $\alpha_4\beta_7$ form it is expressed by mucosal lymphocytes[12], NK cells and eosinophils[13]. In contrast, $\alpha_E\beta_7$ is highly expressed by intraepithelial lymphocytes[12] and only by a small minority of peripheral blood lymphocytes. Endothelial cells[14].

Disease association

OMIM 147559

Knockout

MGI:96616

Amino acid sequence of human integrin β₇

```
  1 MVALPMVLVL LLVLSRGESE LDAKIPSTGD ATEWRNPHLS MLGSCQPAPS CQKCILSHPS
 61 CAWCKQLNFT ASGEAEARRC ARREELLARG CPLEELEEPR GQQEVLQDQP LSQGARGEGA
121 TQLAPQRVRV TLRPGEPQQL QVRFLRAEGY PVDLYYLMDL SYSMKDDLER VRQLGHALLV
181 RLQEVTHSVR IGFGSFVDKT VLPFVSTVPS KLRHPCPTRL ERCQSPFSFH HVLSLTGDAQ
241 AFEREVGRQS VSGNLDSPEG GFDAILQAAL CQEQIGWRNV SRLLVFTSDD TFHTAGDGKL
301 GGIFMPSDGH CHLDSNGLYS RSTEFDYPSV GQVAQALSAA NIQPIFAVTS AALPVYQELS
361 KLIPKSAVGE LSEDSSNVVQ LIMDAYNSLS STVTLEHSSL PPGVHISYES QCEGPEKREG
421 KAEDRGQCNH VRINQTVTFW VSLQATHCLP EPHLLRLRAL GFSEELIVEL HTLCDCNCSD
481 TQPQAPHCSD GQGHLQCGVC SCAPGRLGRL CECSVAELSS PDLESGCRAP NGTGPLCSGK
541 GHCQCGRCSC SGQSSGHLCE CDDASCERHE GILCGGFGRC QCGVCHCHAN RTGRACECSG
601 DMDSCISPEG GLCSGHGRCK CNRCQCLDGY YGALCDQCPG CKTPCERHRD CAECGAFRTG
661 PLATNCSTAC AHTNVTLALA PILDDGWCKE RTLDNQLFFF LVEDDARGTV VLRVRPQEKG
721 ADHTQAIVLG CVGGIVAVGL GLVLAYRLSV EIYDRREYSR FEKEQQQLNW KQDSNPLYKS
781 AITTTINPRF QEADSPTL
```

Database accession

EMBL/GenBank   S80335
SwissProt   P26010

References

[1] Erle, D.J. et al. (1991) J. Biol. Chem. 266, 11009–11016.
[2] Yaun, Q. et al. (1990) Int. Immunol. 2, 1097–1108.
[3] Micklem, K.J. et al. (1991) Am. J. Pathol. 139,1297–1301.
[4] Lobb, R.R. and Hemler, H.E. (1994) J. Clin. Invest. 94, 1722–1728.
[5] Kilger, G. et al. (1995) J. Biol. Chem. 270, 5979–5984.
[6] Berlin, C. et al. (1993) Cell 74, 185–195.
[7] Rott, L.S. et al. (1996) J. Immunol. 156, 3727–3736.
[8] Cepek, K.L. et al. (1994) Nature 372, 190–193.
[9] Higgins, J.M. et al. (1998) J. Cell Biol. 140, 197–210.
[10] Karecla, P. et al. (1996) J. Biol. Chem. 271, 30909–30915.
[11] Cepek, K.L. et al. (1993) J. Immunol. 150, 3459–3470.
[12] Parker, C.M. et al. (1992) Proc. Natl Acad. Sci. USA 89, 1924–1928.
[13] Erle, D.J. et al. (1994) J. Immunol. 153, 517–528.
[14] Brezinschek, R. et al. (1996) Am. J. Pathol. 149, 1651–1660.

Integrin β₈

Family

Integrin

Structure

Molecular weights

Amino acids	769
Polypeptide	85 631

SDS-PAGE reduced   95 kDa

Carbohydrate

N-linked sites	7
O-linked sites	

Gene location

Gene structure

Alternative forms

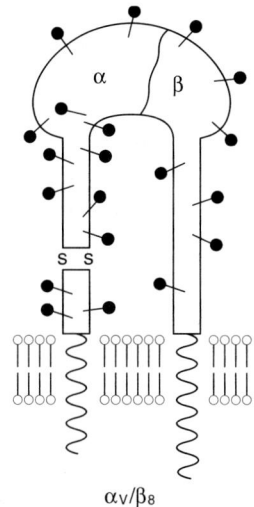

αv/β₈

Structure

Integrin β₈ dimerizes with the integrin α_V chain to form $\alpha_V\beta_8$. The β₈ chain sequence is divergent from other β chains[1] especially in the cytoplasmic tail, which does not interact with cytoskeleton[2].

Ligands

$\alpha_V\beta_8$ binds to vitronectin in addition to fibronectin, laminin and Type IV collagen[2,3].

Function

There is increasing evidence for a role for $\alpha_V\beta_8$ in neural function[4,5].

Distribution

Placenta, kidney, brain, ovary, uterus; transformed cell lines. Early oligodendrocyte precursors express $\alpha_V\beta_8$ at high levels, reducing during differentiation[4]; it is also expressed in synapses and glia of selected areas of the adult brain[5].

Disease association

OMIM   604160

Knockout

MGI:1338035

Amino acid sequence of human integrin β_8

```
  1 MCGSALAFFT AAFVCLQNDR RGPASFLWAA WVFSLVLGLG QGEDNRCASS NAASCARCLA
 61 LGPECGWCVQ EDFISGGSRS ERCDIVSNLI SKGCSVDSIE YPSVHVIIPT ENEINTQVTP
121 GEVSIQLRPG AEANFMLKVH PLKKYPVDLY YLVDVSASMH NNIEKLNSVG NDLSRKMAFF
181 SRDFRLGFGS YVDKTVSPYI SIHPERIHNQ CSDYNLDCMP PHGYIHVLSL TENITEFEKA
241 VHRQKISGNI DTPEGGFDAM LQAAVCESHI GWRKEAKRLL LVMTDQTSHL ALDSKLAGIV
301 VPNDGNCHLK NNVYVKSTTM EHPSLGQLSE KLIDNNINVI FAVQGKQFHW YKDLLPLLPG
361 TIAGEIESKA ANLNNLVVEA YQKLISEVKV QVENQVQGIY FNITAICPDG SRKPGMEGCR
421 NVTSNDEVLF NVTVTMKKCD VTGGKNYAII KPIGFNETAK IHIHRNCSCQ CEDNRGPKGK
481 CVDETFLDSK CFQCDENKCH FDEDQFSSES CKSHKDQPVC SGRGVCVCGK CSCHKIKLGK
541 VYGKYCEKDD FSCPYHHGNL CAGHGECEAG RCQCFSGWEG DRCQCPSAAA QHCVNSKGQV
601 CSGRGTCVCG RCECTDPRSI GRFCEHCPTC YTACKENWNC MQCLHPHNLS QAILDQCKTS
661 CALMEQQHYV DQTSECFSSP SYLRIFFIIF IVTFLIGLLK VLIIRQVILQ WNSNKIKSSS
721 DYRVSASKKD KLILQSVCTR AVTYRREKPE EIKMDISKLN AHETFRCNF
```

Database accession

EMBL/GenBank   M73780
SwissProt   P26012

References

[1] Moyle, M. et al. (1991) J. Biol. Chem. 266, 19650–19658.
[2] Nishimura, S.L. et al. (1994) J. Biol. Chem. 269, 28708–28715.
[3] Venstrom, K. and Reichardt, L. (1995) Mol. Biol. Cell 6, 419–431.
[4] Milner, R. et al. (1997) Glia 21, 350–360.
[5] Nishimura, S.L. et al. (1998) Brain Res. 791, 271–282.

Selectins

E-Selectin

Family

Selectin (Ca^{2+}-dependent, C-type lectin)

Structure

Molecular weights

Amino acids	610
Polypeptide	66 655

SDS-PAGE reduced   97, 107, 115 kDa

Carbohydrate

N-linked sites	11
O-linked sites	

Gene location   1q23–q25

Gene structure   approximately 13 kb, 14 exons

Alternative forms

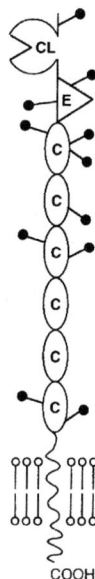

Structure

The extracellular domain of E-selectin consists of an N-terminal C-type lectin domain, an EGF-like domain and six complement control protein domains. The crystal structure of the C-type lectin and EGF domain has been solved[1]. Activation of endothelial cells leads to association of E-selectin with the actin cytoskeleton[2].

Ligands

E-selectin displays calcium-dependent, low-affinity binding to oligosaccharide sequences related to sialylated Lewis x (sLe^x, CD15s), particularly 3-sialyl di-Le^x, via the C-type lectin domain[1,3,4]. Well established ligands are ESL-1 on myeloid cells, PSGL-1 on neutrophils, monocytes and most lymphocytes, and leucocyte L-selectin[2,5,6].

Function

E-selectin mediates the rolling and tethering of leucocytes on activated endothelial surfaces.

Distribution

E-selectin is mainly expressed on activated endothelial cells and on endothelium in chronic inflammatory lesions of the skin and synovium[7]. In addition, E-selectin is detected on non-inflamed endothelia in the skin, placenta and bone marrow. Soluble E-selectin forms are found in plasma[8].

Disease association

OMIM 131210
In addition to its role in leucocyte–endothelial interactions, E-selectin may play a role in tumour metastasis since carcinoma cells have been shown to express E-selectin ligands and bind activated endothelium[9]. Patients with LAD-2 syndrome, who lack the sialyl Lewis x component on their selectins, suffer pyogenic infections (see Knockout).

Knockout

MGI:98278
The Gene Knockout FactsBook[10], p. 935.
E-selectin-deficient mice display no gross abnormalities or overt inflammatory or immunological deficiencies despite circulating leucocytes showing a decreased adhesion to activated endothelia[11-13]. However, mice deficient in both E- and P-selectin have severe defects in leucocyte recruitment into inflammatory sites and are highly susceptible to opportunistic bacterial infections, suggesting overlapping roles for these two family members.

Amino acid sequence of human E-selectin

```
  1 MIASQFLSAL TLVLLIKESG AWSYNTSTEA MTYDEASAYC QQRYTHLVAI QNKEEIEYLN
 61 SILSYSPSYY WIGIRKVNNV WVWVGTQKPL TEEAKNWAPG EPNNRQKDED CVEIYIKREK
121 DVGMWNDERC SKKKLALCYT AACTNTSCSG HGECVETINN YTCKCDPGFS GLKCEQIVNC
181 TALESPEHGS LVCSHPLGNF SYNSSCSISC DRGYLPSSME TMQCMSSGEW SAPIPACNVV
241 ECDAVTNPAN GFVECFQNPG SFPWNTTCTF DCEEGFELMG AQSLQCTSSG NWDNEKPTCK
301 AVTCRAVRQP QNGSVRCSHS PAGEFTFKSS CNFTCEEGFM LQGPAQVECT TQGQWTQQIP
361 VCEAFQCTAL SNPERGYMNC LPSASGSFRY GSSCEFSCEQ GFVLKGSKRL QCGPTGEWDN
421 EKPTCEAVRC DAVHQPPKGL VRCAHSPIGE FTYKSSCAFS CEEGFELHGS TQLECTSQGQ
481 WTEEVPSCQV VKCSSLAVPG KINMSCSGEP VFGTVCKFAC PEGWTLNGSA ARTCGATGHW
541 SGLLPTCEAP TESNIPLVAG LSAAGLSLLT LAPFLLWLRK CLRKAKKFVP ASSCQSLESD
601 GSYQKPSYIL
```

Database accession

EMBL/GenBank   M24736
SwissProt   P16581

References

[1] Graves, B.J. et al. (1994) Nature 367, 532–538.
[2] Yoshida, M. et al. (1996) J. Cell Biol. 133, 445–455.
[3] Varki, A. et al., (1994) Proc. Natl Acad. Sci. USA 91, 7390–7397.
[4] Patel, T.P. et al. (1994) Biochemistry 33, 14815–14824.
[5] McEver, R.P. et al. (1995) J. Biol. Chem. 270, 11025–11028.
[6] Vestweber, D. and Blanks, J.E. (1999) Physiol. Rev. 79, 181–213.
[7] Tedder, T.F. et al. (1995) FASEB J. 9, 866–873.
[8] Gearing, A.J.H. and Newman, W. (1993) Immunol. Today 14, 506–512.

[9] Aruffo, A. et al. (1992) Proc. Natl Acad. Sci. USA 82, 2292–2296.
[10] Mak, T.W. (1998) The Gene Knockout FactsBook. Academic Press, London.
[11] Frenette, P.S. et al. (1996) Cell 84, 563–574.
[12] Bevilacqua, M.P. et al. (1989) Science 243, 1160–1165.
[13] Labow, M.A. et al. (1994) Immunity 1, 709–720.

L-Selectin

Family

Selectin (Ca^{2+}-dependent, C-type lectin)

Structure

Molecular weights

Amino acids	372
Polypeptide	42 187

SDS-PAGE reduced   74 kDa, 90–100 kDa

Carbohydrate

N-linked sites	7
O-linked sites	0

Gene location   1q23–q25

Gene structure   >30 kb, at least 10 exons

Alternative forms

Structure

The extracellular domain consists of an N-terminal lectin C-type domain, an EGF-like domain, two complement control protein domains and a 15 amino acid spacer containing a proteolytic cleavage site to generate a soluble L-selectin[1]. The cytoplasmic domain mediates association with a complex of cytoskeletal proteins including α-actinin[2] and calmodulin[3].

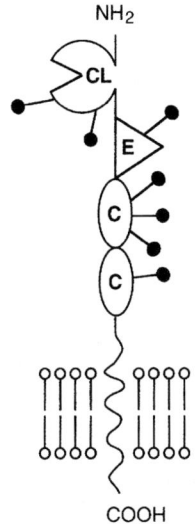

Ligands

L-selectin binds with low affinity to oligosaccharide sequences related to sialylated Lewis x (sLe^x, CD15s) via the C-type lectin domain[4], which are presented on the major ligands CD34, GlyCAM-1, MAdCAM-1 and PSGL-1[5-8]. In addition, L-selectin binds PLCP1[9] unrelated anionic carbohydrates such as heparan sulphate[4].

Function

L-selectin mediates the rolling and tethering of leucocytes on endothelial surfaces, which is a prerequisite for leucocyte adhesion and extravasation[5]. In particular, L-selectin mediates homing of naive lymphocytes via high endothelial venules to peripheral lymph nodes and Peyer's patches. L-selectin also plays a role in recruitment of leucocytes to inflammatory sites[5,10]. In vitro, association of L-selectin with GlyCAM-1 can activate β$_2$ integrins[11].

Distribution

L-selectin is expresesd by most haematopoietic cells at some stage of differentiation[5]. In these cells the localization of L-selectin of the tips of the microvilli is required for optimal adhesion[12]. Activation of leucocytes causes shedding of L-selectin[5].

Disease association

OMIM 153240

Knockout

MGI:98279
The Gene Knockout FactsBook[13], p. 938.
L-selectin-deficient mice are viable and fertile, but exhibit no leucocyte migration across the high endothelial venules of peripheral lymph nodes and a reduced leucocyte emigration into tissues at sites of inflammation[14-16].

Amino acid sequence of human L-selectin

```
  1 MIFPWKCQST QRDLWNIFKL WGWTMLCCDF LAHHGTDCWT YHYSEKPMNW QRARRFCRDN
 61 YTDLVAIQNK AEIEYLEKTL PFSRSYYWIG IRKIGGIWTW VGTNKSLTEE AENWGDGEPN
121 NKKNKEDCVE IYIKRNKDAG KWNDDACHKL KAALCYTASC QPWSCSGHGE CVEIINNYTC
181 NCDVGYYGPQ CQFVIQCEPL EAPELGTMDC THPLGNFSFS SQCAFSCSEG TNLTGIEETT
241 CGPFGNWSSP EPTCQVIQCE PLSAPDLGIM NCSHPLASFS FTSACTFICS EGTELIGKKK
301 TICESSGIWS NPSPICQKLD KSFSMIKEGD YNPLFIPVAV MVTAFSGLAF IIWLARRLKK
361 GKKSKRSMND PY
```

Database accession

EMBL/GenBank   M25280
SwissProt   P14151

References
[1] Chen, A. et al. (1995) J. Exp. Med. 182, 519–530.
[2] Pavalko, F.M. et al. (1995) J. Cell Biol. 129, 1155–1164.
[3] Kahn, J. et al. (1998) Cell 92, 809–818.
[4] Varki, A. (1994) Proc. Natl Acad. Sci. USA 91, 7390–7397.
[5] Tedder, T.F. et al. (1995) FASEB J. 9, 866–873.
[6] Lasky, L.A. (1995) Annu. Rev. Biochem. 64, 113–139.
[7] Rosen, S.D. and Bertozzi, C.R. (1996) Curr. Biol. 6, 261–264.
[8] Walcheck, B. et al. (1996) J. Clin. Invest. 98, 1081–1087.
[9] Sassetti, C. et al. (1998) J. Exp. Med. 187, 1965–1975.
[10] Hemmerich, S. and Rosen, S.D. (1994) Biochemistry 33, 4830–4835.
[11] Hwang, S.T. et al., (1996) J. Exp. Med. 184, 1343–1348.
[12] von Andrian, U.H. et al. (1995) Cell 82, 989–999.
[13] Mak, T.W. (1998) The Gene Knockout FactsBook. Academic Press, London.
[14] Arbones, M.L. et al. (1994) Immunity 1, 247–260.
[15] Xu, J. et al. (1996) J. Exp. Med. 183, 589–598.
[16] Catalina, M.D. et al. (1996) J. Exp. Med. 184, 2341–2351.

P-Selectin

Family

Selectin (Ca^{2+}-dependent, C-type lectin)

Structure

Molecular weights

Amino acids	830
Polypeptide	90 844
SDS-PAGE reduced	140 kDa

Carbohydrate

N-linked sites	12
O-linked sites	

Gene location

1q21–q24

Gene structure

>50 kb, 17 exons

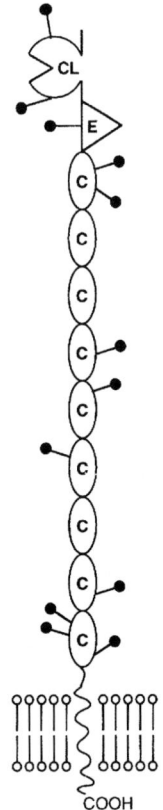

Alternative forms

Alternative splicing results in variants lacking exon 11, which encodes the seventh complement control protein domain, or lacking exon 14, which results in a soluble P-selectin form[1,2].

Structure

The extracellular domain consists of an N-terminal C-type lectin domain, an EGF-like domain and nine complement control protein domains. The cytoplasmic domain is phosphorylated on Ser, Thr, Tyr and His residues following platelet activation[3]. Cys807 in the cytoplasmic domain is acylated with palmitic and stearic acid.

Ligands

P-selectin binds with low affinity to oligosaccharide sequences related to sialylated Lewis x (sLe^x, CD15s) via the C-type lectin domain[4], which are presented on the major P-selectin ligands, PSGL-1[5] and CD24[6].

Function

P-selectin mediates leucocyte and platelet rolling on activated endothelium. Platelet P-selectin promotes leucocyte accumulation in thrombi[7].

Distribution

P-selectin is found on megakaryocytes, activated platelets and activated endothelial cells. P-selectin is stored in secretory granules (Weibel–Palade bodies) and is rapidly transported to the plasma membrane upon activation[2]. Rapid internalization from the plasma membrane attenuates the transient expression.

Disease association

OMIM 173610
Constitutive P-selectin expression in inflammation may contribute to tissue destruction, atherogenesis and thrombosis. Patients with the lifelong bleeding disorder, grey platelet syndrome, have reduced P-selectin expression[8].

Knockout

MGI:98280
The Gene Knockout FactsBook[9], p. 942.
Mice deficient in P-selectin develop normally and are fertile, but show reduced leucocyte rolling and delayed recruitment into inflammatory sites[10]. Mice deficient in both P- and E-selectin have much more severe defects in leucocyte extravasation and a high susceptibility to opportunistic infections[11] suggesting an overlapping role for these family members. Mice deficient in both P- and L-selectin are viable and fertile with a phenotype similar to that observed in the P-selectin −/− mice[12]. Mice deficient in all three selectins are viable and fertile, but with severe leucocytosis and high susceptibility to opportunistic infections, similar to that observed in the P-selectin/E-selectin −/− mice[12].

Amino acid sequence of human P-selectin

```
  1 MANCQIAILY QRFQRVVFGI SQLLCFSALI SELTNQKEVA AWTYHYSTKA YSWNISRKYC
 61 QNRYTDLVAI QNKNEIDYLN KVLPYYSSYY WIGIRKNNKT WTWVGTKKAL TNEAENWADN
121 EPNNKRNNED CVEIYIKSPS APGKWNDEHC LKKKHALCYT ASCQDMSCSK QGECLETIGN
181 YTCSCYPGFY GPECEYVREC GELELPQHVL MNCSHPLGNF SFNSQCSFHC TDGYQVNGPS
241 KLECLASGIW TNKPPQCLAA QCPPLKIPER GNMICLHSAK AFQHQSSCSF SCEEGFALVG
301 PEVVQCTASG VWTAPAPVCK AVQCQHLEAP SEGTMDCVHP LTAFAYGSSC KFECQPGYRV
361 RGLDMLRCID SGHWSAPLPT CEAISCEPLE SPVHGSMDCS PSLRAFQYDT NCSFRCAEGF
421 MLRGADIVRC DNLGQWTAPA PVCQALQCQD LPVPNEARVN CSHPFGAFRY QSVCSFTCNE
481 GLLLVGASVL QCLATGNWNS VPPECQAIPC TPLLSPQNGT MTCVQPLGSS SYKSTCQFIC
541 DEGYSLSGPE RLDCTRSGRW TDSPPMCEAI KCPELFAPEQ GSLDCSDTRG EFNVGSTCHF
601 SCNNGFKLEG PNNVECTTSG RWSATPPTCK GIASLPTPGL QCPALTTPGQ GTMYCRHHPG
661 TFGFNTTCYF GNSTTCYF DSTLSCRPSG QWTAVTPACR AVKCSELHVN KPIAMNCSNL
721 WGNFSYGSIC SFHCLEGQLL NGSAQTACQE NGHWSTTVPT CQAGPLTIQE ALTYFGGAVA
781 STIGLIMGGT LLALLRKRFR QKDDGKCPLN PHSHLGTYGV FTNAAFDPSP
```

Database accession

EMBL/GenBank　　M25322
SwissProt　　　　　P16109

References

1　Johnston, G.I. et al. (1990) J. Biol. Chem. 265, 21381–21385.
2　McEver, R.P. et al. (1995) J. Biol. Chem. 270, 11025–11028.
3　Crovello, C.S. et al. (1995) Cell 82, 279–286.
4　Varki, A. (1994) Proc. Natl Acad. Sci. USA 91, 7390–7397.
5　Rosen, S.D. and Bertozzi, C.R. (1996) Curr. Biol. 6, 261–264.
6　Sammar, M. et al. (1994) Int. Immunol. 6, 1027–1036.
7　Tedder, T.F. et al. (1995) FASEB J. 9, 866–873.
8　Mazurov, A.V. et al. (1996) Eur. J. Haematol. 57, 38–41.
9　Mak, T.W. (1998) The Gene Knockout FactsBook. Academic Press, London.
10　Mayadas, T.N. et al. (1993) Cell 74, 541–554.
11　Frenette, P.S. et al. (1996) Cell 84, 563–574.
12　Robinson, S.D. et al. (1999) Proc. Natl Acad. Sci. USA 96, 11452–11457.

Syndecans

Syndecan-1 CD138

Family

Syndecan

Structure

Molecular weights
Amino acids	310
Polypeptide	32 476

SDS-PAGE reduced	85–95 kDa

Carbohydrate
N-linked sites	1
O-linked sites	probably ++
GAG	5

Gene location   2p24.1

Gene structure   5 exons

Alternative forms

Structure

Syndecan-1, like other members of the syndecan family, is a type I trans-membrane protein with an extracellular domain, which has consensus sequences for the attachment of glycosaminoglycans (GAGs) and O-linked sugars. This extracellular domain can be shed by membrane proximal proteolytic cleavage. The transmembrane and short cytoplasmic domains are highly conserved between the four family members. The cytoplasmic domains of all syndecan family members can be divided into three regions: a membrane proximal constant (C1) region, a variable (V) region, and a carboxyl-terminal constant (C2) region. The C2 region has a consensus sequence for binding proteins with class II PDZ domains and such an association would result in the formation of an adhesive/signalling complex[1-4].

In syndecan-1 the GAG attachment sites occur in two clusters: the three N-terminal sites are usually modified by heparan sulphate, while the membrane proximal sites are usually modified by chondroitin sulphate[5]. The differing molecular weights of syndecan-1 from different tissues suggests that GAG modifications can be cell-type specific[6]. Functional assays indicate that syndecan-1 can associate with the actin cytoskeleton[7], and an association of syndecan-1 with two such proteins, CASK and syntenin, has been demonstrated[8,9].

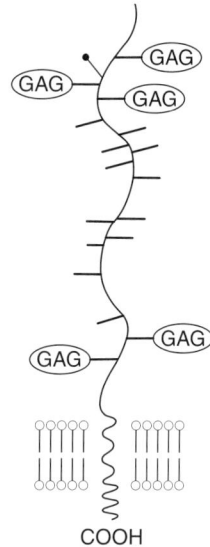

Ligands

Syndecans can bind a wide variety of extracellular components via their heparan sulphate side chains. These include growth factors, extracellular matrix components (fibronectin, laminin, collagens), other adhesion receptors (L-selectin, NCAM, CD31) and enzymes[2].

Function

Syndecans play an important role in modulating the biological activity of heparin binding growth factors. In addition, they can mediate cell–cell and cell–matrix interactions. Binding is strengthened either by clustering of the syndecans and other adhesion receptors, such as integrins, by stabilizing other receptor–ligand interactions and/or the induction of signal transduction cascades[2,4]. *In vitro* studies have demonstrated that loss of syndecan-1 expression results in the loss of a polarized epithelial phenotype and acquisition of anchorage-independent growth[10].

Distribution

Syndecan-1 is predominantly expressed on the basolateral surface of epithelial cells. In addition, expression is found on vascular smooth muscle cells, endothelium, neural cells, pre-B cells, immature B cells and plasma cells[1].

Disease association

OMIM 186355
Syndecan-1 expression is reduced in many epithelial tumours and the amount of loss correlates with tumour progression[11].

Knockout

MGI:1349162

Amino acid sequence of human syndecan-1

```
  1 MRRAALWLWL CALALSLQPA LPQIVATNLP PEDQDGSGDD SDNFSGSGAG ALQDITLSQQ
 61 TPSTWKDTQL LTAIPTSPEP TGLEATAAST STLPAGEGPK EGEAVVLPEV EPGLTAREQE
121 ATPRPRETTQ LPTTHQASTT TATTAQEPAT SHPHRDMQPG HHETSTPAGP SQADLHTPHT
181 EDGGPSATER AAEDGASSQL PAAEGSGEQD FTFETSGENT AVVAVEPDRR NQSPVDQGAT
241 GASQGLLDRK EVLGGVIAGG LVGLIFAVCL VGFMLYRMKK KDEGSYSLEE PKQANGGAYQ
301 KPTKQEEFYA
```

Database accession

EMBL/GenBank   J05392
SwissProt   P18827

References

[1] Bernfield, M. et al. (1992) Annu. Rev. Cell Biol. 8, 365–393.

[2] Carey, D.J. (1997) Biochem. J. 327, 1–16.

[3] Rapraeger, A.C. and Ott, V.L. (1998) Curr. Opin. Cell Biol. 10, 620–628.

[4] Woods, A. and Couchman, J.R. (1998) Trends Cell Biol. 8, 189–192.

[5] Kokenyesi, R. and Bernfield, M. (1994) J. Biol. Chem. 269, 12304–12309.

[6] Kato, M. et al. (1994) J. Biol. Chem. 269, 18881–18890.

[7] Carey, D.J. et al. (1994) J. Cell Biol. 124, 161–170.

[8] Grootjans, J.J. et al. (1997) Proc. Natl Acad. Sci. USA 94, 13683–13688.

[9] Cohen, A.R. et al. (1998) J. Cell Biol. 142, 129–138.

[10] Kato, M. et al. (1995) Mol. Biol. Cell 6, 559–576.

[11] Inki, P. and Jalkanen, M. (1996) Ann. Med. 28, 63–67.

Syndecan-2

Fibroglycan, heparan sulphate proteoglycan, HSPG

Family

Syndecan

Structure

Molecular weights

Amino acids	201
Polypeptide	22 174

SDS-PAGE reduced   46 kDa

Carbohydrate

N-linked sites	0
O-linked sites	probably +
GAG	3

Gene location   8q22–8q24

Gene structure   5 exons

Alternative forms

Structure

Syndecan-2, like other members of the syndecan family, is a type I transmembrane protein with an extracellular domain, which has consensus sequences for the attachment of GAGs and O-linked sugars. This extracellular domain can be shed by membrane proximal proteolytic cleavage. The transmembrane and short cytoplasmic domains are highly conserved between the four family members. The cytoplasmic domains of all syndecan family members can be divided into three regions: a membrane proximal constant (C1) region, a variable (V) region, and a carboxyl-terminal constant (C2) region. The C2 region has a consensus sequence for binding proteins with class II PDZ domains and such an association would result in the formation of an adhesive/signalling complex[1-4].

In syndecan-2 the heparan sulphate attachment sites are clustered at the N-terminus. Within the cytoplasmic domain, an association of syndecan-2 with the PDZ domain containing proteins CASK and syntenin[5-7] and with ezrin[8] has been demonstrated. In addition, syndecan-2 contains a unique serine phosphorylation site, which is a target for protein kinase C[9].

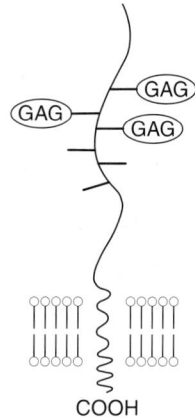

Ligands

Syndecans can bind a wide variety of extracellular components via their heparan sulphate side chains. These include growth factors, extracellular matrix components (fibronectin, laminin, collagens), other adhesion receptors (L-selectin, NCAM, CD31) and enzymes[2].

Function

Syndecans play an important role in modulating the biological activity of heparin binding growth factors. In addition, they can mediate cell–cell and cell–matrix interactions. Binding is strengthened either by clustering of the syndecans and other adhesion receptors, such as integrins, by stabilizing other receptor–ligand interactions and/or the induction of signal transduction cascades[2,4]. Syndecan-2 plays a critical role in the development of dendritic spines whose morphological changes provide a structural basis for learning and memory[10].

Distribution

In adults, syndecan-2 is the major syndecan in fibroblasts. During development syndecan-2 is detected predominantly on mesenchymal cells that represent the hard and connective tissue precursor cells, such as those in prechondrogenic and preosteogenic mesenchymal condensation areas[11]. Syndecan-2 is also induced on primary human monocytes in response to differentiation/activation[12].

Disease association

OMIM 142460

Knockout

MGI:1349165

Amino acid sequence of human syndecan-2

```
  1 MRRAWILLTL GLVACVSAES RAELTSDKDM YLDNSSIEEA SGVYPIDDDD YASASGSGAD
 61 EDVESPELTT TRPLPKILLT SAAPKVETTT LNIQNKIPAQ TKSPEETDKE KVHLSDSERK
121 MDPAEEDTNV YTEKHSDSLF KRTEVLAAVI AGGVIGFLFA IFLILLLVYR MRKKDEGSYD
181 LGERKPSSAA YQKAPTKEFY A
```

Data base accession

EMBL/GenBank   J04621
SwissProt   P34741

References

1 Bernfield, M. et al. (1992) Annu. Rev. Cell Biol. 8, 365–393.
2 Carey, D.J. (1997) Biochem. J. 327, 1–16.
3 Rapraeger, A.C. and Ott, V.L. (1998) Curr. Opin. Cell Biol. 10, 620–628.
4 Woods, A. and Couchman, J.R. (1998) Trends Cell Biol. 8, 189–192.
5 Grootjans, J.J. et al. (1997) Proc. Natl Acad. Sci. USA 94, 13683–13688.
6 Cohen, A.R. et al. (1998) J. Cell Biol. 142, 129–138.
7 Hsueh, Y.P. et al. (1998) J. Cell Biol. 142, 139–151.
8 Granes, F. et al. (2000) J. Cell Sci. 113, 1267–1276.
9 Itano, N.K. et al. (1996) Biochem. J. 315, 925–930.
10 Ethell, I. and Yamaguchi, Y. (1999) J. Cell Biol. 144, 575–586.
11 David, G. et al. (1993) Development 119, 841–854.
12 Clasper, S. (1999) J. Biol. Chem. 274, 24113–24123.

Syndecan-3
N-Syndecan

Family

Syndecan

Structure

Molecular weights

Amino acids	442
Polypeptide	45919

SDS-PAGE reduced   100–120 kDa

Carbohydrate

N-linked sites	0
O-linked sites	probably ++
GAG	6

Gene location   Chromosome 1

Gene structure   5 exons

Alternative forms

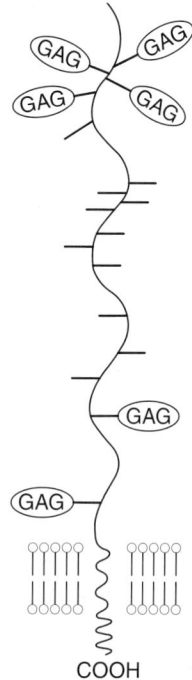

Structure

Syndecan-3, like other members of the syndecan family, is a type I transmembrane protein with an extracellular domain, which has consensus sequences for the attachment of GAGs and O-linked sugars. This extracellular domain can be shed by membrane proximal proteolytic cleavage. The transmembrane and short cytoplasmic domains are highly conserved between the four family members. The cytoplasmic domains of all syndecan family members can be divided into three regions: a membrane proximal constant (C1) region, a variable (V) region, and a carboxyl-terminal constant (C2) region. The C2 region has a consensus sequence for binding proteins with class II PDZ domains and such an association would result in the formation of an adhesive/signalling complex[1-4].

Ligands

Syndecans can bind a wide variety of extracellular components via their heparan sulphate side chains. These include growth factors, extracellular matrix components (fibronectin, laminin, collagens), other adhesion receptors (L-selectin, NCAM, CD31) and enzymes[2].

Function

Syndecans play an important role in modulating the biological activity of heparin binding growth factors. In addition, they can mediate cell–cell and cell–matrix interactions. Binding is strengthened either by clustering of the syndecans and other adhesion receptors, such as integrins, by stabilizing other receptor–ligand interactions and/or the induction of signal transduction cascades[4,5]. A variety of studies have indicate that syndecan-3 is involved in aspects of limb morphogenesis, skeletal development and skeletal muscle differentiation[6,7].

Distribution

In adults, syndecan-3 is the major syndecan in neuronal cells. During development, high levels of syndecan-3 are found in neonatal brain, heart and Schwann cells. Expression in the brain rises rapidly at birth, peaks on postnatal day 7 and then declines in the adult nervous system[2]. Syndecan-3 is also expressed transiently during embryonic limb development[8] and by proliferating immature chondrocytes.

Disease association

OMIM 186357

Knockout

MGI:1349163

Amino acid sequence of rat syndecan-3

```
  1 MKPGPPRRGT AQGQRVDTAT HGPGARGLLL PPLLLLLLAG RAAGAQRWRN ENFERPVDLE
 61 GSGDDDSFPD DELDDLYSGS GSGYFEQESG LETAMRFIPD IALAAPTAPA MLPTTVIQPV
121 DTPFEELLSE HPGPEPVTSP PLVTEVTEVV EEPSQRATTI STTTSTTAAT TTGAPTMATA
181 PATAATTAPS TPAAPPATAT TADIRTTGIQ GLLPLPLTTA ATAKATTPAV PSPPTTVTTL
241 DTEAPTPRLV NTATSRPRAL PRPVTTQEPE VAERSTLPLG TTAPGPTEVA QTPTPESLLT
301 TTQDEPEVPV SGGPSGDFEL QEETTQPDTA NEVVAVEGAA AKPSPPLGTL PKGARPGLGL
361 HDNAIDSGSS AAQLPQKSIL ERKEVLVAVI VGGVVGALFA AFLVTLLIYR MKKKDEGSYT
421 LEEPKQASVT YQKPDKQEEF YA
```

Database accession

EMBL/GenBank   U52825
SwissProt   P33671

References

[1] Bernfield, M. et al. (1992) Annu. Rev. Cell Biol. 8, 365–393.

[2] Carey, D.J. et al. (1997) J. Biol. Chem. 272, 2873–2879.

[3] Rapraeger, A.C. and Ott, V.L. (1998) Curr. Opin. Cell Biol. 10, 620–628.

[4] Woods, A. and Couchman, J.R. (1998) Trends Cell Biol. 8, 189–192.

[5] Carey, D.J. (1997) Biochem. J. 327, 1–16.

[6] Kosher, R.A. (1998) Microsc. Res. Techn. 43, 123–130.

[7] Fuentealba, L. et al. (1999) J. Biol. Chem. 274, 37876–37884.

[8] Gould, S.E. et al. (1995) Dev. Biol. 168, 438–451.

Syndecan-4

Family

Syndecan

Structure

Molecular weights

Amino acids	198
Polypeptide	21 641

SDS-PAGE reduced	35 kDa

Carbohydrate

N-linked sites	0
O-linked sites	probably ++
GAG	4

Gene location   20q12–20q13

Gene structure   5 exons

Alternative forms

Structure

Syndecan-4, like other members of the syndecan family, is a type I transmembrane protein with an extracellular domain, which has consensus sequences for the attachment of GAGs and O-linked sugars. This extracellular domain can be shed by membrane proximal proteolytic cleavage. The transmembrane and short cytoplasmic domains are highly conserved between the four family members. The cytoplasmic domains of all syndecan family members can be divided into three regions: a membrane proximal constant (C1) region, a variable (V) region, and a carboxyl-terminal constant (C2) region. The C2 region has a consensus sequence for binding proteins with class II PDZ domains and such an association would result in the formation of an adhesive/signalling complex[1-4].

In syndecan-4 the heparan sulphate attachment sites are clustered at the N-terminus. Unlike the other family members, the V region in the syndecan-4 cytoplasmic domain can bind activated protein kinase Cα (PKCα) and PIP$_2$[5,6], and, unusually, this binding is via the PKCα catalytic domain[7].The C2 region of the cytoplasmic domain can bind the PDZ-containing protein, syntenin[8,9].

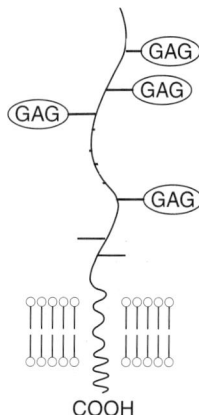

Ligands

Syndecans can bind a wide variety of extracellular components via their heparan sulphate side chains. These include growth factors, extracellular matrix components (fibronectin, laminin, collagens), other adhesion receptors (L-selectin, NCAM, CD31) and enzymes[2].

Function

Syndecans play an important role in modulating the biological activity of heparin binding growth factors. In addition, they can mediate cell–cell and cell–matrix interactions. Binding is strengthened either by clustering of the syndecans and other adhesion receptors, such as integrins, by stabilizing other receptor–ligand interactions and/or the induction of signal transduction cascades[2,4,9].

A number of studies have indicated that syndecan-4 may have a unique function among the syndecan family in regulating focal adhesion formation and hence integrin-based cell adhesion processes[4,9]. However, in recent studies it has been demonstrated that syndecan-4 –/– fibroblasts derived from nockout mice (see below) can form focal adhesions with terminating actin fibres when plated on fibronectin. Defects in focal adhesion formation were only detected when the heparin-binding domain of fibronectin was added in a soluble form[10].

Distribution

Syndecan-4 is usually present in lower amounts than the other syndecans but it is more widespread in its distribution, being found on many epithelial and fibroblastic cells. Syndecan-4 is unique among the family members as it is localized to focal adhesions[11] and this localization is dependent on PKC activation[12].

Disease association

OMIM 600017

Knockout

MGI:1349164

No gross abnormalities are detected in syndecan-4 –/– mice or in syndecan-4 -/- fibroblasts plated onto fibronectin[10].

Amino acid sequence of human syndecan-4

```
  1 MAPARLFALL LLFVGGVAES IRETEVIDPQ DLLEGRYFSG ALPDDEDVVG PGQESDDFEL
 61 SGSGDLDDLE DSMIGPEVVH PLVPLDNHIP ERAGSGSQVP TEPKKLEENE VIPKRISPVE
121 ESEDVSNKVS MSSTVQGSNI FERTEVLAAL IVGGIVGILF AVFLILLLMY RMKKKDEGSY
181 DLGKKPIYKK APTNEFYA
```

Data base accession

EMBL/GenBank   X67016
SwissProt   P31431

References

1 Bernfield, M. et al. (1992) Annu. Rev. Cell Biol. 8, 365–393.
2 Carey, D.J. (1997) Biochem. J. 327, 1–16.
3 Rapraeger, A.C. and Ott, V.L. (1998) Curr. Opin. Cell Biol. 10, 620–628.
4 Woods, A. and Couchman, J.R. (1998) Trends Cell Biol. 8, 189–192.
5 Oh, E.S. et al. (1998) J. Biol. Chem. 273, 10624–10629.
6 Lee, D. et al. (1998) J. Biol. Chem. 273, 13022–13029.
7 Oh, E.S. et al. (1997) J. Biol. Chem. 272, 8133–8136.
8 Grootjans, J.J. et al. (1997) Proc. Natl Acad. Sci. USA 94, 13683–13688.
9 Couchman, J.R. and Woods, A. (1999) J. Cell Sci. 112, 3415–3420.
10 Ishiguro, K. et al. (2000) J. Biol. Chem. 275, 5249–5252.
11 Woods, A. and Couchman, J.R. (1994) Mol. Biol. Cell 5, 183–192.
12 Baciu, P.C. and Goetinck, P.F. (1995) Mol. Biol. Cell 6, 1503–1513.

Other Molecules

AMOG
Adhesion molecule on glia, Na$^+$,K$^+$-ATPase β2 polypeptide

Family
ATPase

Structure

Molecular weights

Amino acids	290
Polypeptide	33 344

SDS-PAGE reduced   45–50 kDa

Carbohydrate

N-linked sites	7
O-linked sites	

Gene location   17p (human), chromosome 11 (mouse)

Gene structure   7 kb

Alternative forms

COOH — β2   NH$_2$ 1   10   NH$_2$ COOH   α$_2$

Structure

AMOG is a type 2 integral membrane glycoprotein of glial cells with homology to the non-catalytic β_2 subunit of Na$^+$,K$^+$-ATPase associated with the α subunit in the active enzyme[1-3]; part of a multigene enzyme family with tissue-specific functions.

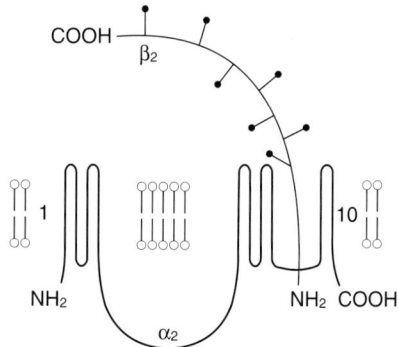

Ligands
Unknown.

Function

Non-catalytic component of plasma membrane ATP-dependent Na$^+$, K$^+$ exchanger. Involved in calcium-independent cell adhesion between neurons and glia, and migration of neurons and astrocytes[4,5].

Distribution

Developmentally regulated expression[6]. Central nervous system neurons and astrocytes[7].

Disease association
OMIM 182331

Knockout

Motor incoordination and paralysis with death by 18 days post-natal[8].

Amino acid sequence of mouse AMOG

```
  1 MVIQKEKKSC GQVVEEWKEF VWNPRTHQFM GRTGTSWAFI LLFYLVFYGF LTAMFSLTMW
 61 VMLQTVSDHT PKYQDRLATP GLMIRPKTEN LDVIVNISDT ESWGQHVQKL NKFLEPYNDS
121 IQAQKNDVCR PGRYYEQPDN GVLNYPKRAC QFNRTQLGDC SGIGDPTHYG YSTGQPCVFI
181 KMNRVINFYA GANQSMNVTC VGKRDEDAEN LGHFVMFPAN GSIDLMYFPY YGKKFHVNYT
241 QPLVAVKFLN VTPNVEVNVE CRINAANIAT DDERDKFAGR VAFKLRINKT
```

Database accession

EMBL/GenBank   X16645 (mouse)
SwissProt   P14231 (mouse)

References

1 Pagliusi, S. et al. (1989) J. Neurosci. Res. 22, 113–119.
2 Martin-Vasallo, P. et al. (1989) J. Biol. Chem. 164, 4613–4618.
3 Gloor, S.M. et al. (1990) J. Cell Biol. 110, 165–174.
4 Antonicek, H. et al. (1987) J. Cell Biol. 104, 1587–1595.
5 Muller-Husmann, G. et al. (1993) J. Biol. Chem. 268, 26260–26267.
6 Lecuona, E. et al. (1996) Brain Res. Bull. 40, 167–174.
7 Pagliusi, S.R. et al. (1990) Eur. J. Neurosci. 2, 471–580.
8 Magyar, J.P. et al. (1994) J. Cell Biol. 127, 835–845.

CD6

T12, Tp120

Family

Scavenger receptor cysteine-rich group B

Structure

Molecular weights

Amino acids	668
Polypeptide	71 783

SDS-PAGE reduced	105–130 kDa

Carbohydrate

N-linked sites	8
O-linked sites	+

Gene location	11q13

Gene structure	> 25 kb, ≥ 13 exons

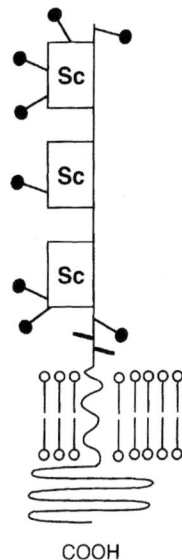

COOH

Alternative forms

The large cytoplasmic domain can be alternatively spliced[1-3].

Structure

The extensively glycosylated CD6 extracellular domain contains three scavenger receptor cysteine-rich domains and an O-glycosylated membrane proximal stalk. The third scavenger receptor domain contains the ALCAM (CD166) binding site[4-6]. The large cytoplasmic domain contains multiple SH2 and SH3 binding sites.

Ligands

CD6 is the counter-receptor for ALCAM (CD166)[7].

Function

Association of CD6 with ALCAM (CD166) expressed in the thymus may play a role in T cell maturation and regulation of T cell activation[8]. In addition, the CD6 cytoplasmic domain becomes phosphorylated following T cell receptor stimulation, suggesting a signalling role for CD6[9].

Distribution

CD6 is expressed on the majority of peripheral blood T cells and a subset of B cells. Expression on immature thymocytes is low but increases with maturation. CD6 expression has also been detected in the brain[2].

Disease association

OMIM 186720
CD6 may be involved in graft versus host disease. In bone marrow transplants, CD6+ T cells are sorted out to reduce graft versus host disease. Expression is found in chronic lymphocytic leukaemia B cells.

Knockout

MGI:103566

Amino acid sequence of human CD6

```
  1 MWLFFGITGL LTAALSGHPS PAPPDQLNTS SAESELWEPG ERLPVRLTNG SSSCSGTVEV
 61 RLEASWEPAC GALWDSRAAE AVCRALGCGG AEAASQLAPP TPELPPPPAA GNTSVAANAT
121 LAGAPALLCS GAEWRLCEVV EHACRSDGRR ARVTCAENRA LRLVDGGGAC AGRVEMLEHG
181 EWGSVCDDTW DLEDAHVVCR QLGCGWAVQA LPGLHFTPGR GPIHRDQVNC SGAEAYLWDC
241 PGLPGQHYCG HKEDAGVVCS EHQSWRLTGG ADRCEGQVEV HFRGVWNTVC DSEWYPSEAK
301 VLCQSLGCGT AVERPKGLPH SLSGRMYYSC NGEELTLSNC SWRFNNSNLC SQSLAARVLC
361 SASRSLHNLS TPEVPASVQT VTIESSVTVK IENKESRELM LLIPSIVLGI LLLGSLIFIA
421 FILLRIKGKY ALPVMVNHQH LPTTIPAGSN SYQPVPITIP KEVFMLPIQV QAPPPEDSDS
481 GSDSDYEHYD FSAQPPVALT TFYNSQRHRV TDEEVQQSRF QMPPLEEGLE ELHASHIPTA
541 NPGHCITDPP SLGPQYHPRS NSESSTSSGE DYCNSPKSKL PPWNPQVFSS ERSSFLEQPP
601 NLELAGTQPA FSAGPPADDS SSTSSGEWYQ NFQPPPQPPS EEQFGCPGSP SPQPDSTDND
661 DYDDISAA
```

Database accession

EMBL/GenBank   U34623
SwissProt   P30203

References

[1] Robinson, W.H. et al. (1995) Eur. J. Immunol. 25, 2765–2769.
[2] Aruffo, A. et al. (1991) J. Exp. Med. 174, 949–952.
[3] Bowen, M.A. et al. (1997) J. Immunol. 158, 949–952.
[4] Bowen, M.A. et al. (1996) J. Biol. Chem. 271, 17390–17396.
[5] Bodian, D.L. et al. (1997) Biochemistry 36, 2637–2641.
[6] Skonier, J.E. et al. (1997) Protein Eng. 10, 943–947.
[7] Aruffo, A. et al. (1997) Immunol. Today 18, 498–504.
[8] Osorio, L.M. (1998) Immunology 93, 358–365.
[9] Kobarg, J. et al. (1997) Eur. J. Immunol. 27, 2971–2980.

CD23

FcεRII, Blast-2

Family

Calcium-dependent C-type lectin

Structure

Molecular weights

Amino acids	321
Polypeptide	36 468

SDS-PAGE reduced   45 kDa (45 kDa unreduced)

Carbohydrate

N-linked sites	1
O-linked sites	Probably +

Gene location   19p13.3

Gene structure   13 kb, 11 exons

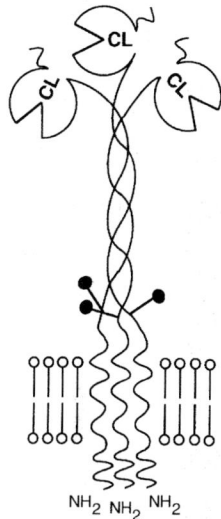

Alternative forms

Multiple soluble forms of molecular weight 37, 33 and 17 kDa produced by proteolytic cleavage[1]; these retain their lectin domains intact. Two further forms exist as a result of alternate RNA splicing resulting in differing cytoplasmic tail sequences in the first nine amino acids (due to 5'-untranslated region splicing)[2].

Structure

CD23 is a type II membrane glycoprotein that exists as a homotrimer[1,3,4]. The extracellular C-terminal part of CD23 contains one C-type lectin domain[5] that mediates ligand interaction. Three membrane proximal repeats of 21 amino acids form a stalk region that is involved in non-covalent trimerization.

Ligands

Low-affinity receptor for IgE[1]. CD23 is also a counter-receptor for the integrins $\alpha_M\beta_1$ and $\alpha_X\beta_1$ but not $\alpha_L\beta_1$[6-8]; this occurs in part via carbohydrate–lectin interaction. CD23 also binds CD21 (complement receptor CR2, C3d receptor)[9].

Function

It has a major role in the negative feedback regulation of IgE synthesis[10], and B lymphocyte proliferation, differentiation and survival; upregulated by IL-4 and many other cytokines (see ref.11 for detailed discussion). The interaction between CD23 and integrins is thought to play a role in monocyte activation[6,8].

239

Distribution

Widespread expression by the majority of leucocytes, including B lymphocytes[12], monocytes, tissue macrophages and follicular dendritic cells, and more weakly on others[11]. The CD23a splice form is restricted to resting B cells.

Disease association

OMIM 151445
Surface expression and plasma levels of CD23 are a prognostic marker for reduced survival in chronic lymphocytic leukaemia[13,14].

Knockout

The Gene Knockout FactsBook[15], p156
CD23 knockout mice have low circulating levels of IgE and reduced IgE-mediated immune responses[16,17].

Amino acid sequence of human CD23

```
  1 MEEGQYSEIE ELPRRRCCRR GTQIVLLGLV TAALWAGLLT LLLLWHWDTT QSLKQLEERA
 61 ARNVSQVSKN LESHHGDQMA QKSQSTQISQ ELEELRAEQQ RLKSQDLELS WNLNGLQADL
121 SSFKSQELNE RNEASDLLER LREEVTKLRM ELQVSSGFVC NTCPEKWINF QRKCYYFGKG
181 TKQWVHARYA CDDMEGQLVS IHSPEEQDFL TKHASHTGSW IGLRNLDLKG EFIWVDGSHV
241 DYSNWAPGEP TSRSQGEDCV MMRGSGRWND AFCDRKLGAW VCDRLATCTP PASEGSAESM
301 GPDSRPDPDG RLPTPSAPLH S
```

Database accession
EMBL/GenBank   M15059
SwissProt   P06734

References

  [1] Sutton, B.J. and Gould, H.J. (1993) Nature 366, 421–428.
  [2] Stengelin, S. et al. (1988) EMBO J. 7, 1053–1059.
  [3] Kikutani, H. et al. (1986) Cell 47, 657–665.
  [4] Ludin, C. et al. (1987) EMBO J. 6, 109–114.
  [5] Ikuta, K. et al. (1987) Proc. Natl Acad. Sci. USA 84, 819–823.
  [6] Dugas, B. et al. (1995) Immunol. Today, 16, 574–580.
  [7] Lecoanet-Henchoz, S. et al. (1995) Immunity 3, 119–125.
  [8] Bonnefoy, J.-Y. et al. (1997) Int. Rev. Immunol. 16, 113–128.
  [9] Aubry, J.-P. et al. (1992) Nature 358, 505–507.
  [10] Yu, P. et al. (1994) Nature 369, 753–756.
  [11] Fridman, W.H. (ed.) (1989) Chemical Immunology, 47, 21–78.
  [12] Sarfati, M. et al. (1995) Leucocyte Typing V, 530–553.
  [13] Keating, M.J. (1999) Semin. Oncol. 26, 107–114.
  [14] D'Arena, G. et al. (2000) Leuk. Lymphoma 36, 225–237.
  [15] Mak, T.W. (1998) The Gene Knockout FactsBook. Academic Press, London.
  [16] Lamers, M.C. and Yu, P. (1995) Immunol. Rev. 148, 71–95.
  [17] Fujiwara, H. et al. (1994) Proc. Natl Acad. Sci. USA 91, 6835–6839.

CD34

Sgp90

Family

Sialomucin

Structure

Molecular weights

Amino acids	385
Polypeptide	40 716

SDS-PAGE reduced   90–120 kDa

Carbohydrate

N-linked sites	9
O-linked sites	+++

Gene location   1q32

Gene structure   ~26 kb, 8 exons

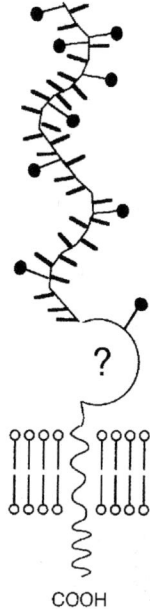

COOH

Alternative forms

Alternative splicing results in an isoform with a truncated 16 amino acid cytoplasmic domain[1,2].

Structure

The N-terminal portion of the CD34 extracellular domain is extensively N- and O-glycosylated. In the membrane proximal region there are six cysteine residues for potential disulphide bond formation. The 73 amino acid full-length cytoplasmic domain becomes serine phosphorylated in response to protein kinase C activation[3].

Ligands

L-selectin binds to sialoglycoconjugates present on CD34; this binding requires sulphation and probably fucosylation[4,5]. Variable CD34 glycosylation results in L-selectin binding to CD34 from high endothelial venules but not to CD34 on other vascular or haematopoietic cells. CD34 can also act as an E-selectin ligand[6].

Function

The ability of endothelial CD34 to mediate attachment and rolling of leucocytes *in vitro* suggests a role in leucocyte trafficking[6]. The similar structure and expression pattern of PCLP (podocalyxin-like protein) and

CD34 suggests that they may have redundant or overlapping physiological roles, and that PCLP may compensate in the CD34 −/− mice, which show no defects in lymphocyte recruitment to secondary sites[7]. Experiments in which CD34 has been ectopically expressed also suggest a function in haematopoietic cells in promoting adhesion to bone marrow stroma[8] and initiation of signal transduction pathways[9]. Like all mucins, CD34 may also have a cytoprotective and/or antiadhesive function.

Distribution

CD34 is expressed by haematopoietic stem and progenitor cells, bone marrow stromal cells, a subset of endothelium and some nervous tissue.

Disease association

OMIM 142230
As discussed by Sutherland et al.[10], there are potential clinical applications in using CD34 to select for haematopoietic stem cell progenitors from tumour contaminated bone marrow in preparation for transplantation and/or gene manipulation in gene therapy.

Knockout

MGI:88329
The Gene Knockout FactsBook[11], p168
CD34 −/− animals develop normally and display no abnormalities in leucocyte trafficking. In one knockout strain, there is evidence for a decrease in number and in colony-forming activity of haematopoietic progenitor cells[12]. The other strain[13] displayed no abnormalities in haematopoiesis but defects in eosinophil trafficking in the lung.

Amino acid sequence of human CD34

```
  1 MLVRRGARAG PRMPRGWTAL CLLSLLPSGF MSLDNNGTAT PELPTQGTFS NVSTNVSYQE
 61 TTTPSTLGST SLHPVSQHGN EATTNITETT VKFTSTSVIT SVYGNTNSSV QSQTSVISTV
121 FTTPANVSTP ETTLKPSLSP GNVSDLSTTS TSLATSPTKP YTSSSPILSD IKAEIKCSGI
181 REVKLTQGIC LEQNKTSSCA EFKKDRGEGL ARVLCGEEQA DADAGAQVCS LLLAQSEVRP
241 QCLLLVLANR TEISSKLQLM KKHQSDLKKL GILDFTEQDV ASHQSYSQKT LIALVTSGAL
301 LAVLGITGYF LMNRRSWSPT GERLGEDPYY TENGGGQGYS SGPGTSPEAQ GKASVNRGAQ
361 ENGTGQATSR NGHSARQHVV ADTEL
```

Database accession
EMBL/GenBank   M81945
SwissProt   P28906

References
1 Satterthwaite, A.B. et al. (1992) Genomics 12, 788–794.
2 Nakamura, Y. et al. (1993) Exp. Hematol. 21, 236–242.
3 Fackler, M.J. et al. (1990) J. Biol. Chem. 265, 11056–11061.
4 Baumhueter, S. et al. (1993) Science 262, 436–438.

5 Rosen, S. and Bertozzi, C.R. (1996) Curr. Biol. 6, 261–264.

6 Puri, K.D. et al. (1995) J. Cell Biol. 131 261–270.

7 Sassetti, C. et al. (1998) J. Exp. Med. 187, 1965–1975.

8 Healy, L. et al. (1995) Proc. Natl Acad. Sci. USA 92, 12240–12244.

9 Tada, J. et al. (1999) Blood 93, 3723–3735.

10 Sutherland, D.R. et al. (1993) Stem Cells 11 (Suppl. 3), 50–57.

11 Mak, T.W. (1998) The Gene Knockout FactsBook. Academic Press, London.

12 Cheng, J. et al. (1996) Blood 87, 479–490.

13 Suzuki, A. et al. (1996) Blood 87, 3550–3562.

CD36

gpIV or gpIIIb, PAS IV, milk fat globule membrane protein

Family

CD36 family

Structure

Molecular weights

Amino acids	471
Polypeptide	52 922

SDS-PAGE reduced   80–90 kDa (80–90 kDa unreduced)

Carbohydrate

N-linked sites	10
O-linked sites	+

Gene location   7q11.2

Gene structure   32 kb, 15 exons

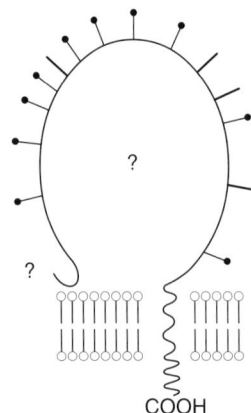

Alternative forms

Two alternately spliced variants, and a short 57 kDa form due to exon skipping[1].

Structure

CD36 is a type I transmembrane heavily glycosylated protein; it is possible that the N-terminus also forms a membrane-spanning domain and/or that the extracellular uncleaved hydrophobic signal sequence interacts with the external face of the plasma membrane[2,3]. The two putative cytoplasmic tails are palmitoylated[4] and the two membrane-spanning model may be incorrect[5]. CD36 is a member of a superfamily that includes a further lipoprotein receptor, SR-BI/CLA-1[4,6]. Platelet glycoprotein IV (gpIV or gpIIIb) and leucocyte CD36 are immunologically related. Threonine 92 is phosphorylated and may be involved with CD36 interaction with intracellular src family protein tyrosine kinase signalling pathways

Ligands

Several extracellular matrix ligands for CD36 have been identified including collagen Types I, IV and V, and thrombospondin[7-10]. Acts as a macrophage scavenger receptor for oxidized low-density lipoproteins[11,12], various lipids and long-chain fatty acids[13,14], and apoptotic neutrophils via interaction with thrombospondin (the latter role in association with integrin $\alpha_V\beta_3$)[15,16].

Function

Major platelet receptor for collagen involved in calcium-dependent platelet adhesion and interaction of platelets with leucocytes and tumour cells. Adhesion of leucocytes to collagen and thrombospondin. Role in atherogenesis via accumulation and differentiation of macrophages in arterial wall. Recognition, and phagocytosis, of apoptotic neutrophils. Mediates cytoadhesion of *Plasmodium falciparum* parasitized erythrocytes to endothelium[8,9]. A role in atherogenesis via CD36 acting as a scavenger receptor on macrophages for oxidized low-density lipoproteins has been suggested[6,17].

Distribution

Platelets, monocytes and macrophages, endothelial cells, adipocytes, developing erythroblasts, activated keratinocytes and some other epithelial cells[18], and small vessel endothelium[7].

Disease association

OMIM 173510

A family with a mutant form of CD36 has been described with a mild bleeding diathesis similar to Bernard–Soulier syndrome[19,20]. Deficiency of CD36 in a transfused patient led to the discovery of the Nak(a) platelet alloantigen[21]. Three per cent of Japanese and Africans have CD36 deficiency – in part this may be due to selection for resistance to malarial infection. There is a strong link in Japanese populations between CD36 deficiency and coronary heart disease and hypertrophic cardiomyopathy[22]. A similar relationship between insulin resistance type II diabetes, hypertension and CD36 (Fat gene) polymorphism is found in the 'spontaneously hypertensive' (SHR) rat[23].

Knockout

MGI:107899

Amino acid sequence of human CD36

```
  1 MGCDRNCGLI AGAVIGAVLA VFGGILMPVG DLLIQKTIKK QVVLEEGTIA FKNWVKTGTE
 61 VYRQFWIFDV QNPQEVMMNS SNIQVKQRGP YTYRVRFLAK ENVTQDAEDN TVSFLQPNGA
121 IFEPSLSVGT EADNFTVLNL AVAAASHIYQ NQFVQMILNS LINKSKSSMF QVRTLRELLW
181 GYRDPFLSLV PYPVTTTVGL FYPYNNTADG VYKVFNGKDN ISKVAIIDTY KGKRNLSYWE
241 SHCDMINGTD AASFPPPVEK SQVLQFFSSD ICRSIYAVFE SDVNLKGIPV YRFVLPSKAF
301 ASPVENPDNY CFCTEKIISK NCTSYGVLDI SKCKEGRPVY ISLPHFLYAS PDVSEPIDGL
361 NPNEEEHRTY LDIEPITGFT LQFAKRLQVN LLVKPSEKIQ VLKNLKRNYI VPILWLNETG
421 TIGDEKANMF RSQVTGKINL LGLIEMILLS VGVVMFVAFM ISYCACRSKT IK
```

Database accession

EMBL/GenBank   M24795
SwissProt   P16671

References

[1] Taylor, K.T. et al. (1993) Gene 133, 205–212.
[2] Tandon, N.N. et al. (1989a) J. Biol. Chem. 264, 7570–7575.
[3] Wyler, B. et al. (1993) Thromb. Haemost. 70, 500–505.
[4] Tao, N. et al. (1996) J. Biol. Chem. 271, 22315–22320.

[5] Pearce, S.F.A. et al. (1994) Blood 84, 384–389.
[6] Platt, N. et al. (1998) Trends Cell Biol. 8, 365–372.
[7] Asch, A.S. et al. (1987) J. Clin. Invest. 79, 1054–1061.
[8] Oquendo, P. et al. (1989) Cell 58, 95–101.
[9] Ockenhouse, C.F. et al. (1989) Science 243, 1469–1471.
[10] Tandon, N.N. et al. (1989b) J. Biol. Chem. 264, 7576–7583.
[11] Endemann, G. et al. (1993) J. Biol. Chem. 268, 11811–11816.
[12] Krieger, M. and Herz, J. (1994) Annu. Rev. Biochem. 63, 601–637.
[13] Nozaki, S. et al. (1995) J. Clin. Invest. 96, 1859–1865.
[14] Ibrahimi, A. et al. (1996) Proc. Natl Acad. Sci. USA 93, 2646–2651.
[15] Savill, J. et al. (1993) Immunol. Today 14, 131–136.
[16] Navazo, M.D.P. et al. (1996) J. Biol. Chem. 271, 15381–15385.
[17] Rigotti, A. et al. (1995) J. Biol. Chem. 270, 16221–16224.
[18] Greenwalt, D.E. et al. (1992) Blood 80, 1105–1115.
[19] Yamamoto, N. et al. (1994) Blood 83, 392–397.
[20] Kashiwagi, H. et al. (1995) J. Clin. Invest. 95, 1040–1046.
[21] Ikeda, H. et al. (1989) Vox Sang. 57, 213–217.
[22] Tanaka, T. et al. (1997) J. Molec. Cell Cardiol. 29, 121–127.
[23] Aitman, T.J. et al. (1999) Nature Genet. 21, 76–83.

CD39 — Vascular ATP diphosphohydrolase, ecto-apyrase (EC 3.6.1.5)

Family

CD39 gene family

Structure

Molecular weights

| Amino acids | 510 |
| Polypeptide | 579 654 |

| SDS-PAGE reduced | 78 kDa |

Carbohydrate

| N-linked sites | 6 |
| O-linked sites | |

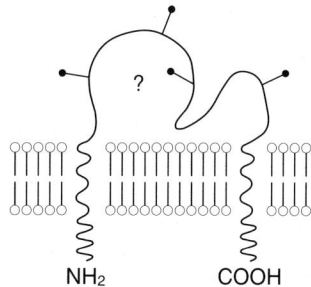

NH$_2$ COOH

Gene location   10q23.1–24.1

Gene structure

Alternative forms

Major 3.2 kb RNA and other tissue forms; at least four structural variants (CD39L1–4) in human and other species indicating a gene family[1].

Structure

CD39 has amino acid homology (four regions) to ecto-apyrase enzymes from animals, lower organisms and plants[2]. Functional studies show that CD39 is a vascular ATP diphosphohydrolase/ecto-apyrase (EC 3.6.1.5) family member[3] with typical enzymatic activity towards ATP and ADP[4,5]. The amino acid sequence of CD39 contains three hydrophobic stretches suggesting a two-membrane span molecule, although the N-terminal segment is unusually short; its exact topology is thus unclear[3].

Ligands

CD39.

Function

CD39 is a mediator of activated B cell homotypic adhesion[6]. Regulation of platelet activation by surface expression of CD39 (endothelial ATP diphosphohydrolase, ecto-apyrase [EC 3.6.1.5]), which hydrolyses extracellular ATP and ADP to AMP, and is further degraded to adenosine by another enzyme 5'-nucleotidase; adenosine, is a stimulant of platelet adhesion to endothelium. A similar mechanism may reduce extracellular ATP by enzymatic hydolysis, which at high levels is toxic to cells.

Distribution

Activated B and NK cells, and subsets of activated T cells (it was first identified on Epstein–Barr virus-transformed lymphocytes[7]), but not resting lymphocytes, and by some other haemopoietic cells. Mantle and paracortical zones of lymph nodes but not germinal centres. Endothelial[6–8] and neural cells also express CD39[9].

Disease association

OMIM 601752

Knockout

MGI:102805

Amino acid sequence of human CD39

```
  1 MEDTKESNVK TFCSKNILAI LGFSSIIAVI ALLAVGLTQN KALPENVKYG IVLDAGSSHT
 61 SLYIYKWPAE KENDTGVVHQ VEECRVKGPG ISKFVQKVNE IGIYLTDCME RAREVIPRSQ
121 HQETPVYLGA TAGMRLLRME SEELADRVLD VVERSLSNYP FDFQGARIIT GQEEGAYGWI
181 TINYLLGKFS QKTRWFSIVP YETNNQETFG ALDLGGASTQ VTFVPQNQTI ESPDNALQFR
241 LYGKDYNVYT HSFLCYGKDQ ALWQKLAKDI QVASNEILRD PCFHPGYKKV VNVSDLYKTP
301 CTKRFEMTLP FQQFEIQGIG NYQQCHQSIL ELFNTSYCPY SQCAFNGIFL PPLQGDFGAF
361 SAFYFVMKFL NLTSEKVSQE KVTEMMKKFC AQPWEEIKTS YAGVKEKYLS EYCFSGTYIL
421 SLLLQGYHFT ADSWEHIHFI GKIQGSDAGW TLGYMLNLTN MIPAEQPLST PLSHSTYVFL
481 MVLFSLVLFT VAIIGLLIFH KPSYFWKDMV
```

Database accession

EMBL/GenBank   S73813
SwissProt   P49961

References

[1] Chadwick, B.P. and Frischauf, A.-M. (1998) Genomics 50, 357–367.

[2] Handa, M. and Guidotti, G. (1996) Biochem. Biophys. Res. Commun. 218, 916–923.

[3] Maliszewski, C.R. et al. (1994) J. Immunol. 153, 3574–3583.

[4] Kaczmarek, E. et al. (1996) J. Biol. Chem. 271, 33116–33122.

[5] Wang, T.F. and Guidotti, G. (1996) J. Biol. Chem. 271, 9898–9901.

[6] Kansas, G.S. et al. (1991) J. Immunol. 146, 2235–2244.

[7] Rowe, M. et al. (1982) Int. J. Cancer 29, 373–381.

[8] Goutefangeas, C. et al. (1992) Eur. J. Immunol. 22, 2681–2685.

[9] Wang, T.F. and Guidotti, G. (1998) Brain Res. 790, 318–322.

CD42a gpIX

Family

Leucine-rich glycoprotein/ CD42 family

Structure

Molecular weights

Amino acids	177
Polypeptide	19046

SDS-PAGE reduced	22 kDa (17–22 kDa unreduced)

Carbohydate

N-linked sites	1
O-linked sites	

Gene location	Chromosome 3
Gene structure	6 kb, 2 exons
Alternative forms	

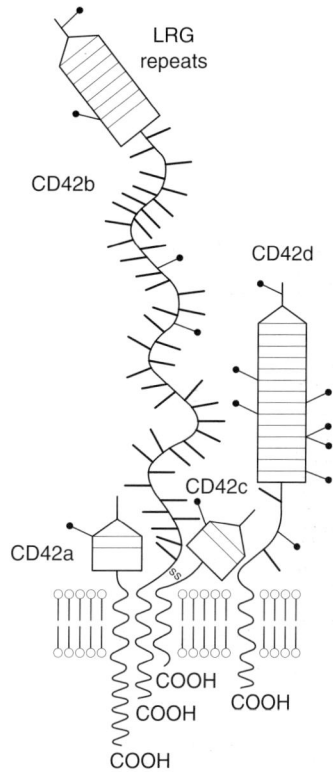

Structure

CD42a is a single-chain type 1 integral membrane glycoprotein[1,2] that forms a non-covalently linked complex with CD42b, CD42c and CD42d[3–5]. The extracellular part of CD42a is characterized by a single 24 amino acid leucine-rich domain[4]; as such it shows homology to CD42c (gpIbβ). CD42a is possibly myristolated[6] and has a single N-glycosylation site (lactosamine chain). CD42a is linked to the platelet actin cytoskeleton (as part of the gpIX–gpIb complex).

Ligands

The CD42 complex forms a receptor for von Willebrand factor[4] and thrombin[7] on platelets (the binding site lies in CD42b/gpIbα).

Function

CD42a mediates adhesion to subendothelial matrix exposed upon endothelial damage at high blood flow rates, and modulates platelet aggregation when activated by thrombin[4]. It is essential for the maintenance of haemostasis in the arterial system.

Distribution

Platelets and megakaryocytes[5].

Disease association

OMIM 173515

Mutations[8,9] and absence of CD42a (along with mutations in CD42b and CD42c) leads to Bernard–Soulier syndrome (OMIM 231200), a bleeding disorder characterized by giant platelets and thrombocytopenia.

Knockout

Amino acid sequence of human CD42a

```
  1 MPAWGALFLL WATAEATKDC PSPCTCRALE TMGLWVDCRG HGLTALPALP ARTRHLLLAN
 61 NSLQSVPPGA FDHLPQLQTL DVTQNPWHCD CSLTYLRLWL EDRTPEALLQ VRCASPSLAA
121 HGPLRLTGYQ LGSCGWQLQA SWVRPGVLWD VALVAVAALG LALLAGLLCA TTEALD
```

Database accession

EMBL/GenBank   X52997
SwissProt   P14770

References

[1] Hickey, M.J. et al. (1989) Proc. Natl Acad. Sci. USA 86, 6773–6777.
[2] Hickey, M.J. et al. (1990) FEBS Lett. 274, 189–192.
[3] Fox, J.E.B. et al. (1988) J. Biol. Chem. 263, 4882–4890.
[4] Roth, G.J. (1992) Immunol. Today 13, 100–105.
[5] Lopez, J.A. et al. (1996) J. Biol. Chem. 269, 23716–23721.
[6] Schick, P.K. and Walker, J. (1996) Blood 87, 1377–1384.
[7] Kawasaki, T. et al. (1996) J. Biol. Chem. 271, 10635–10639.
[8] Wright, S.D. et al. (1993) Blood 81, 2339–2347.
[9] Clemetson, J.M. et al. (1994) Blood 84, 1124–1131.

CD42b gpIbα, glycocalicin

Family
Leucine-rich glycoprotein/CD42 family

Structure
Molecular weights
Amino acids	626
Polypeptide	68 955

SDS-PAGE reduced	145 kDa (160 kDa unreduced)

Carbohydrate
N-linked sites	4
O-linked sites	approx. 43

Gene location
17pter–p12

Gene structure
6 kb, 2 exons

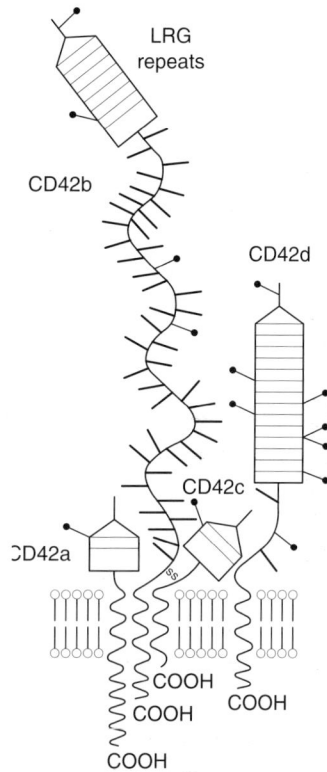

Alternative forms
The cleaved N-terminal part of CD42b is found in serum, named glycocalicin[1].

Structure
CD42b exists as a disulphide-bonded heterodimer with CD42c[2,3], and hence a non-covalent stochiometric complex with CD42a and d, the functional von Willebrand factor/ thrombin receptor on platelets. The N-terminal, 30 amino acid long, and a 60 amino acid stretch C-terminal to the leucine-rich repeat[3] region of CD42b share sequence similarity between members of the CD42 complex members[4]. The extracellular portion of CD42b is contains eight tandem leucine-rich repeats of 24 amino acids flanked by consensus sequences of 22 amino acids. CD42b exists in several size polymorphs, due to the presence of 1–5 copies of a 13 amino acid long, threonine-rich and highly O-glycosylated[5], sequences each containing five O-glycation sites accounting for its extended size[6]. There are three tyrosine sulphation sites (Tyr 276, 278, 279) and CD42b is palmitoylated[7]. The region C-terminal to the leucine-rich repeats contains the von Willebrand factor binding site. The C-terminus interacts with the actin-binding protein, filamin[8], and the

14–3–3 zeta protein[9], which is disrupted upon platelet activation along with dissociation of the CD42 complex on the cell membrane. The N-terminal extracellular part of CD42b is cleaved by calpain during platelet aggregation and released, as glycocalicin, into the serum. The synthesis and assembly of the CD42a–d (gpIb-IX-V) complex has recently been studied[10].

Ligands

Von Willebrand factor and thrombin via the N-terminal part of CD42b[3,11,12].

Function

CD42b mediates adhesion to subendothelial matrix exposed upon endothelial damage at high blood flow rates, and modulates platelet aggregation when activated by thrombin[3]. It is essential for the maintenance of haemostasis in the arterial system.

Platelet activation consequent upon thrombin binding results in dissociation of the CD42 complex and disruption of the CD42b,c (gpIb complex) interaction with the platelet actin cytoskeleton, and hence altered adhesion and spreading on the damaged endothelium/subendothelial matrix.

Distribution

Platelets and megakaryocytes[2].

Disease association

OMIM 231200
Mutations in CD42b lead to Bernard–Soulier syndrome (see CD42 entries; OMIM 231200). Mutations on the double loop region result in increased von Willebrand factor binding – 'platelet type/pseudo von Willebrand's disease' – due to reduced plasma half-life of von Willebrand factor and thrombocytopenia[13,14]. Single amino acid polymorphisms (Met145Thr, and Thr161Met) produce the Ko and Sib(a) alloantigens, respectively, which can result in auto- and allo-immune thrombocytopenia[15,16].

Knockout

Amino acid sequence of human CD42b

```
  1 MPLLLLLLLL PSPLHPHPIC EVSKVASHLE VNCDKRNLTA LPPDLPKDTT ILHLSENLLY
 61 TFSLATLMPY TRLTQLNLDR CELTKLQVDG TLPVLGTLDL SHNQLQSLPL LGQTLPALTV
121 LDVSFNRLTS LPLGALRGLG ELQELYLKGN ELKTLPPGLL TPTPKLEKLS LANNNLTELP
181 AGLLNGLENL DTLLLQENSL YTIPKGFFGS HLLPFAFLHG NPWLCNCEIL YFRRWLQDNA
241 ENVYVWKQGV DVKAMTSNVA SVQCDNSDKF PVYKYPGKGC PTLGDEGDTD LYDYYPEEDT
301 EGDKVRATRT VVKFPTKAHT TPWGLFYSWS TASLDSQMPS SLHPTQESTK EQTTFPPRWT
361 PNFTLHMESI TFSKTPKSTT EPTPSPTTSE PVPEPAPNMT TLEPTPSPTT PEPTSEPAPS
421 PTTPEPTPIP TIATSPTILV SATSLITPKS TFLTTTKPVS LLESTKKTIP ELDQPPKLRG
481 VLQGHLESSR NDPFLHPDFC CLLPLGFYVL GLFWLLFASV VLILLLSWVG HVKPQALDSG
541 QGAALTTATQ TTHLELQRGR QVTVPRAWLL FLRGSLPTFR SSLFLWVRPN GRVGPLVAGR
601 RPSALSQGRG QDLLSTVSIR YSGHSL
```

Database accession
EMBL/GenBank   J02940
SwissProt   P07159

References
1 Clemetson K.J. et al. (1982) Proc. Natl Acad. Sci. 78, 2712–2716.
2 Lopez, J.A. et al. (1996) J. Biol. Chem. 269, 23716–23721.
3 Roth, G.J. (1992) Immunol. Today 13, 100–105.
4 Lopez, J.A. et al. (1987) Proc. Natl Acad. Sci. USA 84, 5615–5619.
5 Lopez, L.A. et al. (1992) J. Biol. Chem. 267, 10055–10061.
6 Fox, J.E.B. et al. (1988) J. Biol. Chem. 263, 4882–4890.
7 Schick, P.K. and Walker, J. (1996) Blood 87, 1377–1384.
8 Fox, J.E. B. et al. (1989) J. Biol. Chem. 264, 9716–9719.
9 Andrews, R.K. et al. (1998) Biochemistry 37, 638–647.
10 Dong, J.F. et al. (1998) J. Biol. Chem. 273, 31449–31454.
11 Harmon, J.T. and Jamieson, G.A. (1986) J. Biol. Chem. 261, 13224–13229.
12 Michelson, A.D. et al. (1986) Blood 67, 19–26.
13 Russell, S.D. and Roth, G.J. (1993) Blood 81, 1787–1791.
14 Murata, M. et al. (1993) J. Clin. Invest. 91, 2133–2137.
15 Murata, M. et al. (1992) Blood 79, 3086–3090.
16 Kuijpers, R.W. et al. (1992) Blood 79, 283–288.

CD42c gpIbβ

Family

Leucine-rich glycoprotein/CD42 family

Structure

Molecular weights

Amino acids	206
Polypeptide	21 717

SDS-PAGE reduced	24 kDa

Carbohydrate

N-linked sites	1
O-linked sites	

Gene location   22q11.2

Gene structure   1 kb, 1 exon

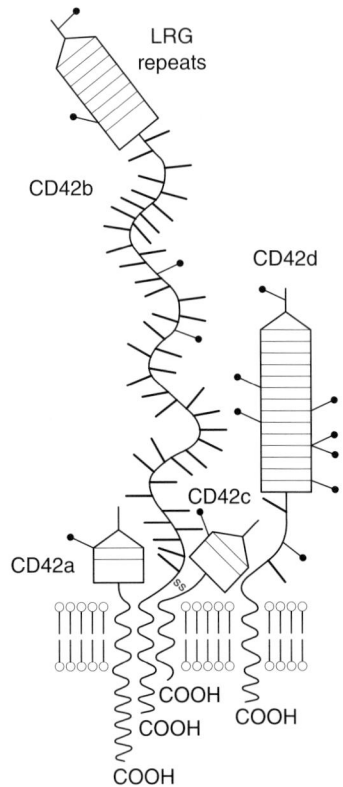

Alternative forms

Larger 411 amino acid form in endothelium, owing to use of alternate upstream start site.

Structure

CD42c forms a 1:1 disulphide-bonded covalent complex with CD42b/gpIbα, and the gpIb complexes with CD42a and d to form the functional platelet receptor for von Willebrand factor and thrombin[1,2]. Contains a single leucine-rich repeat domain[1,3-5] homologous to CD42a/gpIX[4,6,7]. Phosphorylated on intracellular serine 191, suggestive of a role in cell signalling/cytoskeletal interaction[8].

Ligands

Von Willebrand factor and thrombin.

Function

CD42c mediates adhesion to subendothelial matrix exposed upon endothelial damage at high blood flow rates, and modulates platelet aggregation when activated by thrombin (see other CD42 entries). It is essential for the maintenance of haemostasis in the arterial system. Platelet activation consequent upon thrombin binding results in dissociation of the CD42 complex and disruption of the CD42b,c (gpIb complex) interaction with the platelet actin cytoskeleton, and hence altered adhesion and spreading on the damaged endothelium/subendothelial matrix.

Distribution

Platelets, megakaryocytes. Unlike other CD42 components that are tissue restricted, CD42c is expressed in endothelium, heart and brain[9].

Disease association

OMIM 138720

Mutations in CD42c result in Bernard–Soulier syndrome, a platelet haemostatic defect. However, a deletion in 22q11.2 resulted in a complex phenotype that also resembled velocardiofacial syndrome[10] (see claudin 5 entry, whose gene is localized to the same genetic region).

Knockout

Amino acid sequence of human CD42c

```
  1 MGSGPRGALS LLLLLLAPPS RPAAGCPAPC SCAGTLVDCG RRGLTWASLP TAFPVDTTEL
 61 VLTGNNLTAL PPGLLDALPA LRTAHLGANP WRCDCRLVPL RAWLAGRPER APYRDLRCVA
121 PPALRGRLLP YLAEDELRAA CAPGPLCWGA LAAQLALLGL GLLHALLLVL LLCRLRRLRA
181 RARARAAARL SLTDPLVAER AGTDES
```

Database accession

EMBL/GenBank   J03259
SwissProt   P13224

References

1 Roth, G.J. (1992) Immunol. Today 13, 100–105.
2 Lopez, J.A. et al. (1996) J. Biol. Chem. 269, 23716–23721.
3 Lopez, J.A. et al. (1987) Proc. Natl Acad. Sci. USA 84, 5615–5619.
4 Lopez, J.A. et al. (1988) Proc. Natl Acad. Sci. USA 85, 2135–2139.
5 Kelly, M.D. et al. (1994) J. Clin. Invest. 93, 2417–2424.
6 Fox, J.E. et al. (1988) J. Biol. Chem. 263, 4882–4890.
7 Canfield, V.A. et al. (1987) Biochem. Biophys. Res. Commun. 147, 526–534.
8 Wardell, M.R. et al. (1989) J. Biol. Chem. 264, 15656–15661.
9 Rajagopalan, V. et al. (1992) Blood 80, 153–161.
10 Burdarf, M.L. et al. (1995) Hum. Molec. Genet. 4, 763–766.

CD42d gpV

Family

Leucine-rich glycoprotein family

Structure

Molecular weights

Amino acids	560
Polypeptide	60 959

SDS-PAGE reduced	82 kDa (82 kDa unreduced)

Carbohydrate

N-linked sites	8
O-linked sites	2

Gene location

Gene structure

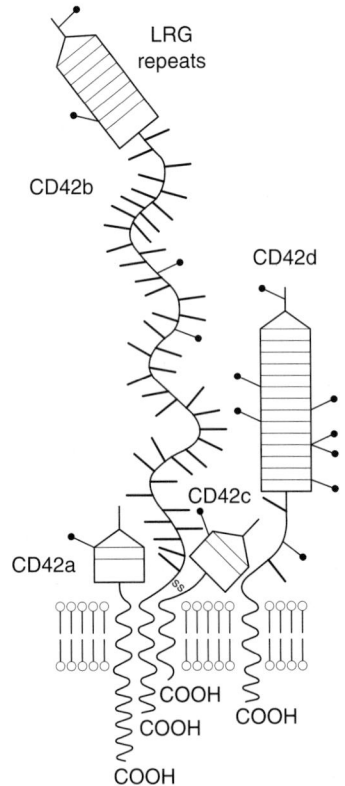

Alternative forms

3 transcripts

Structure

CD42d is a single-chain integral membrane protein[1-4] that forms a non-covalent complex with CD42a,b and c[5-7], and possibly enhances their membrane expression[8]. Contains 15 leucine-rich repeat domains. CD42d contains a thrombin cleavage site that may be functionally relevant.

Ligands

The CD42 complex binds von Willebrand factor and thrombin (see CD42 entries); the CD42d/gpV component may be essential for the formation of a high-affinity receptor for thrombin[9].

Function

Part of the platelet CD42 complex that binds von Willebrand factor and thrombin, and is crucial for maintaining haemostasis in the arterial circulation under high shear rates, and in the modulation of the platelet response to thrombin.

Distribution

Platelets and megakaryocytes.

Disease association

OMIM 173511

Absent from platelets in Bernard–Soulier syndrome, where the entire CD42 complex is lacking; no causative mutations have been found in CD42d.

Knockout

Amino acid sequence of human CD42d

```
  1 MLRGTLLCAV LGLLRAQPFP CPPACKCVFR DAAQCSGGDV ARISALGLPT NLTHILLFGM
 61 GRGVLQSQSF SGMTVLQRLM ISDSHISAVA PGTFSDLIKL KTLRLSRNKI THLPGALLDK
121 MVLLEQLFLD HNALRGIDQN MFQKLVNLQE LALNQNQLDF LPASLFTNLE NLKLLDLSGN
181 NLTHLPKGLL GAQAKLERLL LHSNRLVSLD SGLLNSLGAL TELQFHRNHI RSIAPGAFDR
241 LPNLSSLTLS RNHLAFLPSA LFLHSHNLTL LTLFENPLAE LPGVLFGEMG GLQELWLNRT
301 QLRTLPAAAF RNLSRLRYLG VTLSPRLSAL PQGAFQGLGE LQVLALHSNG LTALPDGLLR
361 GLGKLRQVSL RRNRLRALPR ALFRNLSSLE SVQLDHNQLE TLPGDVFGAL PRLTEVLLGH
421 NSWRCDCGLG PFLGWLRQHL GLVGGEEPPR CAGPGAHAGL PLWALPGGDA ECPGPRGPPP
481 RPAADSSSEA PVHPALAPNS SEPWVWAQPV TTGKGQDHSP FWGFYFLLLA VQAMITVIIV
541 FAMIKIGQLF RKLIRERALG
```

Database accession

EMBL/GenBank   L11238
SwissProt   P40197

References

[1] Shimomura, T. et al. (1990) Blood 75, 2349–2356.
[2] Roth, G.J. et al. (1990) Biochem. Biophys. Res. Commun. 170, 153–161.
[3] Lanza, F. et al. (1993) J. Biol. Chem. 268, 20801–20807.
[4] Hickey, M.J. et al. (1993) Proc. Natl Acad. Sci. 90, 8327–8331.
[5] Lopez, J.A. et al. (1996) J. Biol. Chem. 269, 23716–23721.
[6] Roth, G.J. (1992) Immunol. Today 13, 100–105.
[7] Modderman, P.W. et al. (1992) J. Biol. Chem. 267, 364–369.
[8] Calverley, D.C. et al. (1995) Blood 86, 1361–1367.
[9] Dong, J.F. et al. (1997) Blood 89, 4355–4363.

Family

Sialomucin

Structure

Molecular weights

| Amino acids | 400 |
| Polypeptide | 40 322 |

| SDS-PAGE reduced | neutrophils and platelets | 115–135 kDa |
| | T cells and thymocytes | 95–115 kDa |

Carbohydrate

| N-linked sites | 1 |
| O-linked sites | +++ |

| Gene location | 16q11.2 |

| Gene structure | 4.6 kb, 2 exons |

Alternative forms

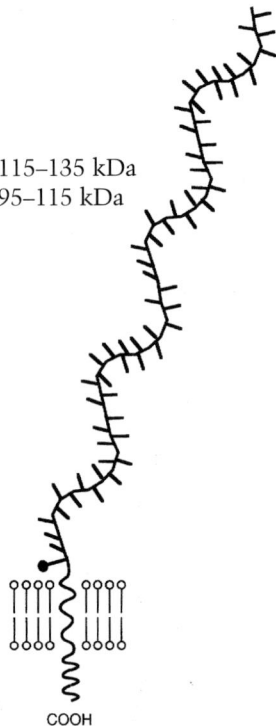

COOH

Structure

The extracellular domain contains five repeats of an 18 amino acid sequence decorated with 70–85 sialylated O-linked carbohydrates[1], resulting in an extended 45 nm rod-like structure that could protrude above the glycocalyx[2]. The 124 amino acid cytoplasmic domain is constitutively phosphorylated, probably by protein kinase C[3] and has been reported to bind members of the ERM family of membrane-cytoskeleton linker proteins[4,5]. The CD43 gene is unusual in that it contains a single 378 bp intron in the 5'-untranslated region and therefore the CD43 coding sequence is within a single exon[6].

Ligands

CD43 has been reported to bind ICAM-1[7] but it is not known whether this represents a relevant physiological event.

Function

CD43 −/− leucocytes show enhanced homotypic adhesion[8], and enhanced rolling and adhesion after chemotactic stimuli[9], indicating a function for

CD43 as an antiadhesive molecule inhibiting T cell interactions. This includes an inhibition of T cell killing and an increase in the threshold for T cell activation. Like all mucins, CD43 may also have a cytoprotective function. However, neutrophil and monocyte infiltration into the peritoneum is significantly reduced in CD43 −/− mice as is leucocyte emigration from the vasculature[9], indicating that CD43 has an adhesive function *in vivo*.

Distribution

CD43 is a major sialylglycoprotein on all leucocytes except resting B cells[6,10]. In different cell types the molecular size of CD43 varies due to differing degrees of *O*-linked carbohydrate sialylation[11]. CD43 is proteolytically cleaved from the surface of stimulated lymphocytes and granulocytes[12].

Disease association

OMIM 182160
CD43 is defective in T cells from males with Wiskott–Aldrich syndrome. Affected individuals are susceptible to opportunistic infections reflecting defects in cytotoxic and helper T cell functions. As the Wiskott–Aldrich syndrome is X-linked, defects in CD43 are not the primary genetic cause. Circulating anti-CD43 antibodies are found in many AIDS patients and may contribute to the severe immunodeficiency.

Knockout

MGI:98384
The Gene Knockout FactsBook[13], p175
CD43 −/− mice develop normally but CD43 −/− T cells show enhanced proliferation in response to T cell activators and in alloreactivity tests, and enhanced homotypic adhesion[8]. CD43 −/− leucocytes show impairments in trafficking through the vasculature[9].

Amino acid sequence of human CD43

```
  1 MATLLLLLGV LVVSPDALGS TTAVQTPTSG EPLVSTSEPL SSKMYTTSIT SDPKADSTGD
 61 QTSALPPSTS INEGSPLWTS IGASTGSPLP EPTTYQEVSI KMSSVPQETP HATSHPAVPI
121 TANSLGSHTV TGGTITTNSP ETSSRTSGAP VTTAASSLET SRGTSGPPLT MATVSLETSK
181 GTSGPPVTMA TDSLETSTGT TGPPVTMTTG SLEPSSGASG PQVSSVKLST MMSPTTSTNA
241 STVPFRNPDE NSRGMLPVAV LVALLAVIVL VALLLLWRRR QKRRTGALVL SRGGKRNGVV
301 DAWAGPAQVP EEGAVTVTVG GSGGDKGSGF PDGEGSSRRP TLTTFFGRRK SRQGSLAMEE
361 LKSGSGPSLK GEEEPLVASE DGAVDAPAPD EPEGGDGAAP
```

Database accession

EMBL/GenBank   J04168
SwissProt   P16150

References

1 Pallant, A. et al. (1989) Proc. Natl Acad. Sci. USA 86, 1328–1332.
2 Cyster, J.G. et al. (1991) EMBO J. 10, 893–902.
3 Piller, V. et al. (1989) J. Biol. Chem. 264, 18824–18831.
4 Serrador, J.M. et al. (1998) Blood 91, 4632–4644.
5 Yonemura, S. et al. (1998) J. Cell Biol. 140, 885–895.
6 Shelley, C.S. et al. (1990) Biochem. J. 270, 569–576.
7 Rosenstein, Y. et al. (1991) Nature 354, 233–235.
8 Manjunath, N. et al. (1995) Nature 377, 535–538.
9 Woodman, R.C. et al. (1998) J. Exp. Med. 188, 2181–2186.
10 Nathan, C. et al. (1993). J. Cell Biol. 122, 243–256.
11 Carlsson, S.R. et al. (1986) J. Biol. Chem. 261, 12787–12795.
12 Bazil, V. and Strominger, J.L. (1993) Proc. Natl Acad. Sci. USA 90, 3792–3796.
13 Mak, T.W. (1998) The Gene Knockout FactsBook. Academic Press, London.

CD44 Pgp-1, Hermes-3

Family
Hyaluronan receptor

Structure

Molecular weights

Amino acids	CD44s	361
	CD44v	up to 742
Polypeptide	CD44s	39 391
	CD44v	up to 81 609
SDS-PAGE reduced	CD44s	80–100 kDa
	CD44v	100–200 kDa

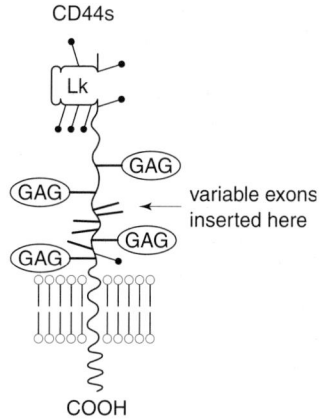

CD44s

Carbohydrate

N-linked sites)	CD44s	7
	CD44v	7–9
O-linked sites	CD44s	++
	CD44v	+++
GAG	CD44s	variable chondroitin sulphate
	CD44v	chondroitin and heparan sulphate

Gene location
11pter–p13

Gene structure
60 kb, 20 exons

Alternative forms
CD44s (standard, also called haematopoietic CD44, CD44H) is generated by exons 1–5, 16–18 and 20. CD44 variant (CD44v) isoforms results from the insertion of various combinations of nine variant exons (exons 7–15 = variant exons 2–10)[1]. An additional variant exon (exon 6 = variant exon 1) can be inserted into murine CD44[2].

Structure
CD44s contains a 341 extracellular domain with a hyaluronan-binding link domain at the N-terminus followed by mucin-like stalk region which is subject to *N*-and *O*-linked glycosylation and variable addition of chondroitin sulphate[3,4]. The cytoplasmic domain is phosphorylated constitutively on a single serine residue[5] and can associate with members of the ERM family of membrane–cytoskeletal linker proteins[6,7]. A large number of different CD44 isoforms have been detected resulting from the insertion of different combinations of variant exons v2–v10. These variant isoforms also encode mucin-like regions and can be extensively *O*-glycosylated. In addition, exon v3 can be modified by heparan sulphate side chains. CD44 can be shed from the cell surface by a proteolytic mechanism and the efficiency of shedding can be modulated by the insertion of variant exons[4,8].

Ligands
CD44 is the principal cell surface receptor for the extracellular matrix glycosaminoglycan, hyaluronan[9–11]. The ability of CD44 to bind hyaluronan

is highly regulated depending on the state of CD44 glycosylation, insertion of variant exons and clustering in the plane of the membrane[4]. CD44 has also been reported to bind collagen, fibronectin and laminin, but these interactions are probably dependent on the chondroitin sulphate side chains rather than the core protein[4]. Variant forms of CD44 containing heparan sulphate modified exon v3 can bind growth factors, such as bFGF and FGF-8[12-14]. A number of other extracellular matrix and signalling molecule ligands have been reported[4,8], although the mechanism and physiological relevance of CD44 binding is not known.

Function

A variety of functions have been ascribed to CD44 including mediating co-stimulatory signalling, leucocyte attachment to, and rolling on, endothelial cells, transmigration through lymphatic tissue, growth factor and cytokine presentation, cell migration and the internalization and degradation of hyaluronan[4,8].

Distribution

CD44 is widely distributed, with CD44s being the most abundant isoform. Expression of variant CD44 isoforms occurs normally in epithelial cells and at other sites in the body. Expression of both CD44s and CD44 variant isoforms is transiently upregulated during lymphocyte and monocyte activation, and extensively in many pathological conditions[4,8].

Disease association

OMIM 107269
CD44–hyaluronan interactions have been implicated to play an important role in the progression of chronic inflammatory conditions, such as rheumatoid arthritis, and in tumour growth and metastatic progression[4,8,15].

Knockout

MGI:88338
The Gene Knockout FactsBook[16], p177
CD44 −/− mice develop normally and exhibit no gross developmental or neurological defects. However, progenitor egress from the bone marrow is defective and mice develop an exaggerated granuloma response to *Cryosporidium parvum* infection[17].

Amino acid sequence of human CD44s

```
  1 MDKFWWHAAW GLCLVPLSLA QIDLNITCRF AGVFHVEKNG RYSISRTEAA DLCKAFNSTL
 61 PTMAQMEKAL SIGFETCRYG FIEGHVVIPR IHPNSICAAN NTGVYILTSN TSQYDTYCFN
121 ASAPPEEDCT SVTDLPNAFD GPITITIVNR DGTRYVQKGE YRTNPEDIYP SNPTDDDVSS
181 GSSSERSSTS GGYIFYTFST VHPIPDEDSP WITDSTDRIP ATRDQDTFHP SGGSHTTHGS
241 ESDGHSHGSQ EGGANTTSGP IRTPQIPEWL IILASLLALA LILAVCIAVN SRRRCGQKKK
301 LVINSGNGAV EDRKPSGLNG EASKSQEMVH LVNKESSETP DQFMTADETR NLQNVDMKIG
361 V
```

CD44v isoforms are generated by the insertion of variants exons after T222 (shown in bold).

Amino acid sequence of human CD44 variant exons 2-10

```
  1 TLMSTSATAT ETATKRQETW DWFSWLFLPS ESKNHLHTTT QMAGTSSNTI SAGWEPNEEN
 61 EDERDRHLSF SGSGIDDDED FISSTISTTP RAFDHTKQNQ DWTQWNPSHS NPEVLLQTTT
121 RMTDVDRNGT TAYEGNWNPE AHPPLIHHEH HEEEETPHST STIQATPSST TEETATQKEQ
181 WFGNRWHEGY RQTPREDSHS TTGTAAASAH TSHPMQGRTT PSPEDSSWTD FFNPISHPMG
241 RGHQAGRRMD MDSSHSTTLQ PTANPNTGLV EDLDRTGPLS MTTQQSNSQS FSTSHEGLEE
301 DKDHPTTSTL TSSNRNDVTG GRRDPNHSEG STTLLEGYTS HYPHTKESRT FIPVTSAKTG
361 SFGVTAVTVG DSNSNVNRSL S
```

Initial residues in each exon are shown in italics and underlined.

Database accession

EMBL/GenBank   M69215
SwissProt   P16070

References

1 Screaton, G.R. et al. (1992) Proc. Natl. Acad. Sci. USA 89, 12160–12164.
2 Screaton, G.R. et al. (1993) J. Biol. Chem. 268, 12235–12238.
3 Stamenkovic, I.S. et al. (1989) Cell 56, 1057–1062.
4 Lesley, J. et. al. (1993) Adv. Immunol. 54, 271–335.
5 Peck, D. and Isacke, C.M. (1998) J. Cell Sci. 111, 1595–1601.
6 Legg, J.W. and Isacke, C.M. (1998) Curr. Biol. 8, 705–708.
7 Yonemura, S. et al. (1998) J. Cell Biol. 140, 885–895.
8 Naor, D. et al. (1997) Adv. Cancer Res. 17, 241–319.
9 Aruffo, A. et al. (1990) Cell 61, 1303–1313.
10 Lesley, J. et al. (1990) Exp. Cell Res. 187, 224–233.
11 Miyake, K. et al., (1990) J. Exp. Med. 172, 69–75.
12 Bennett, K.L. et al. (1995) J. Cell Biol. 128, 687–698.
13 Jackson, D.G. et al. (1995) J. Cell Biol. 128, 673–685.
14 Sherman, L. et al. (1998) Genes Dev. 12, 1058–1071.
15 Mikecz, K. et al., (1999) Arthritis Rheum. 42, 659–668.
16 Mak, T.W. (1998) The Gene Knockout FactsBook. Academic Press, London.
17 Schmits, R. et al. (1997) Blood 90, 2217–2233.

CD57 HNK1, Leu-7

◻ **Family**

Structure
Molecular weights
Carbohydrate epitope

SDS-PAGE reduced   Depends on the protein core on which the CD57
                          epitope is expressed

Carbohydrate   n/a

Gene location   11

Gene structure

Alternative forms
Carbohydrate epitope present on many proteins, including L1, N-CAM,
MAG, P(0) and contactin-1 (see individual entries).

Structure
CD57 is a oligosaccharide determinant expressed on a variety of core
proteins, lipids and proteoglycans in a tissue-specific manner[1,2]. The
associated protein backbone structures have not been studied in detail. The
sulfotransferase enzyme that generates the CD57 carbohydrate epitope has
been cloned[3].

Ligands

Binds to L- and P-selectin; E8 fragment of laminin; involved in homophylic
interactions of P(0) (see entry)[1,2,4].

Function

Unknown, although its expression on myelin-associated glycoprotein
(MAG, see entry) on motor, but not sensory, neurones, and involvement in
homophylic interactions with P(0) suggests a role in peripheral nerve
function[1,2,4]. Its binding to L-selectin (CD62L) and P-selectin (CD62P), but
not E-selectin (CD62E), implies involvement in selective leucocyte
emmigration.

Distribution

Surface antigen detected on natural killer cells and T lymphocytes, but not
other leucocytes, and peripheral nerve axons[1,2,5]. CD57 is found on integrins
in chick[1,2,4].

Disease association
OMIM 151290

Knockout

n/a

Amino acid sequence

n/a

Database accession

n/a

References

[1] Jungalwala, F.B. (1994) Neurochem. Res. 19, 945–957.

[2] Schachner, M. et al. (1995) Prog. Brain Res. 105, 183–188.

[3] Schachner, M. and Martini, R. (1995) Trends Neurosci. 18, 183–191.

[4] Bakker, H. et al. (1997) J. Biol. Chem. 272, 29942–29946.

[5] Schubert, J. et al. (1989) In Leucocyte Typing IV (Knapp, W., ed.) Oxford University Press, Oxford, pp. 711–714.

CD98

Family

Structure
Molecular weights
Amino acids   529
Polypeptide   57 944

SDS-PAGE reduced   80 kDa (+41 kDa light chain) (125 kDa unreduced)

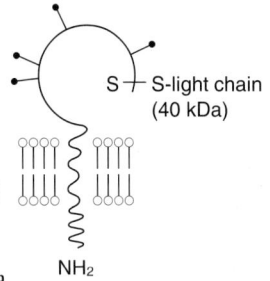

S─┬─ S-light chain
    (40 kDa)

NH$_2$

Carbohydrate
N-linked sites   4 (80 kDa chain; the 41 kDa chain is non-glycosylated)

O-linked site

Gene location   11q13

Gene structure   8 kb, 9 exons (heavy chain)

Alternative forms

Structure
CD98 is a type II membrane protein consisting of a disulphide-linked heterodimer of glycosylated heavy chain (CD98)[1-4], and non-glycosylated light chain[5]. First defined by the 4F2 antibody[6].

Ligands
Unknown.

Function

Potential amino acid[7] and Na$^+$–Ca^{2+} exchanger[8] function is based upon protein homology and functional studies with antibodies. CD98 is part of a family of proteins[9] that associate to form functional amino acid transporters by associating with LAT-1 and LAT-2 proteins[10,11]. CD98 expression on many actively proliferating cells, including epithelial and fibroblastic cells, and by activated leucocytes at sites of inflammation, suggests involvement in cell activation and proliferation[12]. A role in intercellular adhesion and cell fusion is implicated by induction of monocyte fusion by antibodies to CD98[13] and its association with actin.

Distribution

Widespread tissue distribution expressed at high levels by monocytes but low amounts on other leucocytes; upregulated upon leucocyte activation; expressed early in haemopoeiesis[4,12,14].

Disease association

OMIM 158070

Knockout

MGI:106597

Amino acid sequence of human CD98 (heavy chain)

```
  1 MEKKWKYCAV YYIIQIHFVK GVWEKTVNTE ENVYATLGSD VNLTCQTQTV GFFVQMQWSK
 61 VTNKIDLIAV YHPQYGFYCA YGRPCESLVT FTETPENGSK WTLHLRNMSC SVSGRYECML
121 VLYPEGIQTK IYNLLIQTHV TADEWNSNHT IEIEINQTLE IPCFQNSSSK ISSEFTYAWS
181 VEDNGTQETL ISQNHLISNS TLLKDRVKLG TDYRLHLSPV QIFDDGRKFS CHIRVGPNKI
241 LRSSTTVKVF AKPEIPVIVE NNSTDVLVER RFTCLLKNVF PKANITWFID GSFLHDEKEG
301 IYITNEERKG KDGFLELKSV LTRVHSNKPA QSDNLTIWCM ALSPVPGNKV WNISSEKITF
361 LLGSEISSTD PPLSVTESTL DTQPSPASSV SPARYPATSS VTLVDVSALR PNTTPQPSNS
421 SMTTRGFNYP WTSSGTDTKK SVSRIPSETY SSSPSGAGST LHDNVFTSTA RAFSEVPTTA
481 NGSTKTNHVH ITGIVVNKPK DGMSWPVIVA ALLFCCMILF GLGVRKWCQY QKEIMERPPP
541 FKPPPPPIKY TCIQEPNESD LPYHEMETL
```

Database accession

	EMBL/GenBank	SwissProt
CD98 (heavy chain – human)	J02939	P08195
CD98 (light chain – mouse)	AB01789	Q9Z127

References

1 Teixeira, S. et al. (1987) J. Biol. Chem. 262, 9574–9580.
2 Lumadue, J.A. et al. (1987) Proc. Natl Acad. Sci. USA 84, 9204–9208.
3 Quackenbush, E. et al. (1987) Proc. Natl Acad. Sci. USA 84, 6526–6530.
4 Gottesdiener, K.M. et al. (1988) Mol. Cell Biol. 8, 3809–3819.
5 Nakamura, E. et al. (1999) J. Biol. Chem. 274, 3009–3016.
6 Hemler, M.E. and Strominger, J.L. (1982) J. Immun. 129, 623–628.
7 Mastroberardino, L. et al. (1998) Nature 395, 288–291.
8 Michalak, M. et al. (1986) J. Biol. Chem. 261, 92–95.
9 Deves, R. and Boyd, C.A. (2000) J. Membr. Biol. 173, 165–177.
10 Pineda, M. et al. (1999) J. Biol. Chem. 274, 19738–19744.
11 Verrey, F. et al. (1999) J. Membr. Biol. 172, 181–192.
12 Warren, A.P. et al. (1996) Blood 87, 3676–3687.
13 Ohgimoto, S. et al. (1995) J. Immunol. 155, 3585–3592.
14 Haynes, B.F. et al. (1981) J. Immunol. 126, 1409–1414.

CD164

MGC-24, MGC-24v, endolyn, sialomucin

Family

Sialomucin

Structure

Molecular weights

CD164	Amino acids	197
	Polypeptide	20 903
CD164Δ4	Amino acids	184
	Polypeptide	19 676
CD164Δ5	Amino acids	178
	Polypeptide	19 004

SDS-PAGE reduced	80–100 kDa
non-reduced	160 kDa

Carbohydrate

N-linked sites	9
O-linked sites	+++

Gene location   6q21

Gene structure   > 17 kb, 6 exons

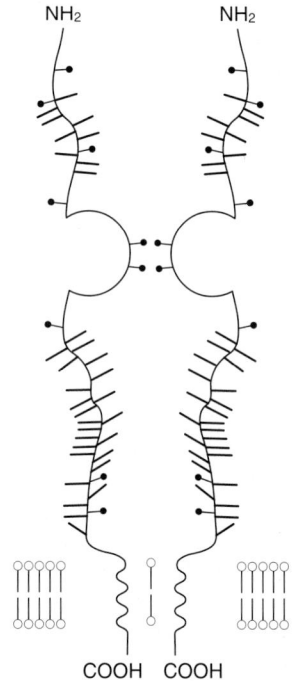

Alternative forms

CD164 isoforms lacking exon 5, CD164(EΔ5)[1]; lacking exon 4, CD164(EΔ4); or containing all six exons, CD164(E1-6)/endolyn[2], have been identified. MGC-24[3] probably represents a splice variant lacking the transmembrane domain and diverging from CD164(E1-6) after L174.

Structure

CD164 is a type I integral transmembrane protein containing two mucin domains, I (encoded by exon 1) and II (encoded by exons 4, 5 and part of 6) interrupted by a cysteine-rich region (encoded by exons 2 and 3), which probably contains a disulphide linkage. The remainder of exon 6 encodes the transmembrane and cytoplasmic domains. The heavy O-glycosylation predicts that the mucin domains of CD164 would have an extended filamentous conformation. Western blotting under non-reducing conditions revealed a 160 kDa form, which resolved at 80 kDa under reducing conditions, suggesting CD164 exists as a homodimer. The cytoplasmic domain of CD164 terminates in the sequence YHTL, which is characteristic of lysosomal glycoproteins[1,2].

Ligands

Predicted to bind a lectin(s) expressed on bone marrow stroma.

Function

CD164 can mediate adhesion of haematopoietic progenitor cells to bone marrow stromal cells *in vitro*. In addition, antibody ligation results in the reduction of cells entering the cell cycle, suggesting that CD164 can function as a signalling molecule with the capacity to suppress haematopoietic cell proliferation[1,4]. Like all mucins, CD164 may also have a cytoprotective and/or antiadhesive function. A large proportion of CD164 is localized intracellularly but the function of this intracellular pool is not know.

Distribution

CD164 is expressed in all tissues with high levels detected in liver. CD164 is expressed on CD34+ haematopoietic progenitor cells throughout ontogeny and is down-regulated on mature neutrophils and absent from erythrocytes, but maintained on monocytes and lymphoid subsets in peripheral blood. Monoclonal antibodies that define three different classes of epitopes on CD164 (class I, II and III) have been identified. Two react with glycosylation-dependent epitopes in mucin domain I, the third reacts with conformation epitopes in the cysteine-rich domain. All three epitopes are expressed on the surface of CD34+ cells but in adult lymphoid tissues the CD164 epitope classes are differentially expressed, with the class I epitope occurring on lymphoid cells, the class II epitope on endothelial and basal epithelia, and the class III epitope on both sets of cells[4-6]. As predicted from the cytoplasmic domain sequence, a substantial proportion of CD164 is found intracellularly in endosomes and lysosomes[2].

Disease association

OMIM 603356

Knockout

Amino acid sequence of human CD164 (E1–6)

```
  1 MSRLSRSLLW AATCLGVLCV LSADKNTTQH PNVTTLAPIS NVTSAPVTSL PLVTTPAPET
 61 CEGRNSCVSC FNVSVVNTTC FWIECKDESY CSHNSTVSDC QVGNTTDFCS VSTATPVPTA
121 NSTAKPTVQP SPSTTSKTVT TSGTTNNTVT PTSQPVRKST FDAASFIGGI VLVLGVQAVI
181 FFLYKFCKSK ERNYHTL
```

Amino acid sequence of human CD164(EΔ4)

```
  1 MSRLSRSLLW AATCLGVLCV LSADKNTTQH PNVTTLAPIS NVTSAPVTSL PLVTTPAPET
 61 CEGRNSCVSC FNVSVVNTTC FWIECKDESY CSHNSTVSDC QVGNTTDFCS AKPTVQPSPS
121 TTSKTVTTSG TTNNTVTPTS QPVRKSTFDA ASFIGGIVLV LGVQAVIFFL YKFCKSKERN
181 YHTL
```

Amino acid sequence of human CD164(EΔ5)

```
  1 MSRLSRSLLW AATCLGVLCV LSADKNTTQH PNVTTLAPIS NVTSAPVTSL PLVTTPAPET
 61 CEGRNSCVSC FNVSVVNTTC FWIECKDESY CSHNSTVSDC QVGNTTDFCS VSTATPVPTA
121 NSTGTTNNTV TPTSQPVRKS TFDAASFIGG IVLVLGVQAV IFFLYKFCKS KERNYHTL
```

Database accession

	EMBL/GenBank	SwissProt
Human CD164(EΔ5)	AF106518	Q04900
Rat CD164(E1–6)	AJ238574	

References

1 Zannettino, A. C. et al. (1998) Blood 92, 2613–2628.
2 Ihrke, G. et al. (2000) Biochem. J. 345, 287–296.
3 Masuzawa, Y. et al. (1992) J. Biochem. 112, 609–615.
4 Watt, S. M. et al. (1998) Blood 932, 849–866.
5 Watt, S. M. and Chan J. Y.-H. (2000) Leuk. Lymphoma 37, 1–25.
6 Watt, S. M. et al. (2000) Blood 95, 3113–3124.

Claudin-1

Senescence-associated epithelial membrane protein 1, SEMP1

Family

Claudin

Structure

Molecular weights
Amino acids	211
Polypeptide	22 881

SDS-PAGE reduced   22 kDa

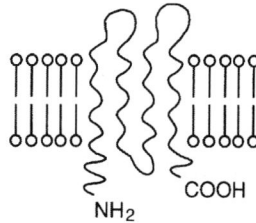

Carbohydrate
N-linked sites
O-linked sites

Gene location

Gene structure   4 kb

Alternative forms

Structure
The exact topology of the protein has not been determined; it has four (possibly three, see senescence-associated epithelial membrane protein 1, SEMP1 below) predicted transmembrane domains. No known structural motifs have been identified in the claudin polypeptide sequence. The cytoplasmic tails of all claudins (except claudin-11) contain a conserved -Tyr-Val sequence at their C-termini. By analogy with the Shaker K+ and neurexin sequences, it is predicted that this is the interaction site for binding to intracellular proteins containing PDZ domains. These include the three ZO proteins of tight junctions[2], which have previously been shown to interact with occludin (see separate entry).

Ligands

It is presumed that claudins interact homophylically or heterophylically with other claudins or occludin in the tight junction of epithelium and endothelium. Fibroblasts transfected with claudin-1, -2, or -3 exhibit calcium-independent cell adhesion[3].

Function

Claudin-1 is a integral membrane protein of epithelial tight junctions, involved in maintaining epithelial barrier function and hence restricting small molecule movement between cells of organized cellular sheets (hence claudin from 'claudere', to close). It is associated, and co-purified, with claudin-2 (38% identical) and occludin (unrelated amino acid sequence) (see relevant sections) from a tight junction membrane fraction from chicken

271

liver[1,4]. The importance of additional tight junction proteins, other than occludin, was shown as a consequence of an *in vitro* knockout of occludin where the cells continued to form functional tight junctional complexes in its absence[4].

A human claudin-1 homologue was also isolated as a differentially expressed sequence in senescing mammary epithelial cells (hence, senescence-associated epithelial membrane protein 1, SEMP1). Sequence analysis predicted only three transmembrane domains[5].

Distribution

Ubiquitous at the RNA level, being particularly abundant in liver and kidney. Epithelial and endothelial tight junctions at protein level.

Disease association

OMIM 603718

Knockout

MGI:1276109

Amino acid sequence of mouse claudin-1

```
  1 MANAGLQLLG FILASLGWIG SIVSTALPQW KIYSYAGDNI VTAQAIYEGL WMSCVSQSTG
 61 QIQCKVFDSL LNLNSTLQAT RALMVIGILL GLIAIFVSTI GMKCMRCLED DEVQKMWMAV
121 IGGIIFLISG LATLVATAWY GNRIVQEFYD PLTPINARYE FGQALFTGWA AASLCLLGGV
181 LLSCSCPRKT TSYPTPRPYP KPTPSSGKDY V
```

Database accession

EMBL/GenBank	AF115546
SwissProt	O95832

References

1 Furuse, M. et al. (1998) J. Cell Biol. 141, 1539–1550.
2 Morita, K. et al.. (1999) Proc. Natl Acad. Sci. 96, 511–516.
3 Kubota, K. et al. (1999) Curr. Biol. 9, 1035–1038.
4 Furuse, M. et al. (1998) J. Cell Biol. 143, 391–401.
5 Swisshelm, K. et al. (1999) Gene 226, 285–295.

Claudin-3

Clostridium perfringens enterotoxin receptor 2, ventral prostate 1 homologue, RVP1 androgen withdrawal apoptosis protein homologue

Family

Claudin

Structure

Molecular weights

Amino acids	154
Polypeptide	18 648

SDS-PAGE reduced   20 kDa

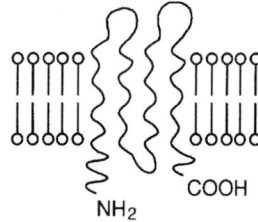

Carbohydrate

N-linked sites
O-linked sites

Gene location   7q11

Gene structure   1 exon

Alternative forms

Structure

Claudin-3 is a four membrane spanning domain structure of the claudin family of tight junction proteins shown to have at least 15 members[1]. Low-affinity version of the *Clostridium perfringens* enterotoxin receptor (*Clostridium perfringens* enterotoxin receptor 1 is the high-affinity form – see claudin-4)[2,3]. It was first identified as a hormonally regulated protein in prostate[4].

Ligands

Clostridium perfringens enterotoxin; the associated tight junction component is yet to be identified. Fibroblasts transfected with claudin-1, -2, or -3 exhibit calcium-independent cell adhesion[5].

Function

Identified as a homologue of *Clostridium perfringens* enterotoxin receptor 2 in human breast cells. Expression in tight junction implies structural or functional role in epithelial barrier.

Distribution

Abundant in intestine, liver and lung by Northern analysis, and widely expressed in other tissues at low levels.

Disease association

OMIM 602910

Knockout

MGI:1329044

Amino acid sequence of rat claudin-3

```
  1 MSMGLEITGT ALAVLGWLGT IVCCALPMWR VSAFIGSNII TSQNIWEGLW MNCVVQSTGQ
 61 MQCKVYDSLL ALPQDLQAAR ALIVVAILLA AFGLLVALVG AQCTNCVQDD TAKAKITIVA
121 GVLFLLAALL TLVPVSWSAN TIIRDFYNPV VPEAQKREMG AGLYVGWAAA ALQLLGGALL
181 CCSCPPREKK YTATKVVYSA PRSTGPGASL GTGYDRKDYV
```

Database accession

EMBL/GenBank　　　AF007189
SwissProt　　　　　O15551

References

[1] Morita, K. et al. (1999) Proc. Natl Acad. Sci. 96, 511–516.
[2] Katahira, J. et al. (1997) J. Biol. Chem. 272, 26652–26658.
[3] Peacock, R. E. et al. (1997) Genomics 46, 443–449.
[4] Ho, K.C. et al. (1989) Biochem. 28, 6367–6373.
[5] Kubota, K. et al. (1999) Curr. Biol. 9, 1035–1038.

Claudin-4

Clostridium perfringens enterotoxin receptor 1

Family

Claudin

Structure

Molecular weights
Amino acids   209
Polypeptide

SDS-PAGE reduced

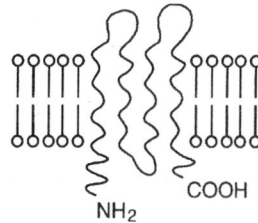

Carbohydrate
N-linked sites
O-linked sites

Gene location   7q11

Gene structure

Alternative forms

Structure
Claudin-4 has four membrane spanning domains. It has no characteristic sequence homologies other than to members of the claudin gene family[1].

Ligands

Clostridium perfringens enterotoxin, high-affinity form of receptor.

Function

Clostridium perfringens enterotoxin is a protein toxin that elicits diarrhoea by altering intestinal permeability. Binds to two forms of *Clostridium perfringens* enterotoxin receptor at high (claudin-4)[1,2] and low-affinity (see claudin-3). Component of tight junctional complex and hence presumed structural or functional role in epithelial barrier integrity.

Distribution

Abundant in intestine and kidney by Northern analysis and lower levels in many other tissues. Expressed in tight junctions of crypt enterocytes.

Disease association

OMIM 602909

Knockout

MGI:1313314

Amino acid sequence of human claudin-4

```
  1 MASMGLQVMG IALAVLGWLA VMLCCALPMW RVTAFIGSNI VTSQTIWEGL WMNCVVQSTG
 61 QMQCKVYDSL LALPQDLQAA RALVIISIIV AALGVLLSVV GGKCTNCLED ESAKAKTMIV
121 AGVVFLLAGL MVIVPVSWTA HNIIQDFYNP LVASGQKREM GASLYVGWAA SGLLLLGGGL
181 LCCNCPPRTD KPYSAKYSAA RSAAASNYV
```

Database accession
EMBL/GenBank   NM001305
SwissProt   O14493

References
[1] Morita, K. et al. (1999) Proc. Natl Acad. Sci. 96, 511–516.
[2] Katahira, J. et al. (1997) J. Biol. Chem. 272, 26652–26658.

Claudin-5

Family
Claudin

Structure
Molecular weights
Amino acids	218
Polypeptide	

SDS-PAGE reduced	23 kDa

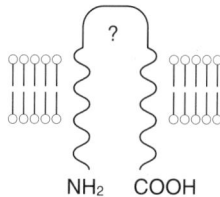

NH₂   COOH

Carbohydrate
N-linked sites
O-linked sites

Gene location
22q11.2

Gene structure
1 exon

Alternative forms

Structure
Claudin-5 is a transmembrane protein of the claudin family[1] with homology to rat RVP1 (rat ventral prostate) protein, a hormonally regulated prostate protein of unknown function[2]. Predicted two membrane spanning domains.

Ligands
Unknown.

Function
Claudin-5 is a member of the claudin family by sequence analysis and is concentrated in tight junctions when expressed *in vitro* in epithelial cells. It is presumed to be involved in tight junction structure or function.

Distribution
Epithelial cells; expression in adult lung, heart, skeletal muscle.

Disease association
OMIM 602101
Commonly deleted in velocardiofacial syndrome (VCFS, OMIM 192430), which is characterized by variable facial and cardiac malformations and cleft palate[3,4].

Knockout
MGI:1276112

Amino acid sequence for human claudin-5

```
  1 MGSAALEILG LVLCLVGWGG LILACGLPMW QVTAFLDHNI VTAQTTWKGL WMSCVVQSTG
 61 HMQCKVYDSV LALSTEVQAA RALTVSAVLL AFVALFVTLA GAQCTTCVAP GPAKARVALT
121 GGVLYLFCGL LALVPLCWFA NIVVREFYDP SVPVSQKYEL GAALYIGWAA TALLMVGGCL
181 LCCGAWVCTG RPDLSFPVKY SAPRRPTATG DYDKKNYV
```

Database accession

EMBL/GenBank   AF000959
SwissProt   O00507

References

[1] Morita, K. et al. (1999) Proc. Natl Acad. Sci. 95, 511–516.
[2] Ho, K.C. et al. (1989) Biochem. 28, 6367–6373.
[3] Puech, A. et al. (1997) Proc. Natl Acad. Sci. 94, 14608–14613.
[4] Sirotkin, H. et al. (1997) Genomics 42, 245–251.

Claudins – other

Family

Claudin gene family

An extensive gene family of at least 20 claudin proteins (from 'claudere', to hold) has recently been identified[1-4]. These show the same overall structure with four predicted membrane spanning segments and a molecular weight in the region 20–25 kDa. They show a variable tissue distribution by Northern analysis and for some there is *in vitro* and electron microscopy evidence for expression in tight junctions (see entries for Claudins-1, -3–5). Claudin-11[3] (mouse oligodendrocyte-specific protein OSP) is somewhat distantly related to the other claudins, and is restricted to myelin sheath interlamellar strands and tight junctions of Sertoli cells in testis[3]. It has a variant cytoplasmic tail C-terminus that suggests interaction with alternative intracellular proteins (see Claudin-1). Mutations in paracellin-1, a distant claudin homologue, have recently been shown to cause inherited renal magnesium loss syndrome (OMIM 603959)[5].

Database accession

	EMBL/GenBank	SwissProt
Mouse sequences		
Claudin-2	AF072128	O88552[1]
Claudin-13	AF124428	Q9Z0S4
Claudin 11 (OSP)	U19582	Q60771
Human sequences		
Claudin-6	AJ249735	P56747
Claudin-7	AJ011497	O95471
Claudin-8	AJ250711	P56748
Claudin-9	AJ130941	O95484
Claudin-10	U89916	P78369
Claudin-11	AF068863	O75508
Claudin-12	AJ250713	P56749
Claudin-14	AJ132445	O95500
Claudin-15	AJ245738	P56746
Claudin 16 (Paracellin-1)	AF152101	Q94517
Claudin-17	AJ250712	P56750
Claudin-18	AF221069	P56856
Claudin-20	AL139101	P56880

References

[1] Furuse, M. et al. (1998) J. Cell Biol. 141, 1539–1550.
[2] Morita, K. et al. (1999a) Proc. Natl Acad. Sci. 96, 511–516.
[3] Morita, K. et al. (1999b) J. Cell Biol. 145, 579–588.
[4] Tsukita, S. and Furuse, M. (1999) Trends Cell Biol. 9, 268-273.
[5] Simon, B.D. et al. (1999) Science 285, 103–106.

Collagen type XIII
(α1 chain)

Family
Short-chain collagen family

Structure

Molecular weights
Amino acids	633
Polypeptide	60 104

SDS-PAGE reduced

Carbohydrate
N-linked sites
O-linked sites

Gene location
10q22

Gene structure
>140 kb, >40 exons

NC1

COL1

NC2

COL2

NC3

COL3

NC4

COOH

Alternative forms
At least 17 alternate splice forms reported with lengths varying from 516 to 633 amino acids[1-5].

Structure
Collagen type XIII is a short-chain, non-fibrillar, transmembrane collagen with three collagenous (COL) domains interspersed by four non-collagenous regions[1-4]. Collagen type XIII forms a homotrimer of α1(XIII) chains[5]. A potential transmembrane region of 21 amino acids is present in the N-terminal NC4 non-collagenous (NC) domain, which results in a predicted type II protein orientation as in collagen Type XVII[6].

Ligands
Interacts with integrin $\alpha_1\beta_1$ but not $\alpha_2\beta_1$[7].

Function
It is an integral membrane form of collagen (see collagen XVII) associated with focal adhesion sites at cell-cell and cell-matrix contacts, but not hemidesmosomes[8]. If the signal sequence is cleaved, then collagen XIII would act as a structural matrix protein.

Distribution
Skin, intestine, placenta, bone, cartilage and striated muscle[4].

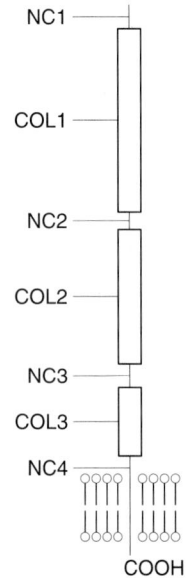

Disease association

OMIM 120350

Knockout

Amino acid sequence of human collagen type XIII

```
  1 METAILGRVN QLLDEKWKLH SRRRREAPKT SPGCNCPPGP PGPTGRPGLP GDKGAIGMPG
 61 RVGSPGDAGL SIIGPRGPPG QPGTRGFPGF PGPIGLDGKP GHPGPKGDMG LTGPPGQPGP
121 QGQKGEKGQC GEYPHRECLS SMPAALRSSQ IIALKLLPLL NSVRLAPPPV IKRRTFQGEQ
181 SQASIQGPPG PPGPPGPSGP LGHPGLPGPM GPPGLPGPPG PKGDPGIQGY HGRKGERGMP
241 GMPGKHGAKG APGIAVAGMK GEPGIPGTKG EKGAEGSPGL PGLLGQKGEK GDAGNSIGGG
301 RGEPGPPGLP GPPGPKGEAG VDGQVGPPGQ PGDKGERGAA GEQGPDGPKG SKGEPGKGEM
361 VDYNGNINEA LQEIRTLALM GPPGLPGQIG PPGAPGIPGQ KGEIGLLGPL GHDGKGPRGK
421 PGDMGPPGPQ GPPGKDGPPG VKGENGHPGS PGEKGEKGET GQAGSPGEKG EAGEKGNPGA
481 EVPGLLGPEG PPGPPGLQGV PGPKGEAGLD GAKGEKGFQG EKGDRGPLGL PGASGLDGRP
541 GPPGTPGPIG VPGPAGPKGE RGSKGDPGMT GPTGAAGLPG LHGPPGDKGN RGERGKKGSR
601 GPKGDKGDQG APGLDAPCPL GQDGYPVQGC WNK
```

Amino acids 1–21 form a possible membrane attachment site or are cleaved off to form the matrix version of the molecule.

Database accession

EMBL/GenBank   U30292 (mouse)
TrEMBL   O70575 (mouse)

References

[1] Pihlajaniemi, T. and Tamminen, M. (1990) J. Biol. Chem. 265, 16922–16928.
[2] Tikka, L. et al. (1991) J. Biol. Chem. 266, 17713–17719.
[3] Juvonen, M. and Pihlajaniemi, T. (1992) J. Biol. Chem. 267, 24693–24699.
[4] Juvonen, M. et al. (1992) J. Biol. Chem. 267, 24700–24707.
[5] Snellman, A. et al. (2000) J. Biol. Chem. 275, 8936–8944.
[6] Peltonen, S. et al. (1997) Cell Biol. 16, 227–234.
[7] Nykvist, P. et al. (2000) J. Biol. Chem. 275, 8255–8261.
[8] Peltonen, S. et al. (1999) J. Invest. Dermatol. 113, 635–642.

Collagen type XVII (α1 chain)

Bullous pemphigoid antigen 2, BPAG2, BP180

Family

Transmembrane collagen family

Structure

Molecular weights

Amino acids	1497
Polypeptide	150 459

SDS-PAGE reduced   180 kDa

Carbohydrate

N-linked sites
O-linked sites

Gene location   10q24.3

Gene structure   52 kb, 56 exons

Alternative forms   2 mRNA spliced forms have been defined in keratinocytes[1].

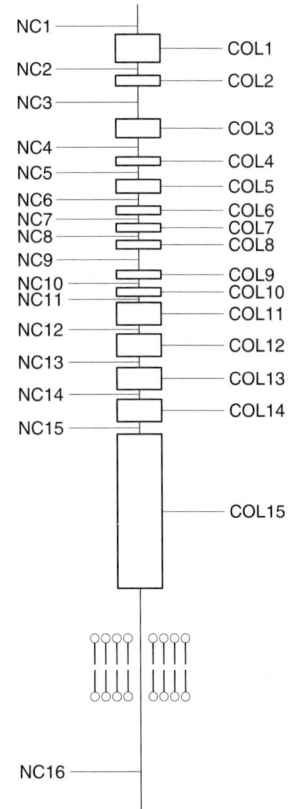

Structure

Collagen type XVII is a homotrimeric transmembrane collagen[1,2] composed of 15 collagenous (COL) domains varying in size from 15 to 242 amino acids long and 16 non-collagenous domains; the N-terminal non-collagenous (NC) domain, NC16, is large (566 amino acids) and globular[1,3-5]. It has a type II membrane protein orientation with a large intracellular segment (NC16)[2]. The N-terminal part of collagen type XVII is involved in targeting of the protein to the hemidesmosome. A region adjacent to the transmembrane segment specifies interaction with the epithelial integrin $\alpha_6\beta_4$[4] and is proteolytically cleaved to form a 120 kDa soluble form. This is the site of the major epitope recognized by autoantibodies in some forms of acquired blistering skin diseases (bullous pemphigoid, etc.)[6].

Ligands

Binds $\alpha_6\beta_4$ integrin at its extracellular face.

Function

It was originally identified as a 180 kDa autoantigen[7,8] associated with blistering skin disease. Collagen type XVII is a transmembrane glycoprotein of epithelial hemidesmosomes which is essential for their structural integrity and hence the adhesion of stratified epithelium to basement membrane (see also integrin $\alpha_6\beta_4$, desmocollins, desmogleins, etc.).

Distribution

Epithelial hemidesmosomes.

Disease association

OMIM 113811

Several mutations have been reported in the extracellular part of collagen type XVII, which result in the absence of protein[9,10], owing to premature termination of protein translation. These result in the relatively benign condition of atrophic epidermolysis bullosa where the epithelium separates from the basement membrane at the level of the hemidesmosome[5,7,10]. An epitope on collagen type XVII is recognized by autoantibodies in bullous pemphigoid and herpes gestationis[6–8].

Knockout

Amino acid sequence of human collagen type XVII

```
   1 MDVTKKNKRD GTEVTERIVT ETVTTRLTSL PPKGGTSNGY AKTASLGGGS RLEKQSLTHG
  61 SSGYINSTGS TRGHASTSSY RRAHSPASTL PNSPGSTFER KTHVTRHAYE GSSSGNSSPE
 121 YPRKEFASSS TRGRSQTRES EIRVRLQSAS PSTRWTELDD VKRLLKGSRS ASVSPTRNSS
 181 NTLPIPKKGT VETKIVTASS QSVSGTYDAT ILDANLPSHV WSSTLPAGSS IGTYHNNMTT
 241 QSSSLLNTNA YSAGSVFGVP NNMASCSPTL HPGLSTSSSV FGMQNNLAPS LTTLSHGTTT
 301 TSTAYGVKKN MPQSPAAVNT GVSTSAACTT SVQSDDLLHK DCKFLILEKD NTPAKKEMEL
 361 LIMTKDSGKV FTASPASIAA TSFSEDTLKK EKQAAYNADS GLKAEANGDL KTVSTKGKTT
 421 TADIHSYSSS GGGGSGGGGG VGGAGGGPWG PAPAWCPCGS CCSWWKWLLG LLLTWLLLLG
 481 LLFGLIALAE EVRRKLKARVD ELERIRRSIL PYGDSMDRIE KDRLQGMAPA AGADLDKIGL
 541 HSDSQEELWM FVRKKLMMEQ ENGNLRGSPG PKGDMGSPGP KGDRGFPGTP GIPGPLGHPG
 601 PQGPKGQKGS VGDPGMEGPM GQRGREGPMG PRGEAGPPGS GEKGERGAAG EPGPHGPPGV
 661 PGSVGPKGSS GSPGPQGPPG PVGLQGLRGE VGLPGVKGDK GPVGPPGPKG DQGEKGPRGL
 721 TGEPGMRGLP GAVGEPGAKG AMGPAGPDGH QGPRGEQGLT GMPGIRGPPG PSGDPGKPGL
 781 TGPQGPQGLP GTPGRPGIKG EPGAPGKIVT SEGSSMLTVP GPPGPPGAMG PPGPPGAPGP
 841 AGPAGLPGHQ EVLNLQGPPG PPGPRGPPGP SIPGPPGPRG PPGEGLPGPP GPPGSFLSNS
 901 ETFLFGPPGP PGPPGPKGDQ GPPGPRGHQG EQGLPGFSTS GSSSFGLNLQ GPPGPPGPQG
 961 PKGDKGDPGV PGALGIPSGP SEGGSSSTMY VSGPPGPPGP PGPPGSISSS GQEIQQYISE
1021 YMQSDSIRSY LSGVQGPPGP PGPPGPVTTI TGETFDYSEL ASHVVSYLRT SGYGVSLFSS
1081 SISSEDILAV LQRDDVRQYL RQYLMGPRGP PGPPGASGDG SLLSLDYAEL SSRILSYMSS
1141 SGISIGLPGP PGPPGLPGTS YEELLSLLRG SEFRGIVGPP GPPGPPGIPG NVWSSISVED
1201 LSSYLHTAGL SFIPGPPGPP GPPGPRGPPG VSGALATYAA ENSDSFRSEL ISYLTSPDVR
1261 SFIVGPPGPP GPQGPPGDSR LLSTDASHSR GSSSSSHSSS VRRGSSYSSS MSTGGGGAGS
1321 LGAGGAFGEA AGDRGPYGTD IGPGGGYGAA AEGGMYAGNG GLLGADFAGD LDYNELAVRV
1381 SESMQRQGLL QGMAYTVQGP PGQPGPQGPP GISKVFSAYS NVTADLMDFF QTYGAIQGPP
1441 GQKGEMGTPG PKGDRGPAGP PGHPGPPGPR GHKGEKGDKG DQVYAGRRRR RSIAVKP
```

Database accession

EMBL/GenBank	NP000485
TrEMBL	Q07563 (mouse)

References

1 Molnar, K. et al. (2000) Clin. Exp. Dermatol. 25, 71–76.
2 Hirako, Y. et al. (1996) J. Biol. Chem. 271, 13739–13745.
3 Li, K. et al. (1991) J. Biol. Chem. 266, 24064–24069.
4 Hopkinson, S.B. et al. (1995) J. Cell Biol. 130, 117–125.
5 Galalica, B. et al. (1997) Am. J. Hum. Genet. 60, 352–365.
6 Schumann, H. et al. (2000) Am. J. Pathol. 156, 685–695.
7 Diaz, L.A. et al. (1990) J. Clin. Invest. 86, 1088–1094.
8 Giudice, G.J. et al. (1993) J. Immunol. 151, 5742–5750.
9 Jonkman, M.F. et al. (1996) Arch. Derm. 132, 145–150.
10 McGrath, J.A. et al. (1995) Nature Genet. 11, 83–86.

Dystroglycan

Family

Structure

Molecular weights

Amino acids	895
Polypeptide	97 580

SDS-PAGE reduced	120–156 kDa (α chain) and 43 kDa (β chain)

Carbohydrate

N-linked sites	3
O-linked sites	+++ (sialated)
Other	1–2 glycosoaminoglycan chains predicted

Gene location   3p21

Gene structure

Aternative forms

Single gene gives rise to two glycoproteins, dystroglycan α and β, by post-translational cleavage. There are substantial tissue-specific differences in α dystroglycan size owing to variable glycosylation[1].

Structure

Post-translational cleavage of dystroglycan yields two polypeptides. β dystroglycan is a 43 kDa type I integral membrane glycoprotein. α dystroglycan is membrane-associated, of 120–156 kDa in size depending upon tissue origin, and non-covalently linked to β dystroglycan[1-7]. Dystroglycan is predicted to have two GAG side chains and a mucin-like O-glycosylated region in the C-terminal end of α dystroglycan[8].

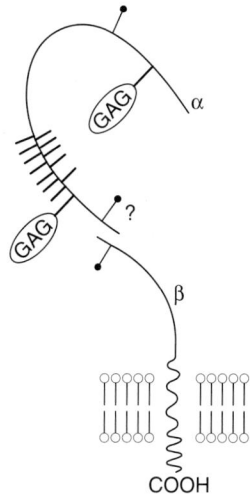

Ligands

In skeletal muscle α dystroglycan (156 kDa form) binds the basement membrane component, laminin-2[7]. Dystroglycan and laminin together act as a receptor for *Mycobacterium leprae* in Schwann cells[9] and may be involved in the pathogenesis of the associated neuropathy. α dystroglycan acts as a receptor for lymphocytic choriomeningitis and Lassa fever viruses[10]. α dystroglycan also functions as a receptor for agrin, an heparan sulphate proteoglycan, which cross-links laminin and dystroglycan at neuromuscular junctions[11,12] and is thus involved in post-synaptic membrane organization and the localization of acetylcholinesterase receptors[13].

Function

The dystroglycan complex is involved in early embryonic development, morphogenesis of muscle, and several other functions in a variety of tissues.

Dystroglycan forms part of the dystrophin-associated protein complex involved in the linkage of basement membrane extracellular matrix to the cellular actin cytoskeleton of skeletal and cardiac muscle[5-7,14]. The cytoplasmic protein, dystrophin, cross-links actin to the cytoplasmic tail of β dystroglycan, which in turn binds α dystroglycan and its matrix ligand, laminin-2. β dystroglycan is also associated with at least five other proteins in the cell membrane (four sarcoglycans and sarcospan). The molecular assembly is somewhat different in neuromuscular junctions, where a different protein, utrophin, acts as the actin linker.

The finding that dystroglycan is more widely expressed than initially reported has led to the demonstration that it combines with other membrane and cytoplasmic proteins in a tissue-specific manner. Thus, dystroglycan is highly expressed in the hemidesmosome of stratified epithelia, but it is not associated with sarcoglycans or sarcospan as in muscle. Here it seems to associate with different membrane molecules, and does not utilize standard dystrophin or utrophin to bind it to the actin cytoskeleton[6,7].

Distribution

Dystroglycan is found in the majority of fetal and adult tissues. There is increasing evidence that expression of dystroglycan in skeletal and cardiac muscle, central and peripheral nervous systems, epithelial hemidesmosomes and placental (Reichert's) membranes is functionally important[7] (see below).

Disease association

OMIM 128239

The involvement of dystroglycan in the pathogenesis of muscular dystrophy is complex.The 156 kDa α dystroglycan protein is dramatically reduced in Duchenne type muscular dystrophy but is not mutated[15,16]. Mutations in other components of the dystrophin complex, such as dystrophin, laminin-2 and sarcoglycans, however, have been shown to be causative[17,18].

Knockout

MGI:101864

The Gene Knockout FactsBook[19], p253

The mouse homologue of dystroglycan, Dag 1, was deleted[20] and resulted in an early embryonic lethal phenotype at 6.5 days owing to the loss of structural integrity of Reichert's membrane, a specialized extra-embryonic basement membrane. Owing to the early lethality, no link to muscular dystrophy pathogenesis could be made.

Amino acid sequence of human dystroglycan

```
  1 MRMSVGLSLL LPLWGRTFLL LLSVVMAQSH WPSEPSEAVR DWENQLEASM HSVLSDLHEA
 61 VPTVVGIPDG TAVVGRSFRV TIPTDLIASS GDIIKVSAAG KEALPSWLHW DSQSHTLEGL
121 PLDTDKGVHY ISVSATRLGA NGSHIPQTSS VFSIEVYPED HSDLQSVRTA SPDPGEVVSS
181 ACAADEPVTV LTVILDADLT KMTPKQRIDL LHRMRSFSEV ELHNMKLVPV VNNRLFDMSA
241 FMAGPGNPKK VVENGALLSW KLGCSLNQNS VPDIHGVEAP AREGAMSAQL GYPVVGWHIA
301 NKKPPLPKRV RRQIHATPTP VTAIGPPTTA IQEPPSRIVP TPTSPAIAPP TETMAPPVRD
```

```
361 PVPGKPTVTI RTRGAIIQTP TLGPIQPTRV SEAGTTVPGQ IRPTMTIPGY VEPTAVATPP
421 TTTTKKPRVS TPKPATPSTD STTTTTRRPT KKPRTPRPVP RVTTKVSITR LETASPPTRI
481 RTTTSGVPRG GEPNQRPELK NHIDRVDAWV GTYFEVKIPS DTFYDHEDTT TDKLKLTLKL
541 REQQLVGEKS WVQFNSNSQL MYGLPDSSHV GKHEYFMHAT DKGGLSAVDA FEIHVHRRPQ
601 GDRAPARFKA KFVGDPALVL NDIHKKIALV KKLAFAFGDR NCSTITLQNI TRGSIVVEWT
661 NNTLPLEPCP KEQIAGLSRR IAEDDGKPRP AFSNALEPDF KATSITVTGS GSCRHLQFIP
721 VVPPRRVPSE APPTEVPDRD PEKSSEDDVY LHTVIPAVVV AAILLIAGII AMICYRKKRK
781 GKLTLEDQAT FIKKGVPIIF ADELDDSKPP PSSSMPLILQ EEKAPLPPPE YPNQSVPETT
841 PLNQDTMGEY TPLRDEDPNA PPYQPPPPFT VPMEGKGSRP KNMTPYRSPP PYVPP
```

Dystroglycan is cleaved into α and β (transmembrane) peptides at amino acid **G**653 (bold).

Database accession

EMBL/GenBank   L19711
SwissProt   Q14118

References

[1] Ibraghimov-Beskrovnaya, O. et al. (1993) Hum. Molec. Genet. 2, 1651-1657.
[2] Ibraghimov-Beskrovnaya, O. et al. (1992) Nature 355, 696-702.
[3] Brancaccio, A. et al. (1995) Matrix Biol. 14, 681-685.
[4] Yotsumoto, S. et al. (1996) Hum. Mol. Genet. 5, 1259-1267.
[5] Tinsley, J.M. et al. (1994) Proc. Natl Acad. Sci. 91, 8307-8313.
[6] Hemler, M.E. (1999) Cell 97, 543-546.
[7] Durbeej, M. et al. (1998) Curr. Opin. Cell Biol. 10, 594-601.
[8] Chiba, A. et al. (1997) J. Biol. Chem. 272, 2156-2162.
[9] Rambukkana, A. et al. (1998) Science 282, 2076-2078.
[10] Cao, W. et al. (1998) Science 282, 2079-2081.
[11] Gee, S.H. et al. (1994) Cell 77, 675-686.
[12] Ma, J. et al. (1993) J. Biol. Chem. 268, 25108-25117.
[13] Campanelli, J.T. et al. (1994) Cell 77, 663-674.
[14] Henry, M.D. and Campbell, K.P. (1998) Cell 95, 859-970.
[15] Matsumura, K. et al. (1992) Nature 359, 320-322.
[16] Matsumura, K. et al. (1993) J. Clin. Invest. 92, 866-871.
[17] Campbell, K.P. (1995) Cell 80, 675-679.
[18] Straub, V et al. (1997) Curr. Opin. Neurol. 10, 168-175.
[19] Mak, T.W. (1998) The Gene Knockout FactsBook. Academic Press, London.
[20] Williamson, R.A. et al. (1997) Hum. Molec. Genet. 6, 831-841.

Family

Ly-6 (mouse) family

Structure

NH₂ — (diagram) L6

Molecular weights

Amino acids	128	
Polypeptide	13 286	
SDS-PAGE reduced	20 kDa	

Carbohydrate

N-linked sites
O-linked sites
Other

Gene location	8q24–qtr
Gene structure	3 exons

Alternative forms

Structure

E48 is the human homologue of the mouse ThB antigen. It is a member of the Ly-6 gene family (containing the Ly-6/UPAR domain) and is GPI-linked to the membrane[1,2].

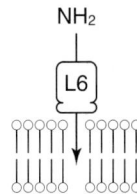

Ligands

Unknown.

Function

E48 is involved in keratinocyte cell–cell adhesion[3], and regulation of keratinocyte activation and function[1,2].

Distribution

Outer layer of transitional epithelia and keratinocytes of stratified squamous epithelia. Human lymphocytes do not express E48, unlike the mouse homologue, ThB, which is found on both lymphocytes and epithelium[2].

Disease association

There is recent evidence that E48 expression may be involved in tumour progression and metastasis[4,5].

Knockout

Amino acid sequence of human E48

```
  1 MRTALLLLAA LAVATGPALT LRCHVCTSSS NCKHSVVCPA SSRFCKTTNT VEPLRGNLVK
 61 KDCAESCTPS YTLQGQVSSG TSSTQCCQED LCNEKLHNAA PTRTALAHSA LSLGLALSLL
141 AVILAPSL
```

The sequences underlined and in italics are cleaved off to form mature E48 and a GPI anchor is added.

Database accession
EMBL/GenBank   X82693
SwissProt   Q14210

References
1 Brakenhoff, R.H. et al. (1995) J. Cell Biol. 129, 1677–1689.
2 Brakenhoff, R.H. et al. (1997) J. Immunol. 159, 4879–4886.
3 Schrijvers, A.H. et al. (1991) Exp. Cell Res. 196, 264–269.
4 Witz, I.P. (1999) J. Cell. Biochem. Suppl. 34, 61–66.
5 Eshel, R. et al. (2000) J. Biol. Chem. 275, 12833–12840.

ESL-1

E-selectin ligand, Golgi sialoglycoprotein MG-160, cysteine-rich FGF receptor

Family

Structure

Molecular weights

Amino acids	1175
Polypeptide	133 733

SDS-PAGE reduced   150 kDa (130 kDa unreduced)

Carbohydrate

N-linked sites	5
O-linked sites	

Gene location   16q22–q23

Gene structure

Alternative forms

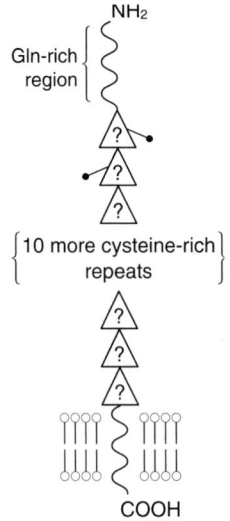

Structure

ESL-1 is a type I membrane sialoglycoprotein consisting of a glutamine-rich N-terminal region followed by 16 repeats of a 50–60 amino acid long cyteine-rich motif. Aside from the N-terminal part of ESL-1, the sequence is highly homologous to the human cysteine-rich FGF receptor, which may be a splice variant of 'human ESL-1'[1], and to the Golgi protein MG-160[2].

Ligands

E-selectin (CD62E) is bound by the isoform of ESL-1 found on myeloid cells[1]. Other variants bind basic FGF (as 'cysteine-rich FGF receptor'[3]) and are distinct from the signal transducing forms of FGF receptors.

Function

The ubiquitous expression of ESL-1 in the Golgi apparatus of most cells from early stages of embryogenesis to adult life suggests that it might be involved in Golgi function[2], although this is not proven. *In vitro* studies have demonstrated that ESL-1 is a major ligand for myeloid E-selectin (CD62E) and is presumably involved in selective leucocyte endothelial transmigration[1].

Distribution

Ubiquitous cellular distribution[1,4]. The carbohydrate variant that binds to E-selectin (CD62E) is only found in myeloid cells[1].

Disease association

OMIM 600753

Knockout

MGI:104967

Amino acid sequence of mouse ESL-1

```
   1 MAVCGRVRGM FRLSAALPLL LLAAAGAQNG HGQGQGPGTN FGPFPGQGGG GSPAGQQPPQ
  61 QPQLSQQQQQ PPPQQQQQQQ QQSLFAAGGL PARRGGAGPG GTGGGWKLAE EESCREDVTR
 121 VCPKHTWSNN LAVLECLQDV REPENEISSD CNHLLWNYKL NLTTDPKFES VAREVCKSTI
 181 SEIKECAEEP VGKGYMVSCL VDHRGNITEY QCHQYITKMT AIIFSDYRLI CGFMDDCKND
 241 INLLKCGSIR LGEKDAHSQG EVVSCLEKGL VKEAEEKEPK IQVSELCKKA ILRVAELSSD
 301 DFHLDRHLYF ACRDDRERFC ENTQAGEGRV YKCLFNHKFE ESMSEKCREA LTTRQKLIAQ
 361 DYKVSYSLAK SCKSDLKKYR CNVENLPRSR EARLSYLLMC LESAVHRGRQ VSSECQGEML
 421 DYRRMLMEDF SLSPEIILSC RGEIEHHCSG LHRKGRTLHC LMKVVRGEKG NLGMNCQQAL
 481 QTLIQETDPG ADYRIDRALN EACESVIQTA CKHIRSGDPM ILSCLMEHLY TEKMVEDCEH
 541 RLLELQYFIS RDWKLDPVLY RKCQGDASRL CHTHGWNETS ELMPPGAVFS CLYRHAYRTE
 601 EQGRRLSREC RAEVQRILHQ RAMDVKLDPA LQDKCLIDLG KWCSEKTETG QELECLQDHL
 661 DDLAVECRDI VGNLTELESE DIQIEALLMR ACEPIIQNFC HDVADNQIDS GDLMECLIQN
 721 KHQKDMNEKC AIGVTHFQLV QMKDFRFSYK FKMACKEDVL KLCPNIKKKV DVVICLSTTV
 781 RNDTLQEAKE HRVSLKCRKQ LRVEELEMTE DIRLEPDLYE ACKSDIKNYC STVQYGNAQI
 841 IECLKENKKQ LSTRCHQKVF KLQETEMMDP ELDYTLMRVC KQMIKRFCPE ADSKTMLQCL
 901 KQNKNSELMD PKCKQMITKR QITQNTDYRL NPVLRKACKA DIPKFCHGIL TKAKDDSELE
 961 GQVISCLKLR YADQRLSSDC EDQIRIIIQE SALDYRLDPQ LQLHCSDEIA NLCAEEAAAQ
1021 EQTGQVEECL KVNLLKIKTE LCKKEVLNML KESKADIFVD PVLHTACALD IKHHCAAITP
1081 GRGRQMSCLM EALEDKRVRL QPECKKRLND RIEMWSYAAK VAPADGFSDL AMQVMTSPSK
1141 NYILSVISGS ICILFLIGLM CGRITKRVTR ELKDR
```

Database accession

EMBL/GenBank   X84037 (mouse)
SwissProt   Q61543

References

[1] Steegmaier, M. et al. (1995) Nature 373, 615–620.
[2] Gonatas, J.O. et al. (1989) J. Biol. Chem. 264, 646–653.
[3] Burrus, L.W. et al. (1992) Mol. Cell Biol. 12, 5600–5609.
[4] Gonatas, J.O. et al. (1995) J. Cell Sci. 108, 457–467.

Galectin 3

Family

C-terminal lectin domain belongs to galaptin/S-lectin family

Structure

Molecular weights

Amino acids	250
Polypeptide	26 057

SDS-PAGE reduced   31 kDa (35, 67 and 80 kDa)

Carbohydrate

N-linked sites	0
O-linked sites	

Gene location   14q21–q22

Gene structure

Alternative forms

Exists as monomeric, dimeric and trimeric forms.

Structure

The C-terminal domain belongs to the galaptin/S-lectin family[1-4]. Towards the Pro/Gly-rich N-terminal end of galectin 3, there are eight nine-amino-acid-long tandem repeats of Y-P-G-xxx-P-G-A. Galectin 3 exists as a mixture of multimers through its N-terminal domain[5-7]. Serine phophoryalated (Ser6 and 12). No signal or transmembrane sequences, and the surface expression of galectin 3 is probably due to secretion and association with cells via carbohydrate binding to other membrane molecules through its lectin domain[6,7].

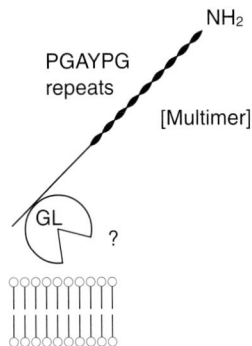

Ligands

Galectin 3 binds β galactosides on laminin and IgE via the C-terminal lectin domain[5,6]. It also binds to FcεR1 (high-affinity IgE receptor) and the secreted glycoprotein, Mac-2 binding protein, which belongs to the scavenger receptor family[7].

Function

It is a membrane and secreted protein of inflammatory macrophages, which binds galactose-containing ligands. It also mediates macrophage adhesion to laminin when cross-linked by transglutaminase. The intracellular distribution of galectin 3/Mac-2 to the apical peripheral membrane of polarized colonic epithelia cells suggests an additional role in targeted cellular protein secretion. A separate intracellular role is postulated due to its localization in the nucleus[6,8]. Recent evidence suggests that galectin 3 is involved in tumour proliferation and apoptosis[9]. It interacts with the neural adhesion molecule, L1, and promotes neurite outgrowth[10].

Distribution

Activated macrophages, basophils and mast cells, sensory neurons[6,7,10], and colonic epithelium and colon cancers[11].

Disease association

OMIM 153619

Knockout

Amino acid sequence of human galectin-3

```
  1 MADNFSLHDA LSGSGNPNPQ GWPGAWGNQP AGAGGYPGAS YPGAYPGQAP PGAYPGQAPP
 61 GAYHGAPGAY PGAPAPGVYP GPPSGPGAYP SSGQPSAPGA YPATGPYGAP AGPLIVPYNL
121 PLPGGVVPRM LITILGTVKP NANRIALDFQ RGNDVAFHFN PRFNENNRRV IVCNTKLDNN
181 WGREERQSVF PFESGKPFKI QVLVEPDHFK VAVNDAHLLQ YNHRVKKLNE ISKLGISGDI
241 DLTSASYTMI
```

Database accession

EMBL/GenBank   M57710
SwissProt   P17931

References

1. Cherayil, B.J. et al. (1990) Proc. Natl Acad. Sci. USA 87, 7324–7328.
2. Robertson, M.W. et al. (1990) Biochem. 29, 8093–8100.
3. Oda, Y. et al. (1991) Gene 99, 279–283.
4. Raz, A. (1991) Cancer Res. 51, 2173–2178.
5. Liu, F.-T. (1993) Immunol. Today 14, 486–490.
6. Barondes, S.H. et al. (1994) J. Biol. Chem. 269, 20807–20810.
7. Hughes, R.C. (1994) Glycobiology 4, 5–12.
8. Lotz, M.M. et al. (1993) Proc. Natl Acad. Sci. USA 90, 3466–3470.
9. Kim, H.R. et al. (1999) Cancer Res. 59, 4148–4154.
10. Pesheva, P. et al. (1998) J. Neurosci. Res. 54, 639–654.
11. Ho, M.-K. and Springer, T.A. (1982) J. Immunol. 128, 1221–1227.
12. Huflejt, M.E. et al. (1997) J. Biol. Chem. 272, 14294–14303.

GlyCAM-1

Glycosylation-dependent cell adhesion molecule 1, SGP50 Proteose-Peptone component 3 (PP3)

Family

Structure

Molecular weights

Amino acids	146 amino acids
Polypeptide	15 440

SDS-PAGE reduced   50 kDa (50 kDa unreduced)

Carbohydrate

N-linked sites	1
O-linked sites	Abundant +++

Gene location   Mouse chromosome 15

Gene structure   2.5 kb

Alternative forms

NH$_2$

? COOH

Structure

Sulphated secreted mucoprotein[1-3] with two extracellular serine/threonine rich regions, which are heavily O-glycosylated[1,4].

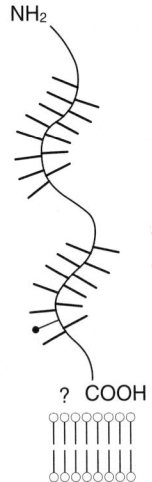

Ligands

Carbohydrate-dependent binding to L-selectin (CD62L) on the luminal surface of high endothelial venules of peripheral lymph nodes, involving O-linked carbohydrates, including sulphated forms of sialyated Lewis[x] carbohydrate sequence.

Function

GlyCAM-1 is involved in leucocyte adhesion via L-selectin (CD62L) and hence migration through high endothelium during inflammation[5] (see L-selectin entry). Being a secreted molecule it is speculated that GlyCAM-1 reassociates with the cell surface and acts to downregulate L-selectin-mediated adhesion[6]. The soluble form of GlyCAM-1 may be involved in regulating T cell activation.

Distribution

Peripheral and mesenteric lymph node high venular endothelium[6], lactating mammary gland[8] and lung[9]. Detectable in blood and milk as secreted form[9]; the milk form does not bind L-selectin. Increased plasma levels of GlyCAM-1 are observed following exposure to inflammatory stimuli[10].

Disease association

Knockout

MGI:95759

Amino acid sequence of rat GlyCAM-1

```
  1 MKFFTVLLFA SLAATSLAAV PGSKDELHLR TQPTDAIPAS QFTPSSHISK ESTSSKDLSK
 61 ESFIFNEELV SEDNVGTEST KPQSQEAQDG LRSGSSQQEE TTSAATSEGK LTMLSQAVQK
141 ELGKVIEGFI SGVEDIISGA SGTVRP
```

Database accession

EMBL/GenBank   L08100 [rat]
SwissProt   Q04807 [rat]

References
1 Lasky, L.A. et al. (1992) Cell 69, 927–938.
2 Dowbenko, D. et al. (1993a) J. Biol. Chem. 268, 4525–4529.
3 Dowbenko, D. et al. (1993b) J. Biol. Chem. 268, 14399–14403.
4 Hemmerich, S. et al. (1995) J. Biol. Chem. 270,12035–12047.
5 Johnson-Leger, C. et al. (2000) J. Cell Sci. 113, 921–933.
6 Brustein, M. et al. (1992) J. Exp. Med. 176, 1415–1419.
7 Onrust, S.V. et al. (1996) J. Clin. Invest. 97, 54–64.
8 Nishimura, T. et al. (1993) J. Biochem. 114, 567–569.
9 Lasky, L.A. (1995) Annu. Rev. Biochem. 64, 113–139.
10 Suguri, T. et al. (1996) J. Leukoc. Biol. 60, 593–597.

Family

Structure

Molecular weights

Amino acids	80
Polypeptide	8 083

SDS-PAGE reduced	35–45 kDa

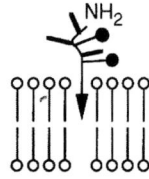

Carbohydrate

N-linked sites	2
O-linked sites	probably +++

Gene location   6q21

Gene structure   2 exons

Alternative forms

Multiple HSA/CD24 genes have been identified in human and mouse[1-3] as have RNA splice variants.

Structure

HSA is a sialoglycoprotein that is GPI-linked to the membrane. The mature protein is predicted to be of 33 amino acids[2,4-6]. Its high content of threonine and serine implies extensive O-glycosylation.

Ligands

HSA is a ligand for P-selectin (CD62P)[7,8].

Function

Possible functional involvement in B cell proliferation and differentiation[9]. HSA/P-selectin interaction may play an ancillary role to the major selectin (CD62) ligands in leucocyte interactions or endothelial transit[7,8]. HSA mediates tumour cell binding to endothelium under static conditions[8]. Signal transduction via HSA may modify integrin $\alpha_5\beta_1$ (CD49d)-mediated adhesive interactions[10]. HSA, decorated with the CD57 epitope (see entry), interacts with L1 in the brain[11].

Distribution

Activated, but not late (antibody forming cells), B lymphocytes and granulocytes, but not T cells and monocytes, express HSA[5,6,12,13]. Tissue distribution differs in the mouse being expressed by thymocytes, erythrocytes, developing epithelium and central nervous system[14,15]. High expression by small cell lung carcinoma[6].

Disease association

OMIM 600074

Knockout

MGI:88323
HSA knockout mice have alterations in numbers of immature bone marrow B-cells[16] and erythrocyte abnormalities.

Amino acid sequence of human HSA

```
 1 MGRAMVARLG LGLLLLALLL PTQIYSSETT TGTSSNSSQS TSNSGLAPNP TNATTKAAGG
61 ALQSTASLFV VSLSLLHLYS
```

The sequences underlined and in italics are cleaved off to form mature HSA and a GPI anchor is added.

Database accession

EMBL/GenBank   M58664
SwissProt   P25063

References

[1] Wenger, R.H. et al. (1991) Eur. J. Immunol. 21, 1039–1046.
[2] Wenger, R.H. et al. (1993) J. Biol. Chem. 268, 23345–23352.
[3] Hough, M. R. et al. (1994) Genomics 22, 154–161.
[4] Kay, R. et al. (1990) J. Immunol. 145, 1952–1959.
[5] Kay, R. et al. (1991) J. Immunol. 147, 1412–1416.
[6] Jackson, D. et al. (1992) Cancer Res. 52, 5264–5270.
[7] Sammar, M. et al. (1994) Int. Immunol. 6, 1027–1036.
[8] Aigner, S. et al. (1997) Blood 899, 3385–3395.
[9] Kadmon, G. et al. (1992) J. Cell Biol. 118, 1245–1258.
[10] Hahne, M. et al. (1994) J. Exp. Med. 179, 1391–1395.
[11] Sammar, M. et al. (1997) Biochem. Biophys. Data 1337, 287–294.
[12] Kemshead, J.T. et al. (1982) Hybridoma 1, 109–123.
[13] Hsu, S.-M. and Jaffe, E.S. (1984) Am. J. Pathol. 114, 387–395.
[14] Takei, F. et al. (1981) Immunology 42, 371–378.
[15] Enk, A.H. and Katz, S.I. (1994) J. Immunol. 152, 3264–3270.
[16] Wenger, R.H. et al. (1995) Transgenic Res. 4, 173–183.
[17] Nielsen, P.J. et al. (1997) Blood 89, 1058–1067.

LYVE-1

Family

Hyaluronan receptor

Structure

Molecular weights

Amino acids	322
Polypeptide	35 158

SDS-PAGE reduced	60 kDa

Carbohydrate

N-linked sites	2
O-linked sites	+

Gene location	11q15

Gene structure	6 exons, approximately 11 kb

Alternative forms

Structure

The N-terminus of LYVE-1 contains an extended link domain generating the presumptive hyaluronan binding site[1].

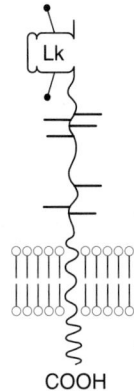

Ligands

LYVE-1 is a cell surface receptor for the extracellular matrix glycosaminoglycan, hyaluronan[1].

Function

LYVE-1 has a putative function in sequestering hyaluronan on lymph vessel wall and/or supporting hyaluronan-mediated rolling of leucocytes in the lymph fluid.

Distribution

LYVE-1 exhibits a restricted pattern of expression being mostly confined to lymphatic endothelial cells[1].

Disease association

Knockout

Amino acid sequence of human LYVE-1

```
  1 MARCFSLVLL LTSIWTTRLL VQGSLRAEEL SIQVSCRIMG ITLVSKKANQ QLNFTEAKEA
 61 CRLLGLSLAG KDQVDTALKA SFETCSYGWV GDGFVVISRI SPNPKCGKNG VGVLIWKVPV
121 SRQFGAYCYN SSDTWTNSRI PEIITTKDPI FNTQTATQTT EFIVSDSTYS VASPYSTIPA
181 PTTTPPAPAS TSIPRRKKLI CVTEVFMETS TMSTETEPFV ENKAAFKNEA AGFGGVPTAL
241 LVLALLFFGA AAGLGFCYVK RYVKAFPFTN KNQQKEMIET KVVKEEKAND SNPNEESKKT
301 DKNPEESKSP SKTTVRCLEA EV
```

Database accession
EMBL/GenBank   AF118108
TrEMBL   Q9Y5Y7

References
[1] Banerji, S. et al. (1999) J. Cell Biol. 144, 789–801.

Occludin

Family

ELL/occludin family

Structure

Molecular weights

Amino acids	522
Polypeptide	59 143

SDS-PAGE reduced   60 kDa

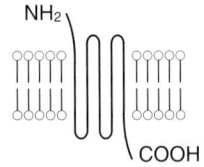

Carbohydrate

N-linked sites	0
O-linked sites	0

Gene location

Gene structure

Alternative forms

Structure

Occludin is an integral membrane protein of epithelial tight junctions[1,2]. The first occludin sequence cloned was from the kangeroo rat and was predicted to have four transmembrane domains with both N- and C-termini oriented cytoplasmically. However, those occludins that have been cloned subsequently from other species (including human) have been found to have five predicted membrane spanning domains and hence would have a different configuration with their N-terminus extracellular (as illustrated). The first extracellular domain, which may be functional in intercellular adhesion, contains a high proportion of tyrosine and glycine amino acid residues. The C-terminal part of the molecule interacts with TJP1/ZO-1 (OMIM 601009), a PDZ domain[3] containing protein of tight junction plaques.

Ligands

Probably interacts with occludin or claudins on opposing cells within the tight junction. It has been shown recently that occludin interacts directly with the gap junction component, Cx32, in polarised hepatocytes[4].

Function

Occludin is involved in the formation and function of cell–cell tight junctional interactions in epithelial and endothelial cells[1,5,6]. Tight junction structure, component proteins and function have been extensively reviewed recently[7-14]. *In vitro* deletion of occludin[15] failed to abrogate tight junctional transcellular resistance suggesting the involvement of other key molecules (see claudins). Occludin is presumed to either play a secondary role in tight junction formation, or to be involved in functional gating of molecules passing through epithelial or endothelial barriers.

Distribution

Tight junctions of epithelium and endothelium in a wide range of tissues.

Disease association

OMIM 602876

Knockout

MGI:106183

In vitro knockout[15] in embryonic stem cells showed that tight junctions formed in the absence of occludin upon epithelial differentiation. This led to the discovery of the claudin family of proteins as being essential components for the structural integrity of epithelial junctions.

Amino acid sequence of human occludin

```
  1 MSSRPLESPP PYRPDEFKPN HYAPSNDIYG GEMHVRPMLS QPAYSFYPED EILHFYKWTS
 61 PPGVIRILSM LIIVMCIAIF ACVASTLAWD RGYGTSLLGG SVGYPYGGSG FGSYGSGYGY
121 GYGYGYGYGG YTDPRAAKGF MLAMAAFCFI AALVIFVTSV IRSEMSRTRR YYLSVIIVSA
181 ILGIMVFIAT IVYIMGVNPT AQSSGSLYGS QIYALCNQFY TPAATGLYVD QYLYHYCVVD
241 PQEAIAIVLG FMIIVAFALI IFFAVKTRRK MDRYDKSNIL WDKEHIYDEQ PPNVEEWVKN
301 VSAGTQDVPS PPSDYVERVD SPMAYSSNGK VNDKRFYPES SYKSTPVPEV VQELPLTSPV
361 DDFRQPRYSS GGNFETPSKR APAKGRAGRS KRTEQDHYET DYTTGGESCD ELEEDWIREY
421 PPITSDQQRQ LYKRNFDTGL QEYKSLQSEL DEINKELSRL DKELDDYREE SEEYMAAADE
481 YNRLKQVKGS ADYKSKKNHC KQLKSKLSHI KKMVGDYDRQ KT
```

Database accession

EMBL/GenBank   U49184
SwissProt   Q16625

References

1 Furuse, M. et al. (1993) J. Cell Biol. 123, 1777–1788.
2 Ando-Akatsuka, Y. et al. (1996) J. Cell Biol. 133, 43–47.
3 Craven, S.E and Bredt, D.S. (1998) Cell 93, 495–498.
4 Kojima, T. et al. (1999) Biochem. Biophys. Res. Comm. 266, 222–229.
5 Furuse, M. et al. (1994) J. Cell Biol. 127, 1617–1626.
6 McCarthy, K.M. et al. (1996) J. Cell Sci. 109, 2287–2298.
7 Tsukita, S and Furuse, M. (1999) Trends Cell Biol. 9, 268–273.
8 Rubin, L.L. and Staddon, J.M. (1999) Annu. Rev. Neurosci, 22, 11–28.
9 Stevenson, B.R. and Keon, B.H. (1998) Annu. Rev. Cell Dev. Biol. 14, 89–109.
10 Citi, S. and Cordenonsi, M. (1998) Biochem. Biophys. Acta 1448, 1–11.
11 Matter, K. and Balda, M.S. (1999) Int. Rev. Cytol. 186, 117–146.
12 Madara, J.L. (1998) Annu. Rev. Physiol. 60, 143–159.
13 Mitic, L.L. and Anderson, J.M. (1998) Annu. Rev. Physiol. 60, 121–142.
14 Denker, B.M. and Nigam, S.K. (1998) Am. J. Physiol. 274, F1-9.
15 Saitou, M. et al. (1998) J. Cell Biol. 141, 397–408.

PCLP1 — Podocalyxin-like protein, Ly102

Family
Sialomucin

Structure

Molecular weights
Amino acids	528
Polypeptide	55 561

SDS-PAGE reduced   160–165 kDa

Carbohydrate
N-linked sites	5
O-linked sites	+++
GAG	4 potential GAG attachment sites

Gene location   7q32–7q33

Gene structure

Alternative forms

Structure
The sequence of podocalyxin-like protein (PCLP) together with the discrepancy between predicted and actual molecular weight indicates extensive O-linked glycosylation between the N-terminus and Thr 300. This suggests that PCLP would have an extended filamentous conformation. Carbohydrate analysis demonstrates that the net negative charge derives from both sulphate and sialic acid residues[1,2]. In addition, in the extracellular domain there are four potential GAG attachment sites and four membrane proximal cysteine residues for potential disulphide bond formation[3]. The cytoplasmic domain is highly acidic with a number of potential phosphorylation sites.

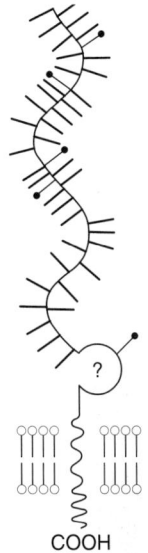

Ligands
PCLP is a ligand for L-selectin[4].

Function
The expression of PCLP in high endothelial venules and the demonstration that it can bind L-selectin, and mediate the tethering and rolling of lymphocytes under physiological flow conditions, indicates a function in the recruitment of lymphocytes to secondary lymphoid organs. The similar structure and expression patterns of PCLP and CD34 suggest that they may have redundant or overlapping physiological roles[4]. It is not known whether PCLP has an adhesive function in the podocytes or whether it has a structural role maintaining the integrity of the filtration slits by contributing to the anionic glycocalyx[1]. Like all mucins, PCLP may also have a cytoprotective and/or antiadhesive function.

Distribution

PCLP is localized to the apical surface of renal glomerular epithelial (podocyte) foot processes, vascular endothelial cells and high endothelial venules[3-6]. In addition PCLP1 is a marker for haemangioblasts that give rise to long-term repopulating haematopoietic stem cells[7].

Disease association

OMIM 602632
Reduced sialylation of PCLP is found in aminonucleoside nephrosis[8]. Urinary levels of PCLP is an indicator of glomerular epithelial cell injury[9].

Knockout

Amino acid sequence of human podocalyxin-like protein

```
  1 MRCALALSAL LLLLSTPPLL PSSPSPSPSP SPSQNATQTT TDSSNKTAPT PASSVTIMAT
 61 DTAQQSTVPT SKANEILASV KATTLGVSSD SPGTTTLAQQ VSGPVNTTVA RGGGSGNPTT
121 TIESPKSTKS ADTTTVATST ATAKPNTTSS QNGAEDTTNS GGKSSHSVTT DLTSTKAEHL
181 TTPHPTSPLS PRQPTLTHPV ATPTSSGHDH LMKISSSSST VAIPGYTFTS PGMTTTLPSS
241 VISQRTQQTS SQMPASSTAP SSQETVQPTS PATALRTPTL PETMSSSPTA ASTTHRYPKT
301 PSPTVAHESN WAKCEDLETQ TQSEKQLVLN LTGNTLCAGG ASDEKLISLI CRAVKATFNP
361 AQDKCGIRLA SVPGSQTVVV KEITIHTKLP AKDVYERLKD KWDELKEAGV SDMKLGDQGP
421 PEEAEDRFSM PLIITIVCMA SFLLLVAALY GCCHQRLSQR KDQQRLTEEL QTVENGYHDN
481 PTLEVMETSS EMQEKKVVSL NGELGDSWIV PLDNLTKDDL DEEEDTHL
```

Database accession

EMBL/GenBank U97519
TrEMBL O00592

References

[1] Kerjaschki, D. et al. (1984) J. Biol. Chem. 98, 1591–1596.
[2] Dekan, G. et al. (1991) Proc. Natl Acad. Sci. USA 88, 5398–5402.
[3] Kershaw, D.B. et al. (1997) J. Biol. Chem. 272, 15708–15714.
[4] Sassetti, C. et al. (1998) J. Exp. Med. 187, 1965–1975.
[5] Horvat, R. et al. (1986) J. Cell Biol. 102, 484–491.
[6] Kershaw, D.B. et al. (1995) J. Biol. Chem. 270, 29439–29446.
[7] Hara, T. et al. (1999) Immunity 11, 567–578.
[8] Kerjaschki, D. et al. (1985) Am. J. Pathol. 118, 343–390.
[9] Hara, M. et al. (1995) Nephron 69, 397–403.

PSGL-1
P-selectin ligand, CD162

Family

Sialomucin

Structure

Molecular weights
Amino acids   412
Polypeptide   41 301

SDS-PAGE reduced   120 kDa (250, 160 kDa (neutrophils)) (220 kDa in HL60 cells)

Carbohydrate
N-linked sites   3
O-linked sites   Abundant +++

Gene location   12q24

Gene structure   9 kb, 1 exon, gene structure resembling that of CD43 and CD42b (gpIbα)

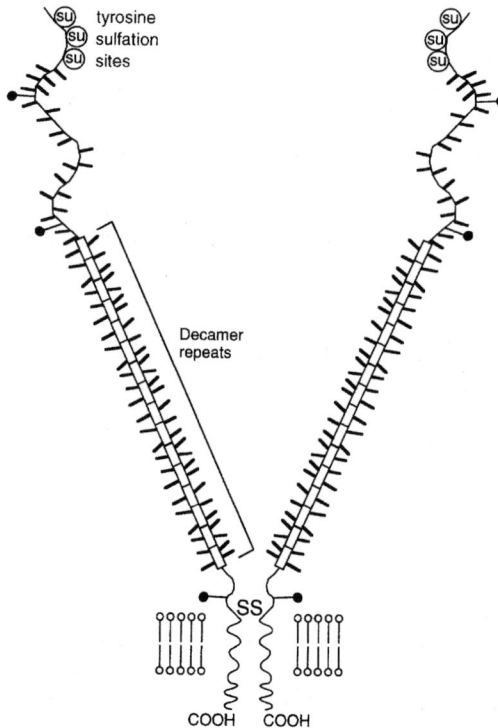

Alternative forms

Two variants (short and long forms, probably by post-translational proteolytic cleavage rather than splice variation as PSGL-1 is encoded by a single exon).

Structure

Disulphide-bonded homodimeric type I transmembrane mucin-like glycoprotein[1,2]. An N-terminal 23 amino acid propeptide sequence is cleaved at a consensus dibasic site (RxRR) by 'PACE' (paired basic amino acid-converting enzyme)[3]. This is followed by a sulphated region containing three tyrosines. A region containing a series of 16 repeats of 10 amino acids is present in its extracellular part in most leucocytes (or 15 in HL60 and other cell lines; and only 10 in murine PSGL-1)[1,4]. These are heavily O-glycosylated forming a mucin-like structure[5]. O-glycan structure has been determined[6], a subset containing 'core-2', sialylated and fucosylated structures being required for functional integrity of PSGL-1 and binding to all three selectins. N-terminal tyrosine sulphation is also required for recognition of P-selectin (CD62P) and L-selectin (CD62L)[7-11]. Binding to E-selctin (CD62E) occurs at lower affinity and does not require tyrosine sulphation[2,10].

Ligands

Binds to P-, L- and E-selectins (CD62P, CD62L and CD62E, respectively)[2,11-14].

Function

High-affinity counter-receptor for P-selectin (CD62P) (hence, PSGL-1, P-selectin glycoprotein ligand) on myeloid cells and activated T lymphocytes[2,15]. PSGL-1 mediates rolling of leucocytes (neutrophils) on activated endothelium under conditons of physiological blood flow[2,14,16,17], leucocyte binding to platelets and leucocyte interactions (neutrophils, T lymphocytes) at sites of inflammation via their interaction with P- and L-selectins.

Distribution

Most peripheral T cells, some B cells, neutrophils, monocytes and platelets[2,11,15,18,19].

Disease association

OMIM 600738

Knockout

MGI:106689
PSGL-1 -/- mice are viable and fertile but leukocyte infiltration into inflammatory sites is significantly delayed suggesting that PSGL-1 is the predominant neutrophil P-selectin ligand but is not required as a counter-receptor for E-selectin in vivo[20].

Amino acid sequence of human PSGL-1

```
  1 MPLQLLLLLI LLGPGNSLQL WDTWADEAEK ALGPLLARDR RQATEYEYLD YDFLPETEPP
 61 EMLRNSTDTT PLTGPGTPES TTVEPAARRS TGLDAGGAVT ELTTELANMG NLSTDSAAME
121 IQTTQPAATE AQTTQPVPTE AQTTPLAATE AQTTRLTATE AQTTPLAATE AQTTPPAATE
181 AQTTQPTGLE AQTTAPAAME AQTTAPAAME AQTTPPAAME AQTTQTTAME AQTTAPEATE
241 AQTTQPTATE AQTTPLAAME ALSTEPSATE ALSMEPTTKR GLFIPFSVSS VTHKGIPMAA
301 SNLSVNYPVG APDHISVKQC LLAILILALV ATIFFVCTVV LAVRLSRKGH MYPVRNYSPT
361 EMVCISSLLP DGGEGPSATA NGGLSKAKSP GLTPEPREDR EGDDLTLHSF LP
```

The propeptide (amino acids 19–41) cleaved by 'PACE' enzyme is underlined and italicized.

Database accession

EMBL/GenBank   U25956
SwissProt   Q14242

References
[1] Sako, D. et al. (1993) Cell 75, 1179–1186.
[2] McEver, R.P. et al. (1995) J. Biol. Chem. 270, 11025–11028.
[3] Li, F. et al. (1996a) J. Biol. Chem. 271, 6342–6348.
[4] Veldman, G.M. et al. (1995) J. Biol. Chem. 270, 16470–16475.
[5] Moore, K.L. et al. (1994) J. Biol. Chem. 269, 23318–23327.
[6] Wilkins, P.P. et al. (1995) J. Biol. Chem. 270, 22677–22680.
[7] Sako, D. et al. (1995) Cell 83, 323–331.
[8] Pouyani, T. and Seed, B. (1995) Cell 83, 333–343.
[9] Wilkins, P.P. et al. (1995) J. Biol. Chem. 270, 22677–22680.
[10] Li, F. et al. (1996b) J. Biol. Chem. 271, 3255–3265.
[11] Spertini, O. et al. (1996) J. Cell Biol. 135, 523–531.
[12] Rosen, S.D. and Bertozzi, C.R. (1996) Curr. Biol. 6, 261–264.
[13] Tu, L. et al. (1996) J. Immunol. 157, 3995–4004.
[14] Walcheck, B. et al. (1996) J. Clin. Invest. 98, 1081–1087.
[15] Vachino, G. et al. (1995) J. Biol. Chem. 270, 21966–21974.
[16] Alon, R. et al. (1994) J. Cell Biol. 127, 1485–1495.
[17] Norman, K.E. et al. (1995) Blood 86, 4417–4421.
[18] Laszik, Z. et al. (1996) Blood 88, 3010–3021.
[19] Frenette, P.S. et al. (2000) J. Exp. Med. 191, 1413–1422.
[20] Yang, J. et al. (1999) J. Exp. Med. 190, 1769–1782.

VAP-1

Vascular adhesion protein-1

Family

Copper/topaquinone oxidase family

Structure

Molecular weights

| Amino acids | 763 |
| Polypeptide | 84 568 |

| SDS-PAGE reduced | 110 kDa |

Carbohydrate

| N-linked sites | 6 |
| O-linked sites | + |

Gene location

Gene structure

14.4 kb, 4 exons in mouse[1]

Alternative forms

Cell-specific isoforms have been described[2]

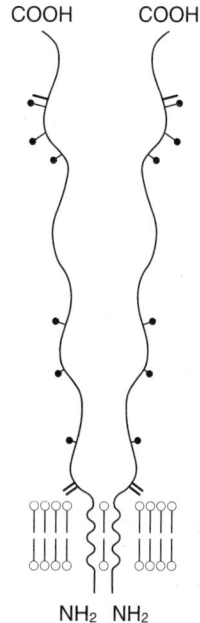

Structure

Unusually for an adhesion receptor, VAP-1 is a type II transmembrane protein with its N-terminus intracellularly. The extracellular domain contains sites for both N- and O-linked glycosylation and enzyme treatment demonstrates that VAP-1 is a sialoglycoprotein. Strikingly, sequence analysis of VAP-1 revealed significant identity to the copper-containing amine oxidase family. Moreover, VAP-1 was shown to have a quinone cofactor and monoamine oxidase activity. Western blot analysis demonstrates that VAP-1 is expressed at the cell surface as a 170–190 kDa homodimer[3-5].

Ligands

VAP-1 ligand(s) have not been identified. However, it is known that VAP-1 mediates lymphocyte binding to peripheral lymph nodes in a selectin-independent manner[6].

Function

Endothelial VAP-1 initiates the contact of a subset of lymphocytes to the peripheral lymph node high endothelial venules[4,6,7]. The relationship between the copper-dependent monoamine oxidase enzyme activity[5] and the role of VAP-1 as an adhesion receptor is not clear. It is possible that different functions are mediated in different VAP-1 expressing tissues[4].

Distribution

VAP-1 is expressed mainly in the high endothelial venules of peripheral lymph node lymphatic tissue and is upregulated in chronic inflammation[8,9]. Expression is also detected in hepatic endothelial cells, smooth muscle and dendritic cells.

Disease association

Knockout

Amino acid sequence of human VAP-1

```
  1 MNQKTILVLL ILAVITIFAL VCVLLVGRGG DGGEPSQLPH CPSVSPSAQP WTHPGQSQLF
 61 ADLSREELTA VMRFLTQRLG PGLVDAAQAR PSDNCVFSVE LQLPPKAAAL AHLDRGSPPP
121 AREALAIVFF GRQPQPNVSE LVVGPLPHPS YMRDVTVERH GGPLPYHRRP VLFQEYLDID
181 QMIFNRELPQ ASGLLHHCCF YKHRGRNLVT MTTAPRGLQS GDRATWFGLY YNISGAGFFL
241 HHVGLELLVN HKALDPARWT IQKVFYQGRY YDSLAQLEAQ FEAGLVNVVL IPDNGTGGSW
301 SLKSPVPPGP APPLQFYPQG PRFSVQGSRV ASSLWTFSFG LGAFSGPRIF DVRFQGERLV
361 YEISLQEALA IYGGNSPAAM TTRYVDGGFG MGKYTTPLTR GVDCPYLATY VDWHFLLESQ
421 APKTIRDAFC VFEQNQGLPL RRHHSDLYSH YFGGLAETVL VVRSMSTLLN YDYVWDTVFH
481 PSGAIEIRFY ATGYISSAFL FGATGKYGNQ VSEHTLGTVH THSAHFKVDL DVAGLENWVW
541 AEDMVFVPMA VPWSPEHQLQ RLQVTRKLLE MEEQAAFLVG SATPRYLYLA SNHSNKWGHP
601 RGYRIQMLSF AGEPLPQNSS MARGFSWERY QLAVTQRKEE EPSSSSVFNQ NDPWAPTVDF
661 SDFINNETIA GKDLVAWVTA GFLHIPHAED IPNTVTVGNG VGFFLRPYNF FDEDPSFYSA
721 DSIYFRGDQD AGACEVNPLA CLPQAAACAP DLPAFSHGGF SHN
```

Database accession

EMBL/GenBank   AF067406
SwissProt   Q16853

References

[1] Bono, P. et al. (1998) J. Immunol. 161, 2953–2960.

[2] Salmi, M. and Jalkanen, S. (1995) Eur. J. Immunol. 25, 2803–2812.

[3] Bono, P. et al. (1998) J. Immunol. 160, 5563–5571.

[4] Smith, D.J. et al. (1998) J. Exp. Med. 188, 17–27.

[5] Zhang, X. and McIntire, W.S. (1996) Gene 179, 279–286.

[6] Salmi, M. et al. (1997) J. Exp. Med. 186, 589–600.

[7] Salmi, M. and Jalkanen, S. (1996) J. Exp. Med. 183, 569–579.

[8] Salmi, M. and Jalkanen, S. (1992) Science 257, 1407–1409.

[9] Salmi, M. et al. (1993) J. Exp. Med. 178, 2255–2260.

Index

GLASGOW UNIVERSITY LIBRARY
WITHDRAWN